MATHEMATICAL MODELING

with Applications for the Managerial, Life, and Social Sciences

Taken from:

College Algebra in Context with Applications for the Managerial, Life, and Social Sciences, Second Edition
by Ronald J. Harshbarger and Lisa S. Yocco

PEARSON Custom Publishing

PEARSON Addison Wesley

Taken from:

College Algebra in Context with Applications for the Managerial, Life, and Social Sciences, Second Edition
by Ronald J. Harshbarger and Lisa S. Yocco

Published by Addison-Wesley, Inc.
Boston, Massachusetts 02116

This special edition published in cooperation with Pearson Custom Publishing.

Printed in the United States of America

10 9 8 7 6 5 4 3 2

ISBN 0-536-43993-1

2007360885

EC

Please visit our web site at *www.pearsoncustom.com*

PEARSON Custom Publishing

PEARSON CUSTOM PUBLISHING
501 Boylston Street, Suite 900, Boston, MA 02116
A Pearson Education Company

Contents

Preface

College Algebra in Context is designed for a course in algebra that is based on real-life applications from the management, life, and social sciences, and on data analysis and modeling. The text is designed to show students how to analyze, solve, and interpret problems in this course, in future courses, and in future careers. The text features a constructive chapter-opening Algebra Toolbox, which reviews previously learned algebra concepts and presents the prerequisite skills needed for successful completion of the chapter. In addition, the authors emphasize how students can use their new knowledge in a variety of calculus courses by devoting a section to calculus preparation at the end of the text.

The text is application driven and uses numerous real-data problems to motivate interest in the skills and concepts of algebra. Modeling is introduced early, in the discussion of linear functions and in the discussion of quadratic and power functions. Additional models are introduced when exponential, logarithmic, logistic, cubic, and quartic functions are discussed. Mathematical concepts are introduced informally with the emphasis on applications. Each chapter contains real-data problems and extended application projects that can be solved by students working collaboratively.

Students can take advantage of available technology (such as graphing calculators or spreadsheet software) to solve applied problems that are drawn from real-life situations. There are examples of how technology can be used throughout the text, and step-by-step solutions using the technology are given in the accompanying technology manual. Technology is used to check answers of solved problems, to study function types, and to solve equations graphically and numerically.

Changes in the Second Edition

Many users of the first edition, and other reviewers, have made suggestions on how to enhance the presentation of material in the text. Below is a summary of some of the changes we have made, based on their suggestions and from our own classroom experiences.

- The organization and exposition are improved throughout the text, and the breadth and depth of topics have been expanded. In addition, Chapters 1 and 2 are reorganized to streamline the discussion.
- The number of Toolbox Exercises and Skills Check problems is increased by more than 25%, to provide more practice and reinforcement of algebra skills.
- New chapter-opening photographs and accompanying applications draw students into the chapter and give immediate examples of how algebra can be used in their future careers and/or studies.
- New Section Previews provide motivation for each section. Examples tied to these previews are then solved in the section.

- The Answers to Selected Exercises section, Appendix D now includes answers to all the Chapter Skills Checks and Chapter Reviews.
- A list of Key Concepts now appears at the beginning of each Algebra Toolbox.
- Mathematical models are covered earlier, in Section 1.1.
- Models of finance from former Section 1.7 are blended in throughout Chapter 1.
- Chapter 1 now contains the section on systems of linear equations in two variables.
- Chapter 1 now contains the discussion of parallel and perpendicular lines and difference quotients.
- The discussion of basic function types is consolidated into a new "Library of Functions" section in Chapter 2.
- The discussion of transformations of functions and symmetry of graphs is strengthened in Chapter 2.
- Complex numbers are introduced in the Chapter 2 Toolbox in Appendix B, and the discussion of quadratic equations with complex solutions is now in Section 2.2.

Continued Features

Features of the text include the following:

- Early on, the text discusses most types of functions, so that students can look at real application data with an understanding of what type of function describes a given data set.

  Basic functions and quadratic functions are discussed after the study of linear functions. Exponential, logarithmic, and logistic functions follow immediately.
- The development of algebra is motivated by, and used to find, the solution of real-data-based applications.

  Real-life problems are used to demonstrate the need for specific algebraic concepts and techniques. Each section begins with a motivational problem that presents a real-life setting for that topic. The problem is solved after the necessary skills are learned in that section. The aim is to prepare students to solve problems of all types by first introducing them to various functions and encouraging them to take advantage of available technology. Special business and finance models are included to demonstrate the applications of functions to the business world. At the heart of this text is its emphasis on problem solving in meaningful contexts.
- The use of technology has been integrated into the text.
- The text ends with a Preparing for Calculus section that shows how algebra skills from the first four chapters are used in a calculus context.

  Many students have difficulty in calculus because they have trouble applying algebra skills to calculus. This section reviews earlier topics and shows how they apply to the development and application of calculus.
- Many problems posed in the text are multilevel.

  Many problems require thoughtful, real-world answers under varying conditions rather than just a numerical answer. Questions like "When will this model no longer be valid?", "What additional limitations must be placed on your answer?", and "Interpret your answer," are commonplace in the text.

- Each chapter has a Chapter Summary, a Chapter Skills Check, and a Chapter Review.

  The Chapter Summary lists the key terms and formulas discussed in the chapter, with section references. Chapter Skills Check and Chapter Review exercises give additional review problems.

- The text encourages collaborative learning.

  Each chapter ends with one or more Group Activity/Extended Application requiring students to solve multilevel problems involving real data or situations, making it desirable that students collaborate in their solutions. These activities provide opportunities for students to work together to solve real problems that involve the use of technology and frequently require modeling.

- The text encourages students to improve communication skills and research skills.

  The Group Activities/Extended Applications require written reports and frequently require use of the Internet or library. Some Extended Applications require students to use the literature or the Internet to find a graph or table of discrete data describing an issue. They are then required to make a scatter plot of the data, determine the function type that is the best fit for the data, create the model, discuss how well the model fits the data, and discuss how it can be used to analyze the issue.

- Appendix A, B and C contains Algebra Toolbox section that provides the prerequisite skills needed for the successful completion of the chapter.

  Each Toolbox covers topics that are prerequisite to a college algebra course (usually found in a Chapter R or appendix of a college algebra text). Key concepts to be learned are listed at the beginning of each Toolbox, and topics are introduced "just in time" to be used in the chapter under consideration.

Acknowledgments

Many individuals contributed to the development of this textbook. We would like to thank the following reviewers, whose comments and suggestions were invaluable in preparing this text.

Jay Abramson, *Arizona State University*
Khadija Ahmed, *Monroe County Community College*
Jamie Ashby, *Texarkana College*
Sohrab Bakhtyari, *St. Petersburg College*
Jean Bevis, *Georgia State University*
Thomas Bird, *Austin Community College*
Len Brin, *Southern Connecticut State University*
Marc Campbell, *Daytona Beach Community College*
Florence Chambers, *Southern Maine Community College*
Floyd Downs, *Arizona State University*
Aniekan Ebiefung, *University of Tennessee at Chattanooga*
Toni W. Fountain, *Chattanooga State Technical Community College*
John Gosselin, *University of Georgia*
Linda Green, *Santa Fe Community College*
Richard Brent Griffin, *Georgia Highlands College*
Deborah Hanus, *Brookhaven College*
Steve Heath, *Southern Utah University*
Todd A. Hendricks, *Georgia Perimeter College*
Suzanne Hill, *New Mexico State University*
Sue Hitchcock, *Palm Beach Community College*
Arlene Kleinstein, *State University of New York—Farmingdale*
Danny Lau, *Kennesaw State University*
Ann H. Lawrence, *Wake Technical Community College*
Kit Lumley, *Columbus State University*
Antonio Magliaro, *Southern Connecticut State University*
Beverly K. Michael, *University of Pittsburgh*
Demetria Neal, *Gwinnett Technical College*
Malissa Peery, *University of Tennessee*
Ingrid Peterson, *University of Kansas*
Beverly Reed, *Kent State University*
Jeri Rogers, *Seminole Community College—Oviedo*
Michael Rosenthal, *Florida International University*
Jacqueline Underwood, *Chandler-Gilbert Community College*
Denise Widup, *University of Wisconsin—Parkside*
Sandi Wilbur, *University of Tennessee—Knoxville*

Many thanks to Jon Booze, Patricia Nelson, and Alicia Gordon for checking the accuracy of this text. Thanks also to Jennifer Albanese for her help with the manuscript and Kathy Diamond for her assistance with production. A special thanks goes to the Addison-Wesley team for their assistance, encouragement, and direction throughout this project: Greg Tobin, Anne Kelly, Becky Anderson, Joanne Ha, Kathy Manley, Dennis Schaefer, Ashley O'Shaughnessy, and Maureen McLaughlin.

Ronald J. Harshbarger
Lisa S. Yocco

CHAPTER 1

Functions, Graphs, and Models; Linear Functions

BMW, the most profitable car company in the world, introduced the new Mini Cooper in 2002 and predicted that it would break even on its development costs in 7 years. It quickly became very popular, and is now very profitable. Government officials predicted a budget surplus in 2000 (and there was) and that the Social Security Trust Fund would be empty by 2038. These predictions are made by collecting real-world data and creating mathematical models that can be used to make predictions and solve problems.

		TOPICS	APPLICATIONS
1.1	Functions and Models	Determining graphs, tables, and equations that represent functions; finding domains and ranges; evaluating functions and mathematical models	Body temperature, personal computers, stock market, men in the workforce, General Electric revenues
1.2	Graphs of Functions	Graphing and evaluating functions with technology; graphing mathematical models; aligning data; graphing data points	Cost-benefit, crime rates, bankruptcies, U.S. executions, voting
1.3	Linear Functions	Identifying and graphing linear functions; finding and interpreting intercepts and slopes; finding constant rates of change	Ritalin use, loan balances, marginal profit
1.4	Equations of Lines	Writing equations of lines; finding average rates of change; approximately linear data	Depreciation, inmate population, service call charges, Toyota sales, high school enrollment
1.5	Algebraic and Graphical Solutions of Linear Equations	Relating solutions, zeros, and x-intercepts; solving linear equations; finding functional forms of equations in two variables	Prison sentence length, credit card debt, simple interest, stock market
1.6	Fitting Lines to Data Points; Modeling Linear Functions	Linear regression; modeling with linear functions; applying linear models; goodness of fit	Earnings and gender, health service employment, U.S. population, temperature measurements
1.7	Systems of Linear Equations in Two Variables	Solving systems of linear equations in two variables with graphing, substitution, and elimination; finding break-even and market equilibrium; nonunique solutions	Concerta, break-even, market equilibrium, investments, medication

1.1 Functions and Models

Key Concepts
- Scatter plot of data
- Function definition
- Domain and range
- Independent and dependent variables
- Tests for functions
- Vertical line test
- Functional notation
- Modeling
- Mathematical model

Section Preview Body Temperatures

One indication of illness in children is elevated body temperature. There are two common measures of temperature, Fahrenheit (°F) and Celsius (°C), with temperature measured in Fahrenheit degrees in the United States and in Celsius degrees in many other countries of the world. Suppose you know that a child's normal body temperature is 98.6°F and his or her temperature is now 37°C. Does this indicate that the child is ill? To help decide this, we could find the Fahrenheit temperature that corresponds to 37°C. We could do this easily if we knew the relationship between Fahrenheit and Celsius temperature scales. In this section, we will see that the relationship between these measurements can be defined by a **function** and that functions can be defined numerically, graphically, or by an equation. We also explore how this and other functions can be applied to help solve problems that occur in real-world situations.

Function Definitions

There are several techniques to show how Fahrenheit degree measurements are related to Celsius degree measurements.

One way to show the relationship between Celsius and Fahrenheit degree measurements is by listing some Celsius measurements and the corresponding Fahrenheit measurements. These measurements, and any other real-world information collected in numerical form, are called **data**. These temperature measurements can be shown in a table (Table 1.1).

TABLE 1.1

Celsius Degrees (°C)	−20	−10	−5	0	25	50	100
Fahrenheit Degrees (°F)	−4	14	23	32	77	122	212

This relationship is also defined by the set of ordered pairs

$$\{(-20, -4), (-10, 14), (-5, 23), (0, 32), (25, 77), (50, 122), (100, 212)\}$$

We can picture the relationship between the measurements with a graph. Figure 1.1(a) shows a **scatter plot** of the data, that is, a graph of the ordered pairs as points.

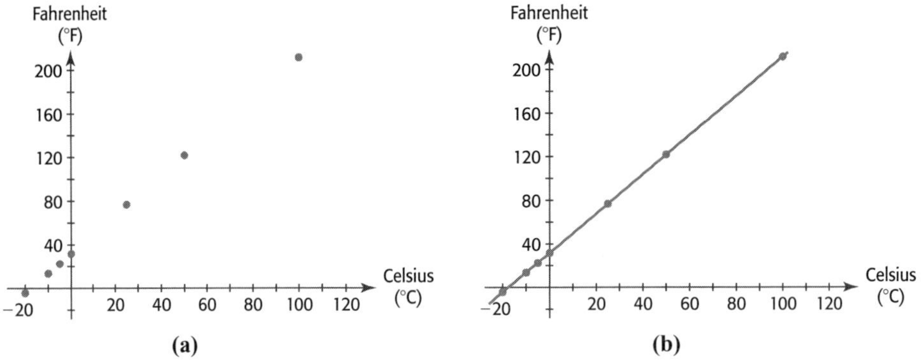

FIGURE 1.1

The points on the graph in Figure 1.1(a) can be connected by the line whose equation is

$$F = \frac{9}{5}C + 32$$

where F is the temperature measured in degrees Fahrenheit and C is the temperature measured in degrees Celsius. This line, shown in Figure 1.1(b), contains the points on the scatter plot and other points that satisfy the relationship between the measures. This relationship defines Fahrenheit (F) as a **function** of Celsius (C), with **input** C and **output** F. It is called a function because each input C results in exactly one output F.

EXAMPLE 1 BODY TEMPERATURE

Suppose a child's normal temperature is 98.6°F and the only thermometer available is Celsius and indicates that the child's temperature is 37°C. Does this reading indicate that the child's temperature is normal?

Solution

We now have a function that will give the output temperature F that corresponds to the input temperature C. Substituting 37 for C in $F = \frac{9}{5}C + 32$ gives $F = \frac{9}{5}(37) + 32 = 98.6$. This indicates that the child's temperature is normal. ■

> **FUNCTION**
>
> A function is a rule or correspondence that assigns to each element of one set (called the **domain**) exactly one element of a second set (called the **range**).
>
> The function may be defined by a set of ordered pairs, a table, a graph, or an equation.

As the definition indicates, functions can be represented by tables and graphs as well as by equations. Arrow diagrams that show how each individual input results in exactly one output can also represent functions. Each arrow in Figure 1.2(a) and in Figure 1.2(c) goes from an input to exactly one output, so each of these diagrams defines a function. On the other hand, the arrow diagram in Figure 1.2(b) does not define a function because one input, 8, results in two different outputs, 6 and 9.

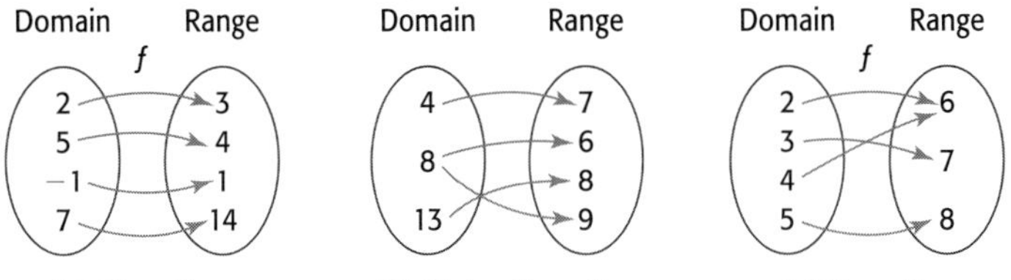

(a) Function **(b) Not a Function** **(c) Function**

FIGURE 1.2

DOMAIN

The domain of a function is the set of all inputs.

RANGE

The range of a function is the set of all outputs.

How a function is defined determines its domain and range. For instance, the domain of the function defined by Table 1.1 or by the scatter plot in Figure 1.1(a) is the finite set $\{-20, -10, -5, 0, 25, 50, 100\}$ with all values measured in degrees Celsius, and the range is the set $\{-4, 14, 23, 32, 77, 122, 212\}$ with all values measured in degrees Fahrenheit. This function has a finite number of inputs in its domain.

The function $F = \frac{9}{5}C + 32$ does not have specific inputs and outputs stated, so we may use any real number that makes sense for the input if the resulting output is a real number. Thus the domain for the function $F = \frac{9}{5}C + 32$ is the set of all real numbers. Some functions defined by equations are restricted by the context in which they are used. For example, in measuring the temperature of water, the domain of this function is limited to the real numbers from 0 to 100 and the range is limited to the real numbers from 32 to 212, because water changes state for other temperatures.

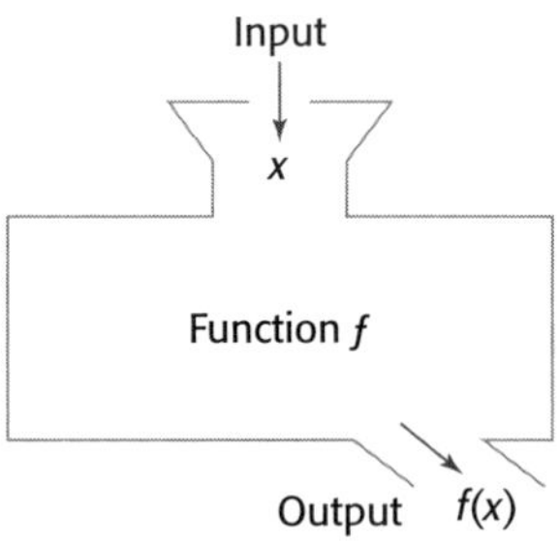

FIGURE 1.3

If x represents any element in the domain, then x is called the **independent variable** and if y represents an output of the function from an input x, then y is called the **dependent variable**. Figure 1.3 shows a general "function machine" in which the input is called x, the rule is denoted by f, and the output is symbolized by $f(x)$. The symbol $f(x)$ is read "f of x."

Tests for Functions

Functions play an important role in the solution of mathematical problems. Also, to graph a relationship using graphing calculators and computer software, it is usually necessary to express the association between the variables in the form of a function. It is therefore essential for you to recognize when a relationship is a function. Recall that a function is a rule or correspondence that determines exactly one output for each input.

EXAMPLE 2 Recognizing Functions

For each of the following, determine whether or not the indicated relationship represents a function. Explain your reasoning. For each function that is defined, give the domain and range.

a. The number N of personal computers in use worldwide determined by the year x, as defined in Table 1.2.

b. The daily profit P (in dollars) from the sale of x pounds of candy as shown in Figure 1.4.

c. The number of tons x of coal sold determined by the profit P that is made from the sale of the product, as shown in Table 1.3. (P is the input.)

d. W is a person's weight in pounds during the nth week of a diet for $n = 1$ and $n = 2$.

e. $y^2 = 3x - 3$ (x is the input).

TABLE 1.2

x Year	N Worldwide Personal Computers (millions)
1991	129.4
1992	150.8
1993	177.4
1994	208.0
1995	245.0
2000	535.6

(Source: *The Time Almanac*)

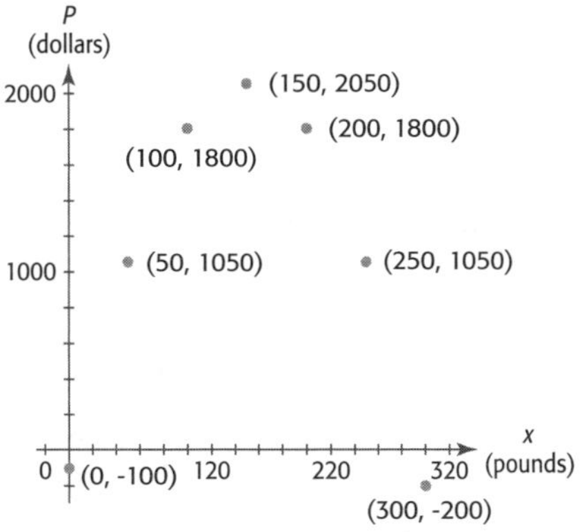

FIGURE 1.4

TABLE 1.3

x (tons)	P ($)
0	$-100{,}000$
500	109,000
1000	480,000
1500	505,000
2000	480,000
2500	109,000
3000	$-100{,}000$

Solution

a. For each year (input) listed in Table 1.2, only one value is given for the number of computers used worldwide (output), so Table 1.2 represents N as a function of x. The set $\{1991, 1992, 1993, 1994, 1995, 2000\}$ is the domain and the range is the set $\{129.4, 150.8, 177.4, 208, 245, 535.6\}$ million.

b. Each input x corresponds to only one daily profit P, so this scatter plot represents P as a function of x. The domain is $\{0, 50, 100, 150, 200, 250, 300\}$ pounds, and the range is $\{-100, -200, 1050, 1800, 2050\}$ dollars. For this function, x is the independent variable and P is the dependent variable.

c. The number of tons x of coal sold is not a function of the profit P that is made, because some values of P result in two values of x. For example, a profit of $480,000 corresponds to both 1000 tons of coal and 2000 tons of coal.

d. A person's weight varies during any week; for example, a woman may weigh 122 lb on Tuesday and 121 lb on Friday. Thus there is more than one output (weight) for each input (week) and this relationship is not a function.

e. This indicated relationship between x and y is not a function because there can be more than one output for each input. For instance, the rule $y^2 = 3x - 3$ determines both $y = 3$ and $y = -3$ for the input $x = 4$. Note that if we solve this equation for y, we get $y = \pm\sqrt{3x - 3}$, so two values of y will result for any value of $x > 1$. ■

Another way to determine whether an equation defines a function is to inspect its graph. If y is a function of x, no two distinct points (x, y) on the graph of $y = f(x)$ can have the same first coordinate. There are two points, (4, 3) and (4, −3), on the graph of

$$y^2 = 3x - 3$$

shown in Figure 1.5(a), so the equation does not represent y as a function of x (as we concluded in Example 2). In general, no two points of the graph can lie on the same vertical line if the relationship is a function.

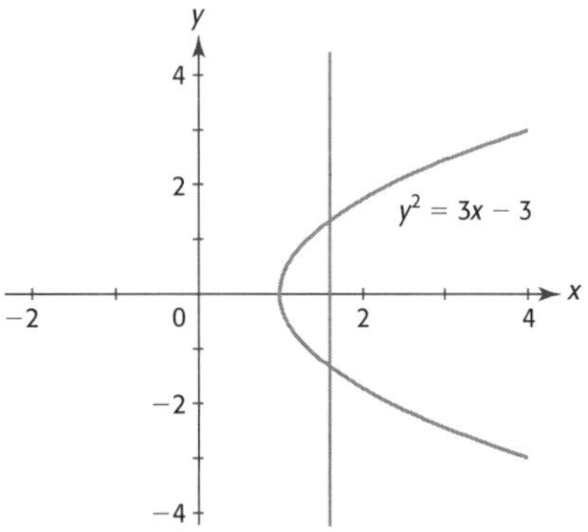

(a) Not a Function

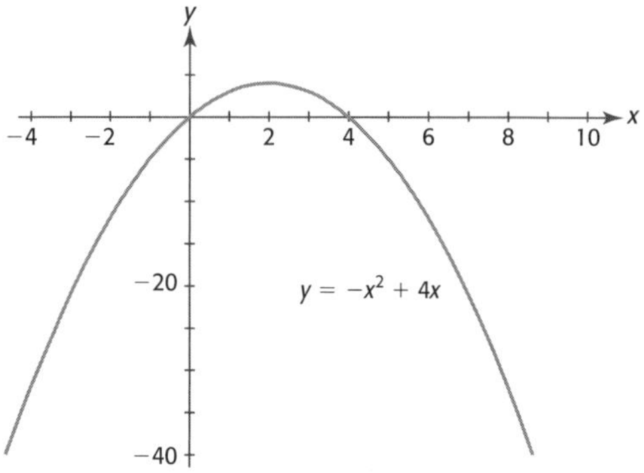

(b) Function

FIGURE 1.5

> **VERTICAL LINE TEST**
>
> A set of points in a coordinate plane is the graph of a function if and only if no vertical line intersects the graph in more than one point.

When we look at the graph of

$$y = -x^2 + 4x$$

in Figure 1.5(b), we can see that any vertical line will intersect the graph in at most one point for the portion of the graph that is visible. If we know that the graph will extend indefinitely in the same pattern, we can conclude that no vertical line will intersect the graph in two points and so the equation $y = -x^2 + 4x$ represents y as a function of x.

The graph in Figure 1.6 gives the stock prices for the first 10 months of 2000 for Coca-Cola Enterprises. The graph shows that the price of a share of Coca-Cola Enterprises stock during each of the first 10 months of 2000 is not a function. The vertical bar above each month shows that the stock has many prices between its monthly high and low. Because of the fluctuations in price during each month, the price of Coca-Cola Enterprises stock during these months is not a function of the month.

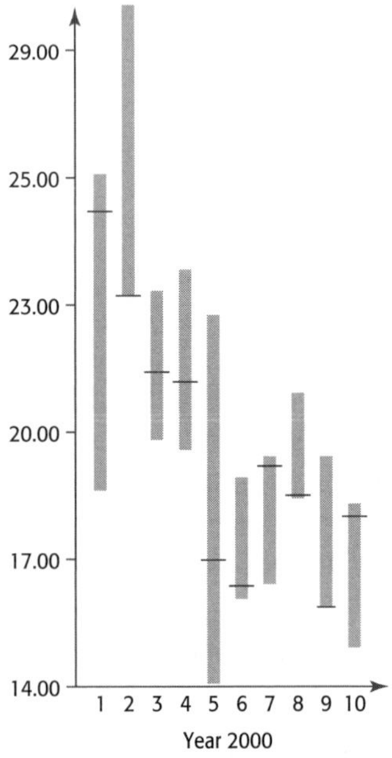

FIGURE 1.6 Coca-Cola Stock

(Source: 2000 Annual Report, Coca-Cola Enterprises)

Functional Notation

We can use the functional notation $y = f(x)$, read "y equals f of x," to indicate that the variable y is a function of the variable x. For specific values of x, $f(x)$ represents the resulting outputs, or y-values. In particular, the point $(a, f(a))$ lies on the graph of $y = f(x)$ for any number a in the domain of the function.

Thus, if

$$f(x) = 4x^2 - 2x + 3$$

then

$$f(3) = 4(3)^2 - 2(3) + 3 = 33$$
$$f(-1) = 4(-1)^2 - 2(-1) + 3 = 9$$

We can also use the functional notation $F(C) = \frac{9}{5}C + 32$ to write the Fahrenheit temperature measurement as a function of the Celsius temperature measurement. From the equation, values from Table 1.4, or points on Figure 1.7, we see that

$$F(50) = 122,\ F(0) = 32, \text{ and } F(25) = 77$$

TABLE 1.4

Celsius Degrees (°C)	−20	−10	−5	0	25	50	100
Fahrenheit Degrees (°F)	−4	14	23	32	77	122	212

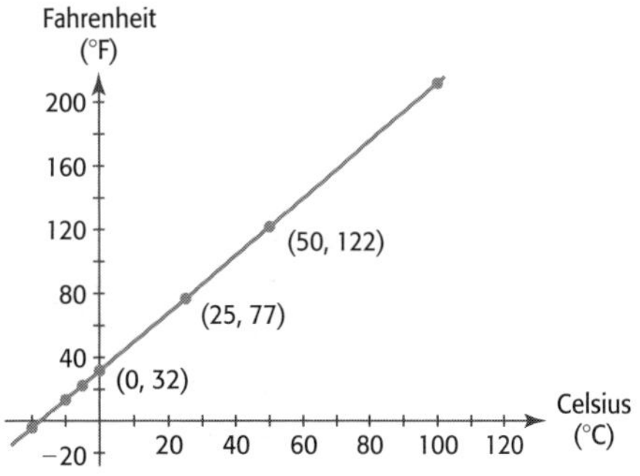

FIGURE 1.7

EXAMPLE 3 Functional Notation

Figure 1.8 shows the graph of

$$f(x) = 2x^3 + 5x^2 - 28x - 15$$

a. Use the points shown on the graph to find $f(-2)$ and $f(4)$.

b. Use the equation to find $f(-2)$ and $f(4)$.

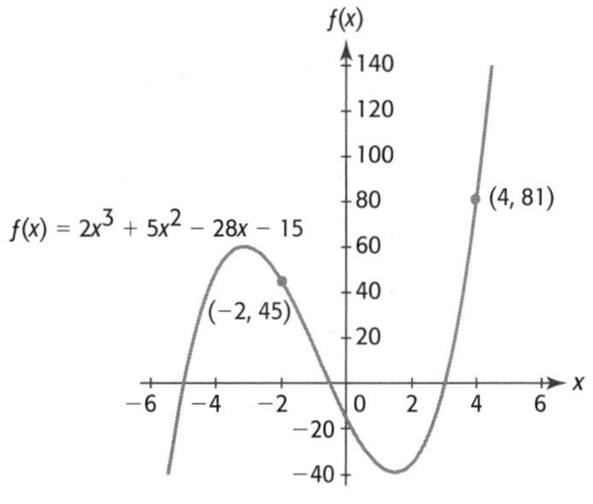

FIGURE 1.8

Solution

a. By observing Figure 1.8, we see that the point $(-2, 45)$ is on the graph of $f(x) = 2x^3 + 5x^2 - 28x - 15$, so $f(-2) = 45$. We also see that the point $(4, 81)$ is on the graph, so $f(4) = 81$.

b. $f(-2) = 2(-2)^3 + 5(-2)^2 - 28(-2) - 15 = 2(-8) + 5(4) - 28(-2) - 15 = 45.$

$$f(4) = 2(4)^3 + 5(4)^2 - 28(4) - 15 = 2(64) + 5(16) - 28(4) - 15 = 81.$$

Note that the values found by substitution agree with the y-coordinates of the points on the graph. ■

EXAMPLE 4 Men in the Workforce

The points on the graph Figure 1.9 give the number of men in the workforce (in millions) as a function g of the year for selected years t from 1890 to 1990.

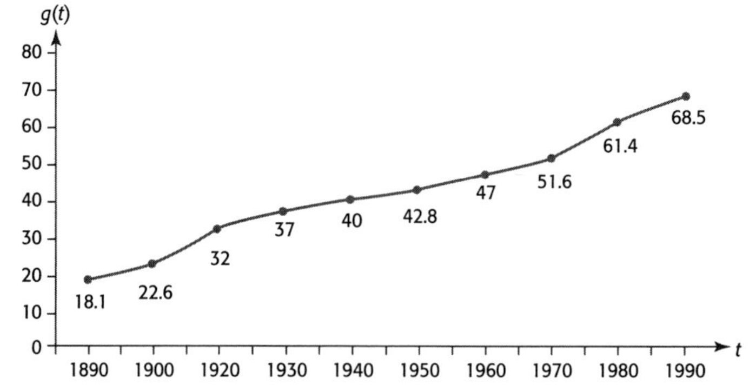

FIGURE 1.9

(Source: 1998 *World Almanac*)

a. Find and interpret $g(1940)$.

b. What is the input t if the output is $g(t) = 51.6$ million men?

c. What can be said about the number of men in the workforce during 1890–1990?

d. What is the maximum number of men in the workforce during the period shown in the graph?

Solution

a. The point above 1940 has coordinates (1940, 40), so $g(1940) = 40$. This means that there were 40 million men in the workforce in 1940.

b. The point (1970, 51.6) occurs on the graph, so $g(1970) = 51.6$ and $t = 1970$ is the input when the output is $g(t) = 51.6$.

c. The number increases during the period 1890–1990.

d. The maximum number of men in the workforce during the period is 68.5 million, in 1990. ■

Mathematical Models

The process of translating real-world information into a mathematical form so that it can be applied and then interpreted in the real-world setting is called **modeling.** As we most often use it in this text, a **mathematical model** is a functional relationship (usually in the form of an equation) that includes not only the function rule but also descriptions of all involved variables and their units of measure. For example, the function $F = \frac{9}{5}C + 32$ was used to describe the relationship between temperature scales at the beginning of the section. The **model** that describes how to convert from one measuring scale to another must include the equation ($F = \frac{9}{5}C + 32$) and a description of the variables (F is the temperature measure in degrees Fahrenheit and C is the temperature measure in degrees Celsius). A mathematical model can sometimes provide an exact description of a real situation (such as the Celsius/Fahrenheit model), but a model frequently provides only an approximate description of a real-world situation.

Suppose the General Electric Company had published only the equation

$$R(x) = 10.027x - 19{,}935.193$$

in its 1998 annual report to inform its stockholders regarding its annual revenues. The stockholders would have had no idea what the equation meant or how it could be used; that is, the equation is meaningless without further explanation. A statement such as: "For the years from 1994 through 1998, General Electric's annual revenue is given by

$$R = 10.03x - 19{,}935.19 \text{ billion dollars in year } x\text{"}$$

allows the reader to understand what the equation means and how it can be used. This statement is a **model** of General Electric's revenue for this period. (Source: Hoover's Online Capsules.)

EXAMPLE 5 General Electric Revenue

Use the model $R(x) = 10.03x - 19{,}935.19$, where R is General Electric's revenue in billions of dollars for year x from 1994 through 1998, to find General Electric's revenue corresponding to the years 1995, 1996, and 1998. (Source: Hoover's Online Capsules.)

Solution

The revenue for each of these years can be found by evaluating the function at $x = 1995$, 1996, and 1998. Table 1.5 gives the year inputs and the revenue outputs that result from substituting each of the years for x in the given equation for R. Note that the annual revenue value for year x can be denoted by $R(x)$, so that $R(1995) = 74.66$.

TABLE 1.5

Year	1995	1996	1998
G.E. Revenue ($ billions)	74.66	84.69	104.75

Thus, the revenue is *estimated* to be $74.66 billion in 1995, $84.69 billion in 1996, and $104.75 billion in 1998. Unless we know that the function is an exact fit to the data it is modeling, the outputs are only estimates of the actual data. ■

Domains and Ranges

Recall that if the domain and range of a function are not specified, it is assumed that the domain consists of all real number inputs that result in real number outputs in the range, and that the range is a subset of the real numbers. For the functions that we will be studying in this chapter, we have the following.

DOMAIN

If the domain of a function is not specified, we assume that it includes all real numbers except those that give nonreal or undefined outputs. Examples of values that are not in the domain of a function are

- Values that result in a denominator of 0
- Values that result in an even root of a negative number.

We can use the graph of a function to find or to verify its domain and range. We can usually see the interval(s) on which the graph exists and therefore agree on the subset of the real numbers for which the function is defined. This set is the domain of the function. We can also usually determine if the outputs of the graph form the set of all real numbers or some subset of real numbers. This set is the range of the function.

EXAMPLE 6 Domains and Ranges

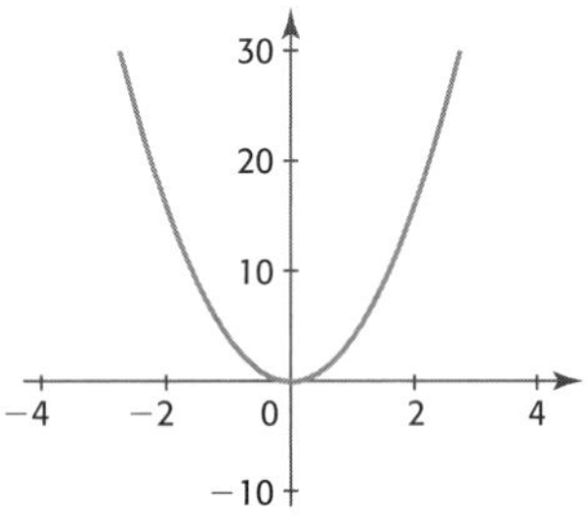

FIGURE 1.10

x	y
2	$\sqrt{2}$
3	1
4	0
4.001	Undefined (not real)

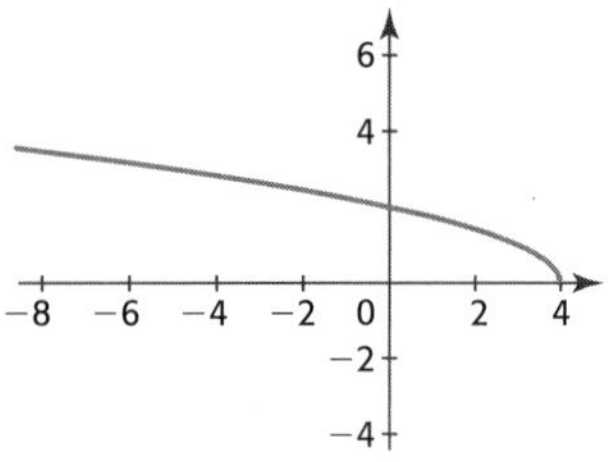

FIGURE 1.11(a)

For each of the following functions determine the domain. Determine the range of the function in parts (a) and (b).

a. $y = 4x^2$ **b.** $y = \sqrt{4 - x}$ **c.** $y = 1 + \dfrac{1}{x - 2}$

Solution

a. Because any real number input for x, when squared and multiplied by 4, results in a real number output for y, we conclude that the domain is the set of all real numbers. Because

$$y = 4x^2$$

cannot be negative for any value of x that is input, the range is the set of all nonnegative real numbers ($y \geq 0$). This can be confirmed by looking at the graph of this function, shown in Figure 1.10.

b. Because

$$y = \sqrt{4 - x}$$

cannot be a real number if $4 - x$ is negative, the only values of x that give real outputs to the function are values that satisfy $4 - x \geq 0$, or $4 \geq x$, so the domain is $x \leq 4$. Because $\sqrt{4 - x}$ (the principal square root) can never be negative, the range is $y \geq 0$. A table of selected inputs and their outputs, and the graph of this function are shown in Figure 1.11(a). They confirm that the domain is $(-\infty, 4]$ and the range is $[0, \infty)$.

c. Because the denominator of the fractional part of this function will be 0 when $x = 2$, and the outputs for every other value of x are real numbers, the domain of this function contains all real numbers except 2. The graph in Figure 1.11(b) confirms this conclusion. ■

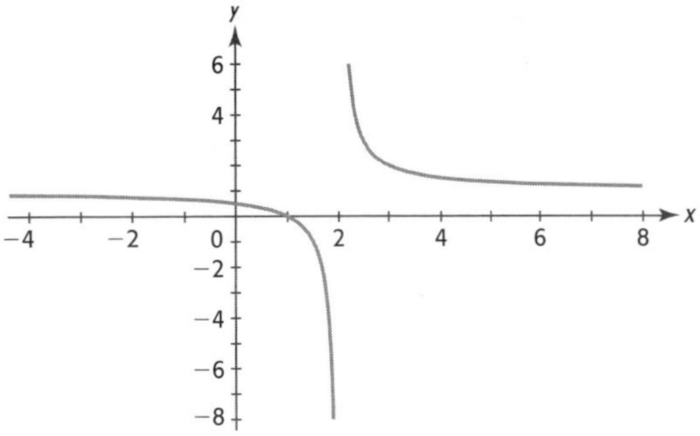

FIGURE 1.11(b)

1.1 SKILLS CHECK

Use the tables below in Exercises 1–6.

TABLE A

x	-9	-5	-7	6	12	17	20
$y = f(x)$	5	6	7	4	9	9	10

TABLE B

x	-4	-1	0	1	3	7	12
$y = g(x)$	5	7	3	15	8	9	10

1. Table A gives y as a function of x, with $y = f(x)$.
  a. Is -5 an input or an output of this function?
  b. Is $f(-5)$ an input or an output of this function?
  c. State the domain and range of this function.
  d. Explain why this relationship describes y as a function of x.

2. Table B gives y as a function of x, with $y = g(x)$.
  a. Is 0 an input or an output of this function?
  b. Is $f(7)$ an input or an output of this function?
  c. State the domain and range of this function.
  d. Explain why this relationship describes y as a function of x.

3. Use Table A to find $y = f(-9)$ and $y = f(17)$.

4. Use Table B to find $y = g(-4)$ and $y = g(3)$.

5. Does Table A describe x as a function of y? Why or why not?

6. Does Table B describe x as a function of y? Why or why not?

7. For each of the functions $y = f(x)$ described below, find $f(2)$.

  a.

x	-1	0	1	2	3
$f(x)$	5	7	2	-1	-8

  b. $y = 10 - 3x^2$

  c.

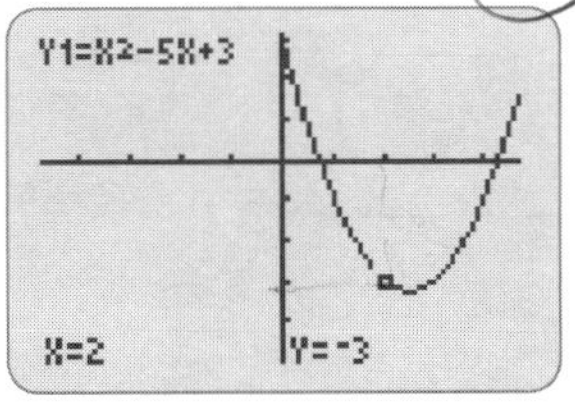

8. For each of the functions $y = f(x)$ described below, find $f(-1)$.

  a.

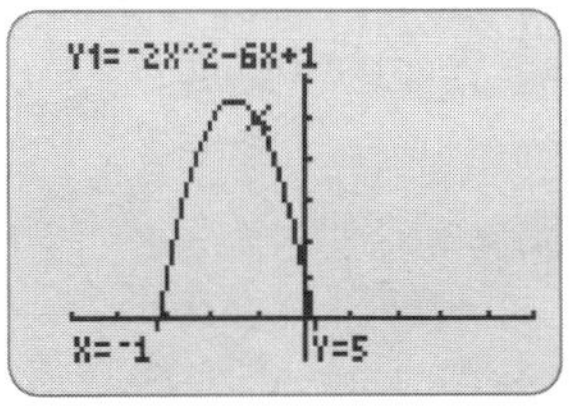

  b.

x	-3	-1	0	3
$f(x)$	12	-8	5	16

  c. $f(x) = x^2 + 3x + 8$

9. If $R(x) = 5x + 8$, find (a) $R(-3)$, (b) $R(-1)$, and (c) $R(2)$.

10. If $C(s) = 16 - 2s^2$, find (a) $C(3)$, (b) $C(-2)$, and (c) $C(-1)$.

11. Does the table below define y as a function of x? If so, give the domain and range of f. If not, state why not.

x	-1	0	1	2	3
y	5	7	2	-1	-8

12. Does the table below define y as a function of x? If so, give the domain and range of f. If not, state why not.

x	−3	2	0	3	2
$f(x)$	12	−3	5	16	4

Determine if each graph in Exercises 13 and 14 indicates that y is a function of x.

13.

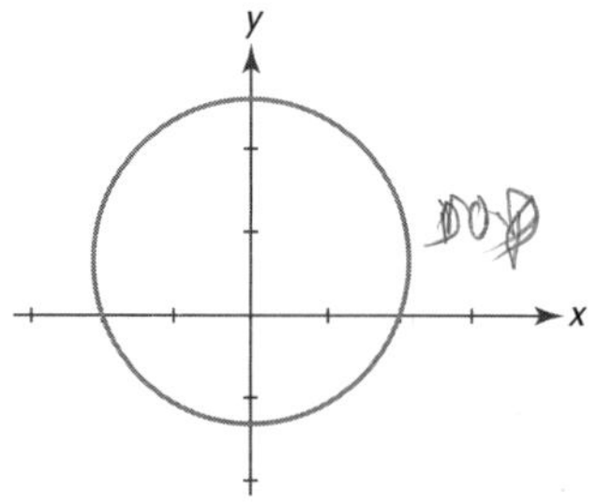

14.

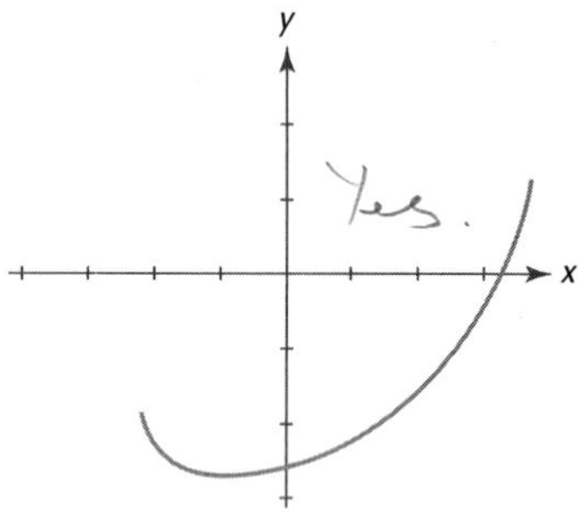

15. Determine if the graph in the figure represents y as a function of x. Explain your reasoning.

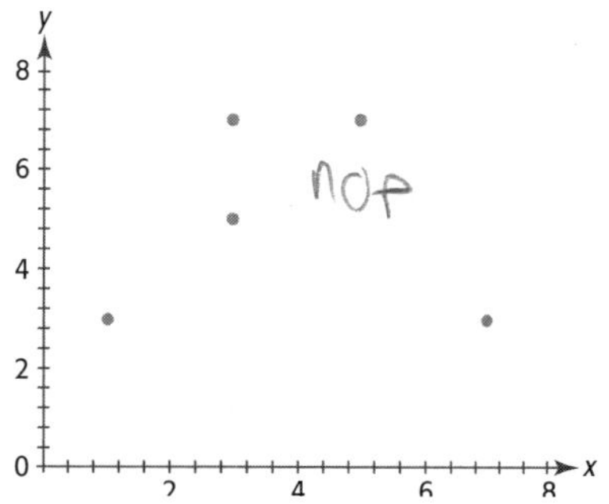

16. Determine if the graph below represents y as a function of x. Explain your reasoning.

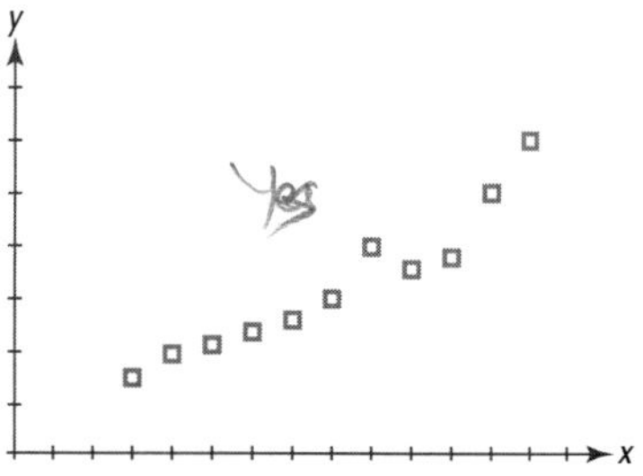

17. Which of the following sets of ordered pairs defines a function?

a. $\{(1, 6), (4, 12), (4, 8), (3, 3)\}$
b. $\{(2, 4), (3, -2), (1, -2), (7, 7)\}$

18. Which of the following sets of ordered pairs defines a function?

a. $\{(1, 3), (-2, 4), (3, 5), (4, 3)\}$
b. $\{(3, 4), (-2, 5), (4, 6), (3, 6)\}$

19. Which of the following arrow diagrams defines a function?

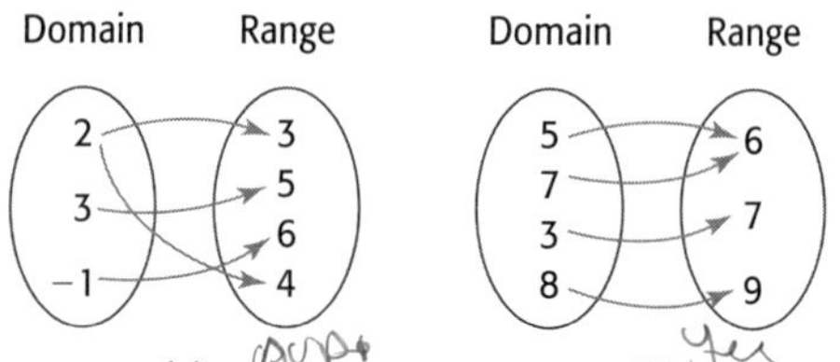

(a) **(b)**

20. Which of the following arrow diagrams defines a function?

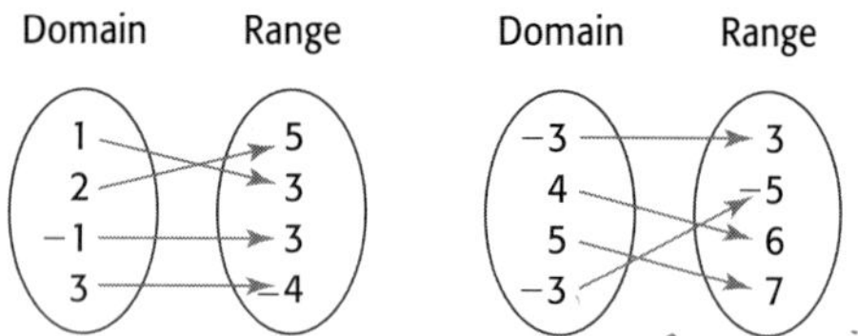

(a) **(b)**

21. Does $x^2 + y^2 = 4$ describe y as a function of x?

22. Does $x^2 + 2y = 9$ describe y as a function of x?

23. Write an equation to represent the function described by the statement: "The circumference C of a circle is found by multiplying 2π times the radius r."

24. Write a verbal statement to represent the function $D = 3E^2 - 5$.

Define, in your own words, the terms in Exercises 25–28.

25. Function

26. Domain of a function

27. Range of a function

28. Vertical-line test

1.1 EXERCISES

In Exercises 29–33, determine whether the given relationship defines a function.

29. *Stock Prices*

a. The price p at which IBM stock can be bought on a given day x.

b. The closing price p of IBM stock on a given day x.

30. *Weight*

a. The weight of a man each time he steps on a scale.

b. The weight of a man in year x.

31. *Groceries* The average price of iceberg lettuce in U.S. cities, in cents per pound, for selected months during 1999, defined by the table.

Month	Average Price (cents per pound)
February	66
March	77
April	75
May	69
June	65

(Source: Bureau of Labor Statistics)

32. *Life Insurance*

a. The monthly premium for a \$100,000 life insurance policy determined by age for males aged 26–32 (as shown in the table.)

b. Ages that get a \$100,000 life insurance policy for a monthly premium of \$11.81 (from the table).

Age (years)	26	27	28	29	30	31	32
Premium (dollars per month)	11.72	11.81	11.81	11.81	11.81	11.81	11.81

33. *Income* The average income of male workers with given years of education, as defined by the graph in the figure.

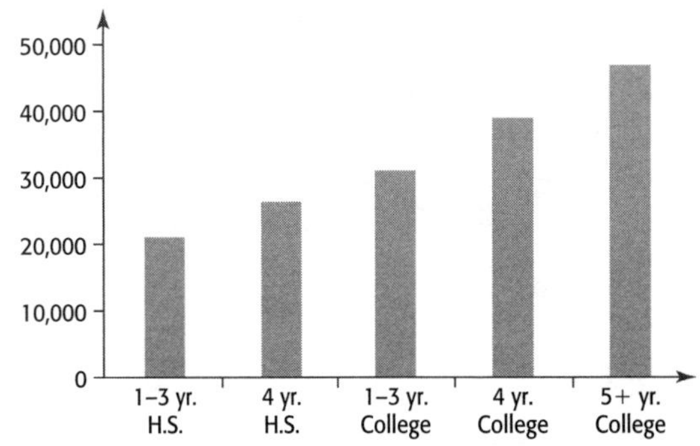

(Source: South Carolina Statistical Abstract, 1998)

34. *Temperature* The figure below shows the graph of $T = 0.43m + 76.8$, which gives the temperature T (in degrees Fahrenheit) inside a concert hall m minutes after a 40-minute power outage during a summer rock concert. Is the temperature a function of the time?

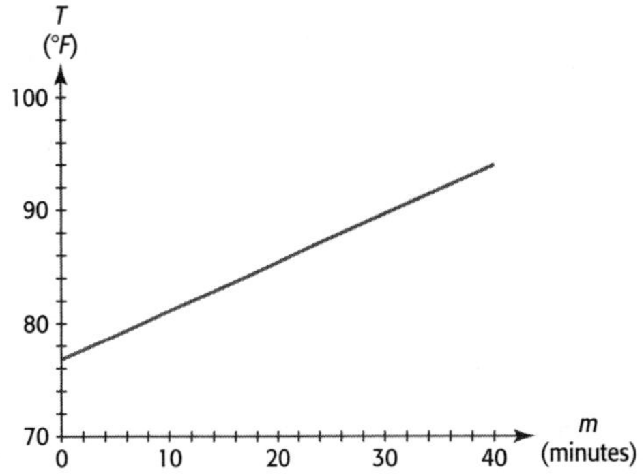

35. *Seawater Pressure* In seawater, the pressure p is related to the depth d according to the model

$$p = \frac{18d + 496}{33}$$

where d is the depth in feet and p is in pounds per square inch. Is p a function of d?

36. *Depreciation* A business property valued at \$300,000 is depreciated over 30 years by the straight-line method, so that its value x years after the depreciation began is

$$V = 300{,}000 - 10{,}000x$$

Is the value of the property a function of the number of years?

37. *Weight* During the first two weeks in May, a man weighs himself daily (in pounds) and records the data in the table below.

a. Does the table define weight as a function of the day in May?
b. What is the domain of this function?
c. What is the range?
d. During what day(s) did he weigh most?
e. During what day(s) did he weigh least?
f. He claimed to be on a diet. What is the longest period of time during which his weight decreased?

May	1	2	3	4	5	6	7
Weight (lbs)	178	177	178	177	176	176	175

May	8	9	10	11	12	13	14
Weight (lbs)	176	175	174	173	173	172	171

38. *Test Scores*

a. Is the average score on the final exam in a course a function of the average score on a placement test for the course, if the table below defines the relationship?
b. Is the average score on a placement test a function of the average score on the final exam in a course, if the table below defines the relationship?

Average Score on Math Placement Test (%)	81	75	60	90	75
Average Score on Final Exam in Algebra Course (%)	86	70	58	95	81

39. *Car Financing* If a couple finances an automobile costing \$35,000 by making equal monthly payments over 5 years at an interest rate of 10%, the balance owed at the end of t years after purchase is given in the table below.

t (years)	$B(t)$ Balance (\$)
1	29,321
2	23,047
3	16,115
4	8458.60
5	0

(Source: Sky Financial Mortgage Tables)

a. What is the balance owed by the couple at the end of 3 years?
b. What is $B(2)$ if we represent the balance at the end of t years as $B(t)$? Write a sentence that explains its meaning.
c. At the end of what year does the couple owe \$23,047?
d. What is t if $B(t) = 8458.60$?

40. *Mortgage* A couple can afford \$800 per month to purchase a home. As indicated in the table below, if they can get an interest rate of 7.5%, the number of years t that it will take to pay off the mortgage is a function of the dollar amount A of the mortgage for the home they purchase.

Amount A (\$)	t (years)
40,000	5
69,000	10
89,000	15
103,000	20
120,000	30

(Source: Comprehensive Mortgage Tables (Publication No. 492). Financial Publishing Co.)

a. If the couple wishes to finance \$103,000, for how long must they make payments? Write this correspondence in functional form if $t = f(A)$.
b. What is $f(89{,}000)$? Write a sentence that explains its meaning.
c. What value of A makes $f(A) = 30$ true?
d. Does $f(3 \cdot 40{,}000) = 3 \cdot f(40{,}000)$? Explain your reasoning.

41. *Working Age* The projected ratio of the working-age population (25- to 64-year-olds) to the elderly shown in the figure below defines the ratio as a function of the year shown. If this function is defined as $y = f(t)$ where t is the year, use the graph to answer the following:

a. What is the projected ratio of the working-age population to the elderly population in 2005?
b. Estimate $f(2005)$ and write a sentence that explains its meaning.
c. What is the domain of this function?

d. Is the projected ratio of the working-age population to the elderly increasing or decreasing over the domain shown in the figure?

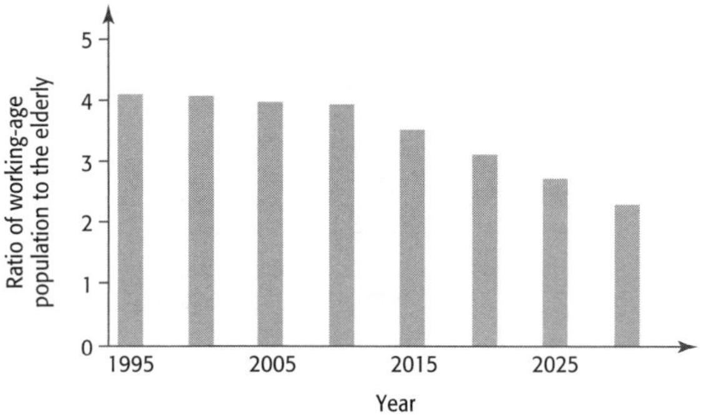

(Source: *Newsweek,* December 1999)

42. *Women in the Workforce* The points on the figure below give the number (in millions) of women in the workforce as a function f of the year for selected years from 1890 to 1990.

a. Approximately how many women were in the workforce in 1960?

b. Estimate $f(1890)$ and write a sentence that explains its meaning.

c. What is the domain of this function if we consider only the indicated points?

d. Is the number of women in the workforce increasing or decreasing over the domain shown in the graph?

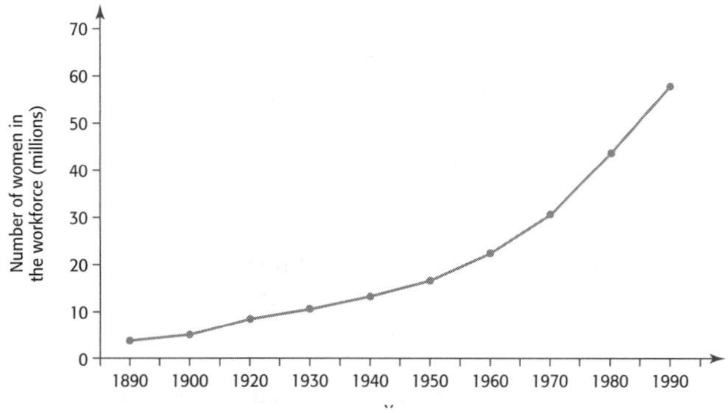

(Source: 1998 *World Almanac*)

43. *Gun Crime* The table below gives the total number of firearm crimes as a function f of the year t.

a. What is $f(1990)$?

b. What is the domain of this function?

c. What is the maximum number of firearm crimes during this period of time? In what year did it occur?

Year	Total Firearm Crimes	Year	Total Firearm Crimes
1985	340,942	1992	565,575
1986	376,064	1993	581,697
1987	365,709	1994	542,529
1988	385,934	1995	504,421
1989	410,039	1996	458,458
1990	492,671	1997	414,530
1991	548,667	1998	364,776

(Source: U.S. Department of Justice)

44. *Movie Attendance* The movie admissions (in billions) in the United States for selected years is given in the table below as a function of the number of years from 1980.

a. What is the domain of this function?

b. What is the range of this function?

c. What is the output when the input is 10, and what does this mean?

d. What can be said about the number of movie admissions over this period of time?

Years from 1980	Movie Admissions (billions of people)
0	1.02
5	1.06
10	1.10
15	1.26
18	1.48

(Source: Motion Picture Association of America)

45. *Farm Workers* The points on the following figure show the percent p of U.S. workers in farm occupations for selected years t.

a. Is the percent of U.S. workers in farm occupations a function of the year?

b. What is $f(1840)$ if $p = f(t)$?

c. What is t if $f(t) = 27$?

d. Write a sentence explaining the meaning of $f(1960) = 6.1$.

e. Describe the change in the percent of U.S. workers in farm occupations.

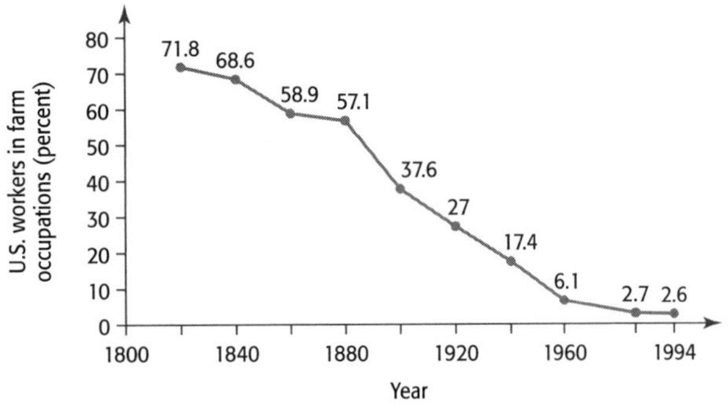

(Source: U.S. Department of Agriculture)

46. *Internet Use* The figure below gives the number of millions of U.S. homes using the Internet for the years 1995–1998. If the number of millions of U.S. homes is the function $f(x)$, where x is years,

a. How many homes used the Internet during 1995?
b. Find $f(1997)$ and explain its meaning.
c. In what year did 31.7 million homes use the Internet?
d. Is this function increasing or decreasing? What do you think has happened to home use of the Internet since 1998?

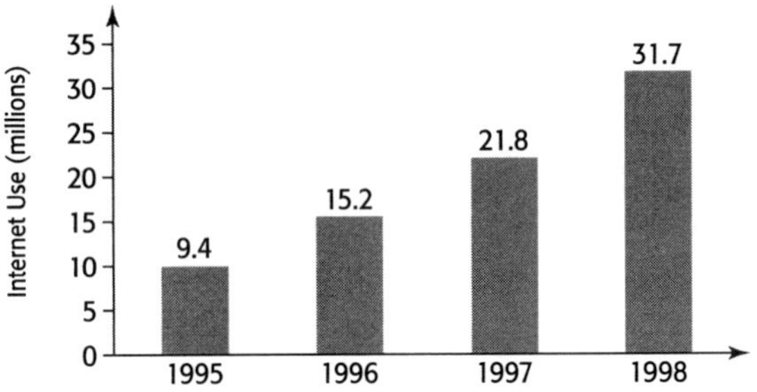

(Source: *Cyber Atlas*)

47. *Social Security Funding* Social Security benefits are funded by individuals who are currently employed. The following graph, based on known data and projections into the future, defines a function that gives the number of workers n supporting each retiree as a function of time t (given by the calendar year). Denote this function by $n = f(t)$.

a. Find $f(1990)$ and explain its meaning.
b. In what year will the number of workers supporting each retiree equal 2?
c. What does this function tell us about Social Security in the future?

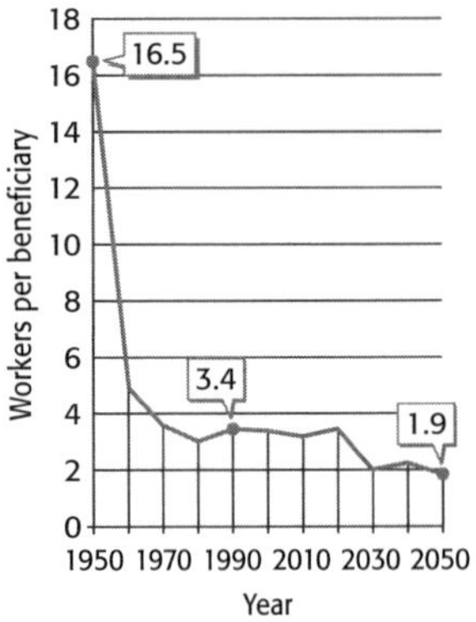

(Source: Social Security Administration)

48. *Teen Pregnancy* The points on the figure below give the number of pregnancies per 1000 U.S. girls (ages 15–19) for the years 1980–1996.

a. Find the output when the input is 1995 and explain its meaning.
b. For what years was the rate 113?
c. During what years did the pregnancy rate increase?
d. For what year is the pregnancy rate at its maximum?

Teen Pregnancy Rate per 1000 Girls

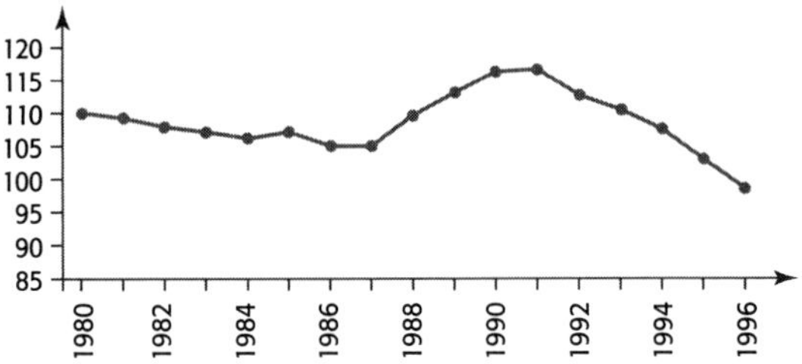

(Source: The National Campaign to Prevent Teen Pregnancy)

49. *Revenue* The revenue from the sale of specialty golf hats is give by the function $R(x) = 32x$, where x is the number of hats sold.

a. What is $R(200)$? Interpret this result.
b. What is the revenue from the sale of 2500 hats? Write this in functional notation.

50. *Cost* The cost from the production of specialty golf hats is given by the function $C(x) = 4000 + 12x$, where x is the number of hats produced.

a. What is $C(200)$? Interpret this result.
b. What is the cost from the production of 2500 hats? Write this in functional notation.

51. *Profit* The profit from the production and sale of ipod players is given by the function $P(x) = 450x - 0.1x^2 - 2000$, where x is the number of units produced and sold.
   a. What is $P(500)$? Interpret this result.
   b. What is the profit from the production of 4000 units? Write this in functional notation.

52. *Profit* The profit from the production and sale of specialty golf hats is given by the function $P(x) = 20x - 4000$, where x is the number of hats produced and sold.
   a. What is $P(200)$? Interpret this result.
   b. What is the profit from the production of 2500 hats? Write this in functional notation.

53. *Utilities* An electric utility company determines the monthly bill by charging 10.5 cents per kilowatt-hour (KWh) used plus a base charge of \$5.80 per month. Thus the monthly charge is given by the function

$$f(W) = 0.105W + 5.80 \text{ dollars}$$

where W is the number of kilowatt-hours.
   a. Find $f(1000)$ and explain what it means.
   b. What is the monthly charge if 1500 KWh are used?

54. *Profit* The daily profit from producing and selling Blue Chief bicycles is given by

$$P(x) = 32x - 0.1x^2 - 1000$$

where x is the number produced and sold and $P(x)$ is in dollars.
   a. Find $P(100)$ and explain what it means.
   b. Find the daily profit from producing and selling 160 bicycles.

55. *Projectiles* Suppose a ball thrown into the air has its height (in feet) given by the function

$$h(t) = 6 + 96t - 16t^2$$

where t is in seconds after the ball is thrown.
   a. Find $h(1)$ and explain what it means.
   b. Find the height of the ball 3 seconds after it is thrown.
   c. Test other values of $h(t)$ to decide if the ball eventually falls. When does the ball stop climbing?

56. *Educational Attainment* The table below shows the percentage of 25- to 29-year-old U.S. white, black, and Hispanic males who are identified in selected years between 1971 and 1999 as having completed at least some college.
   a. If the function f is the percent of white males who are identified in year x as having completed at least some college, find $f(1995)$ and $f(1999)$.
   b. If the function g is the percent of Hispanic males who are identified in year x as having completed at least some college, find $g(1995)$ and explain what it means.
   c. If the function h is the percent of black males who are identified in year x as having completed at least some college, find $h(1983)$ and $h(1999)$.
   d. For what value of x is $f(x) = 51.5$? Write the result in a sentence.

Year	White Male	Black Male	Hispanic Male
1971	50.2	29.0	38.3
1973	53.0	33.5	39.4
1975	57.3	41.0	50.4
1977	59.9	44.2	42.6
1979	59.4	40.7	50.7
1981	54.1	43.0	41.7
1983	53.4	42.0	41.1
1985	52.5	42.4	45.9
1987	51.5	38.4	46.3
1989	53.4	42.2	44.8
1991	54.7	38.3	40.9
1993	60.3	43.6	46.1
1995	62.6	51.2	48.0
1997	66.9	50.2	51.9
1999	66.1	52.1	47.7

(Source: *March Current Population Surveys*, U.S. Department of Commerce, Bureau of the Census)

57. *Body-Heat Loss* The description of body-heat loss due to convection involves a coefficient of convection K_c, which depends on wind speed s according to the equation $K_c = 4\sqrt{4s + 1}$.
   a. Is K_c a function of s?
   b. What is the domain of the function defined by this equation?
   c. What restrictions do the physical nature of the model put on the domain?

58. *Test Reliability* If a test that has reliability 0.7 has the number of questions increased by a factor n, the reliability R of the new test is given by

$$R(n) = \frac{0.7n}{0.3 + 0.7n}$$

a. What is the domain of the function defined by this equation?
b. If the application used here requires that the size of the test be increased, that values of n make sense in the application?

59. *Cost-Benefit* Suppose that the cost C (in dollars) of removing $p\%$ of the particulate pollution from the smokestack of a power plant is given by

$$C(p) = \frac{237{,}000p}{100 - p}$$

a. Use the fact that percent is measured between 0 and 100 and find the domain of this function.
b. Evaluate $f(60)$ and $f(90)$.

60. *Revenue* Suppose the number of units of a product that is demanded by consumers is given as a function of p by

$$q = \frac{100}{\sqrt{2p + 1}} - 1$$

where p is the price charged per unit.
a. What is the domain of the function defined by this equation?
b. What should the domain and range of this function be to make sense in the application?

61. *Postal Restrictions* Some postal restrictions say that the size of the largest box that can be shipped has its length (longest side) plus its girth (distance around the box in the other two dimensions) equal to 108 inches.

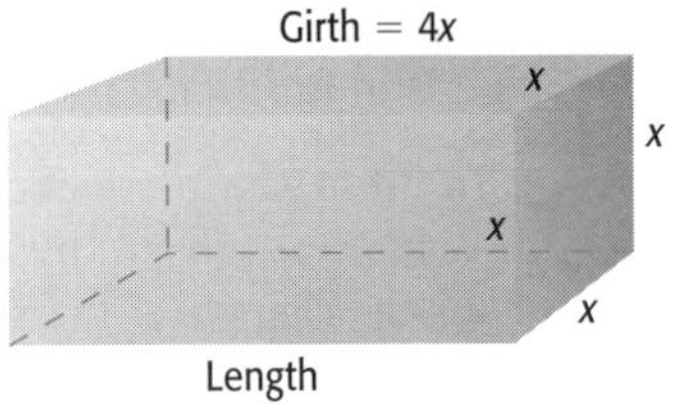

If a box has a square cross section that is x inches on each side, then the volume of the box is given by $V = x^2(108 - 4x)$ cubic inches.
a. Find $V(12)$ and $V(18)$.
b. What restrictions must be placed on x to satisfy the conditions of this model?
c. Create a table of functional values to investigate the value of x that maximizes the volume. What are the dimensions of the box that has the maximum volume?

62. *Height of a Bullet* The height of a bullet shot into the air is given by $S(t) = -4.9t^2 + 98t + 2$, where t is the number of seconds after it is shot and $S(t)$ is in meters.
a. Find $S(0)$ and interpret it.
b. Find $S(9)$, $S(10)$, and $S(11)$.
c. What appears to be happening to the bullet at 10 seconds? Evaluate the function at some additional times near 10 seconds to confirm your conclusion.

1.2 Graphs of Functions

Key Concepts
- Point-plotting method
- Complete graph
- Using a calculator to draw a graph
- Viewing windows
- Spreadsheets
- Aligning data
- Evaluating functions with a calculator
- Graphing data points

Section Preview Bankruptcies

Using data from 1990 through 1999, a model can be created that gives the number of bankruptcies per year of publicly traded companies. If the data is **aligned** so that the input values (x) represent the number of years after 1990, the bankruptcies can be modeled by the function

$$y = 2.84x^2 - 24.40x + 128.94$$

To graph this function for values of x representing 1990 through 1999, we use x-values from 0 through 9 in the function to find the corresponding y-values. These points can be used to sketch the graph of the function. If technology is used to graph the function, the **viewing window** on which the graph is shown can be determined by these values of x and y.

In this section, we will graph functions by point plotting and with technology. We will graph application functions on windows determined by the context of the applications, and we will align data so that smaller inputs can be used in models. We will also graph data points.

Graphs of Functions

If an equation defines y as a function of x, we can sketch the graph of the function by plotting enough points to determine the shape of the graph and then drawing a line or curve through the points. This is called the **point-plotting method** of sketching a graph.

EXAMPLE 1 Graphing an Equation by Hand

a. Graph the equation $y = x^2$ by drawing a smooth curve through points determined by integer values of x between 0 and 3.

b. Graph the equation $y = x^2$ by drawing a smooth curve through points determined by integer values of x between -3 and 3.

Solution

a. We use the values in Table 1.6 to determine the points. The graph of the function drawn through these points is shown in Figure 1.12(a).

TABLE 1.6

x	0	1	2	3
$y = x^2$	0	1	4	9
Points	(0, 0)	(1, 1)	(2, 4)	(3, 9)

b. We use Table 1.7 to find the additional points. The graph of the function drawn through these points is shown in Figure 1.12(b). ■

TABLE 1.7

x	-3	-2	-1	0	1	2	3
$y = x^2$	9	4	1	0	1	4	9
Points	$(-3, 9)$	$(-2, 4)$	$(-1, 1)$	$(0, 0)$	$(1, 1)$	$(2, 4)$	$(3, 9)$

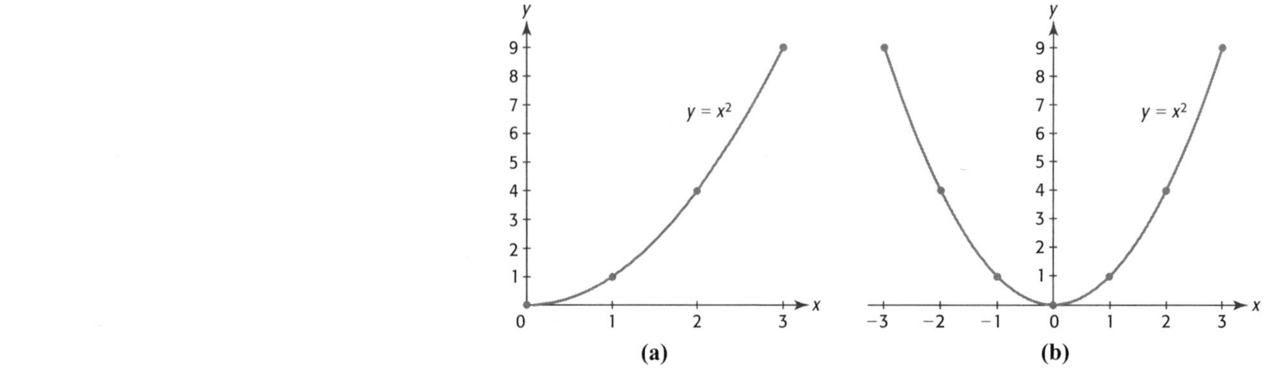

FIGURE 1.12

Although neither graph in Figure 1.12 shows all of the points satisfying the equation $y = x^2$, the graph in Figure 1.12(b) is a much better representation of the function (as we will learn later). When enough points are connected to determine the shape of the graph, the important parts of the graph, and to suggest what the unseen parts of the graph look like, the graph is called **complete**.

COMPLETE GRAPH

A graph is a complete graph if it shows the basic shape of the graph and important points on the graph (including points where the graph crosses the axes and points where the graph turns), and suggests what the unseen portions of the graph will be.

Being able to draw complete graphs, with or without the aid of technology, improves with experience and requires knowledge of the shape of the basic function(s) involved in the graph. As we learn more about types of functions, we will be better able to determine when a graph is complete.

Graphing with Technology

Obtaining a sufficient number of points to sketch a graph by hand can be time consuming. Computers and graphing calculators have graphing utilities that can be used to plot many points quickly, thereby showing the graph with minimal effort. However, graphing with technology involves much more than pushing a few buttons. Unless you know what to enter into the calculator or computer (that is, the domain of the function and sometimes, the range), you cannot see or use the graph that is drawn. We suggest the following steps for graphing most functions with a graphing utility.

USING A GRAPHING CALCULATOR TO DRAW A GRAPH

1. Write the function with x representing the independent variable and y representing the dependent variable. Solve for y, if necessary.
2. Enter the function in the equation editor of the graphing utility. Use parentheses* as needed to ensure the mathematical correctness of the expression.
3. Activate the graph by pressing the ZOOM or GRAPH key. Most graphing utilities have several preset **viewing windows** under ZOOM, including the **standard viewing window** that gives the graph of a function on a coordinate system in which the x-values range from -10 to 10 and the y-values range from -10 to 10.
4. To see parts of the graph of a function other than those shown in a standard window, set the x- and y-boundaries of the viewing window before pressing GRAPH. Viewing window boundaries are discussed below and in the technology supplement. As you gain more knowledge of graphs of functions, determining viewing windows that give complete graphs will be less complicated.

*Parentheses should be placed around numerators and/or denominators of fractions, fractions that are multiplied by a variable, exponents consisting of more than one symbol, and in other situations where the order of operations needs to be indicated.

Although the standard viewing window is a convenient window to use, it may not show the desired graph. To see parts of the graph of a function other than those that might be shown in a standard window, we change the x- and y-boundaries of the viewing window. The values that define the viewing window can be set manually or by using the ZOOM key. The boundaries of a viewing window are:

x_{min}:	the smallest value on the x-axis (the left boundary of the window)	y_{min}:	the smallest value on the y-axis (the bottom boundary of the window)
x_{max}:	the largest value on the x-axis (the right boundary of the window)	y_{max}:	the largest value on the y-axis (the top boundary of the window)
x_{scl}:	the distance between ticks on the x-axis (helps visually find x-intercepts and other points)	y_{scl}:	the distance between ticks on the y-axis (helps visually find y-intercept and other points)

When showing viewing window boundaries on calculator graphs in this text, we write them in the form:

$$[x_{min}, x_{max}] \text{ by } [y_{min}, y_{max}]$$

To illustrate these ideas, consider the graph of $y = x^3 - 3x^2 - 13$. A graph of this function in the standard viewing window appears to be a line (Figure 1.13(a)). By manually setting the viewing window as follows: $x_{min} = -10$, $x_{max} = 10$, $y_{min} = -25$, $y_{max} = 10$ (denoted $[-10, 10]$ by $[-25, 10]$), we obtain the graph in Figure 1.13(b). This window gives a better picture of the actual curvature of the graph of this equation. As we learn more about functions, we will see that the graph in Figure 1.13(b) is a complete graph of this function.

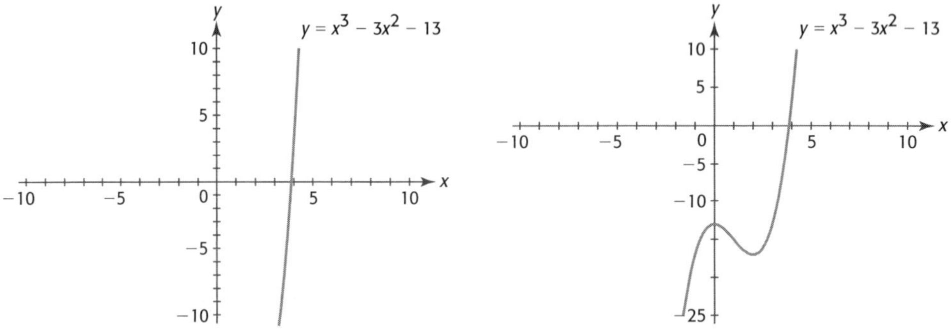

viewing window: [−10, 10] by [−10, 10]
This graph looks like a line
(a)

viewing window: [−10, 10] by [−25, 10]
This is a better graph
(b)

FIGURE 1.13

Different viewing windows give different views of a graph. There are usually many different viewing windows that give complete graphs for a particular function, but some viewing windows do not show all the important parts of a graph.

EXAMPLE 2 Cost-Benefit

Suppose that the cost C of removing $p\%$ of the pollution from drinking water is given by the model

$$C = \frac{5350p}{100 - p} \text{ dollars}$$

a. Use the restriction on p to determine the limitations on the horizontal-axis values (which are the x-values on a calculator).

b. Graph the function on the viewing window [0, 100] by [0, 50,000]. Why is it reasonable to graph this model on a viewing window with the limitation $C \geq 0$?

c. Find the point on the graph that corresponds to $p = 90$. Explain the coordinates of this point.

Solution

a. Because p represents the percent of pollution removed, it is limited to values from 0 to 100. However, 100% of the pollution cannot be removed, so p is restricted to $0 \leq p < 100$ for this model.

b. The graph of the function is shown in Figure 1.14(a), with x representing p and y representing C. The interval [0, 100) contains all the possible values of p. The value of C is bounded below by 0 because it represents the cost of removing the pollution, which cannot be negative.

c. We can find (or estimate) the output of a function $y = f(x)$ at specific inputs with a graphing utility. We do this with [TRACE], [CALC] [VALUE] or [TABLE]. Using [TRACE] or [CALC] [VALUE] with the x-value 90 gives the point (90, 48,150) (see Figure 1.14(b)). Because the values of p are represented by x-values and the values of C are represented by y, the coordinates of the point tell us that the cost of removing

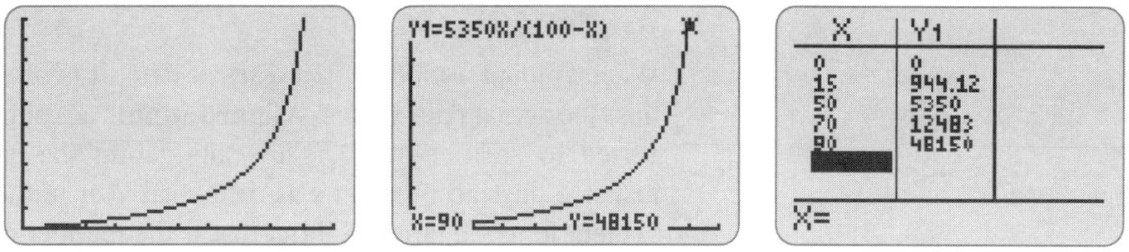

[0, 100] by [0, 50,000] (a)

[0, 100] by [0, 50,000] (b)

(c)

FIGURE 1.14

90% of the pollution from the drinking water is \$48,150. Figure 1.14(c) shows the value of y for $x = 90$ and other values in a calculator table. ■

Spreadsheet Solution

We have briefly described how to use a graphing calculator to graph. Computer software of several types can be used to create more accurate and better-looking graphs. Software such as Scientific Notebook, Maple, and Mathematica can be used to create graphs, and **spreadsheets** like Excel can also be used.

Table 1.8 shows a **spreadsheet** with inputs and outputs for the function $C = \frac{5350p}{100 - p}$ given in Example 2. When evaluating a function with a spreadsheet, we use the cell location of the data to represent the variable. Thus to evaluate $C = \frac{5350p}{100 - p}$ at $p = 0$, we type 0 in cell A2 and $= 5350*A2/(100 - A2)$ in cell B2.
Typing 10 in cell A3 and using the fill-down capacity of the spreadsheet gives the outputs of this function for $p = 0$ to $p = 90$ in increments of 10 (see Table 1.8). These values can be used to create the graph of this function on Excel (Figure 1.15).

TABLE 1.8

	A	B
1	p	$C = 5350p/(100 - p)$
2	0	0
3	10	594.4444
4	20	1337.5
5	30	2292.857
6	40	3566.667
7	50	5350
8	60	8025
9	70	12483.33
10	80	21400
11	90	48150

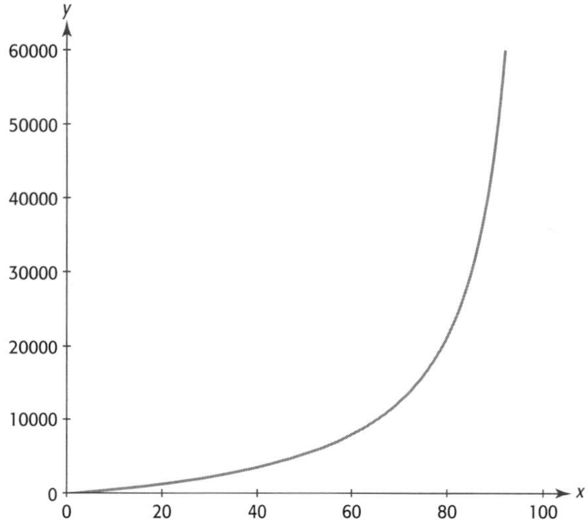

FIGURE 1.15

Aligning Data

When finding a model to fit a set of data, it is often easier to use aligned inputs rather than the actual data values. **Aligned inputs** are simply input values that have been converted to smaller numbers by subtracting the same number from each input. For instance, instead of using x as the actual year in the following example, it is more convenient to use an aligned input that is the number of years after 1950.

EXAMPLE 3 Voting

Between 1950 and 1996, the percent of the voting population who voted in presidential elections (during election years) is given by

$$f(x) = 65.40 - 0.36x$$

where x is the number of years after 1950. (Source: Federal Election Commission.)

a. What are the values of x that correspond to the years 1960 and 1996?

b. Find $f(10)$ and explain its meaning.

c. If this model is accurate for 1984, find the percent of the voting population who voted in the 1984 presidential election.

Solution

a. Because 1960 is 10 years after 1950, $x = 10$ corresponds to the year 1960 in the aligned data. Using similar reasoning, 1996 corresponds to $x = 46$.

b. An input value of 10 represents 10 years after 1950, which is the year 1960. To find $f(10)$, we substitute 10 for x to obtain $f(10) = 65.40 - 0.36(10) = 61.8$. One possible explanation of this answer is this: Approximately 62% of the voting population voted in the 1960 presidential election.

c. To find the percent in 1984, we first calculate the aligned input for the function. Because 1984 is 34 years after 1950, we use $x = 34$. Thus, we find

$$f(34) = 65.40 - 0.36(34) = 53.16$$

So, if this model is accurate for 1984, approximately 53% of the voting population voted in the 1984 presidential election. ■

Determining Viewing Windows

Finding the functional values (y-values) for selected inputs (x-values) can be useful when setting viewing windows for graphing utilities.

> **TECHNOLOGY NOTE**
>
> Once the input values for a viewing window have been selected, [TRACE], [CALC] [VALUE], or [TABLE] can be used to find enough output values to determine a y-view that gives a complete graph.*

*It is occasionally necessary to make more than one attempt to find a window that gives a complete graph.

EXAMPLE 4 Bankruptcies

The number of bankruptcies per year of publicly traded companies during the years 1990 to 1999 can be modeled by the function

$$y = 2.8447x^2 - 24.4023x + 128.9364$$

where x is the number of years after 1990.

a. Choose an appropriate viewing window for the 10-year period and graph the function on a graphing calculator.

b. Use the model to estimate the number of bankruptcies of publicly traded companies in 1996.

Solution

a. The viewing window should include values of x that represent the years 1990 to 1999, so the x-view should include the aligned values $x = 0$ to $x = 9$. The y-view should not include negative numbers and should include the outputs obtained from the inputs between 0 and 9. We can use TRACE, CALC VALUE, or TABLE with some or all of the integers 0 through 9 to find corresponding y-values (regardless of how the y-view is set when we are evaluating).* These evaluations indicate that the y-view should include values from about 76 through about 140. The graph of the function with the boundaries $x_{min} = 0$, $x_{max} = 9$, $y_{min} = 66$ and $y_{max} = 150$ is shown in Figure 1.16(a).

b. The year 1996 is represented by $x = 6$. Evaluating the function with TRACE or CALC VALUE, on the graph (see Figure 1.16(b)) gives an output of approximately 85 when $x = 6$. Figure 1.16(c) shows the values of the function at $x = 0$, 9, and 6 using TABLE. Thus, we estimate that the number of bankruptcies of publicly traded companies in 1996 was 85. ■

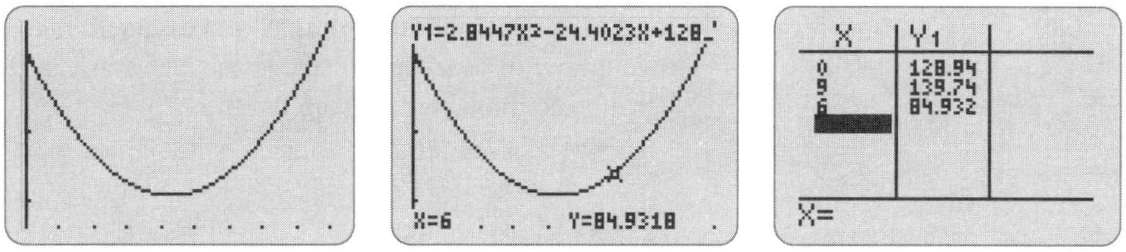

[0, 9] by [66, 150]

(a) (b) (c)

FIGURE 1.16

Spreadsheet Solution

Excel can also be used to graph and evaluate the function $y = 2.8447x^2 - 24.4023x + 128.9364$, discussed in Example 4. The Excel graph is shown in Figure 1.17, and an Excel spreadsheet with values of the function at selected values of x is shown in Table 1.9.

*TRACE may only give an approximate value of a function. On some calculators, a function can be evaluated exactly by pressing TRACE, the x-value, and ENTER or by pressing CALC VALUE, the x-value, and ENTER.

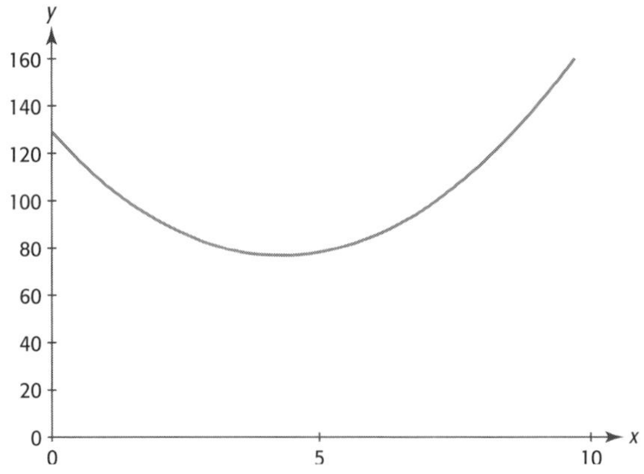

FIGURE 1.17

TABLE 1.9

	A	B
	x	f(x)
1	0	128.9364
2	1	107.3788
3	2	91.5106
4	3	81.3318
5	4	76.8424
6	5	78.0424
7	6	84.9318
8	7	97.5106
9	8	115.7788
10	9	139.7364

Graphing Data Points

Graphing utilities can also be used to create **lists** of numbers and to create graphs of the data stored in the lists. To see how lists and scatter plots are created, consider the following example.

EXAMPLE 5 U.S. Executions

Table 1.10 gives the number of executions in the United States for the decades beginning in 1930. Column 1 contains the decades. The data are aligned in column 2 to give the number of years from 1900 to the *end* of the given decade. Column 3 gives the number of executions in the given decade.

TABLE 1.10

Decade	Years from 1900 to End of Decade	Number of Executions
1930–1939	40	1690
1940–1949	50	1284
1950–1959	60	717
1960–1969	70	191
1970–1979	80	3
1980–1989	90	117
1990–1999	100	518

(Source: "The Death Penalty in the U.S.," www.clarkprosecutor.org)

a. Enter the number of years from 1900 to the end of the decade in list L1 and the number of executions in that decade in list L2.

b. Use a graphing command* to create the graph of these data points (called a scatter plot).

Solution

a. Figure 1.18(a) shows the data from column 2 of Table 1.10 in L1 and the data from column 3 of the table in L2.

b. The scatter plot is shown in Figure 1.18(b), with each entry in L1 represented by an x-coordinate on the graph and the corresponding element in L2 represented by the corresponding y-coordinate on the graph. The window can be set automatically or manually to include x-values from L_1 and y-values from L_2. ■

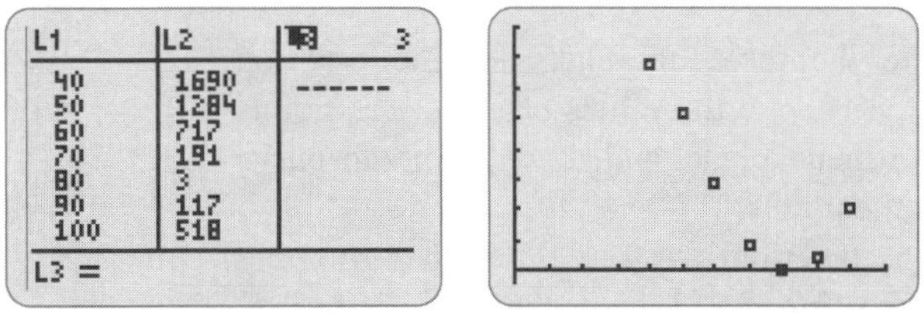

[0, 110] by [−100, 2000]

(a) (b)

FIGURE 1.18

Spreadsheet Solution

Figure 1.19 shows the values from Table 1.10 in an Excel spreadsheet and the scatter plot (graph) of the data.

	A	B
1	40	1690
2	50	1284
3	60	717
4	70	191
5	80	3
6	90	117
7	100	518

(a)

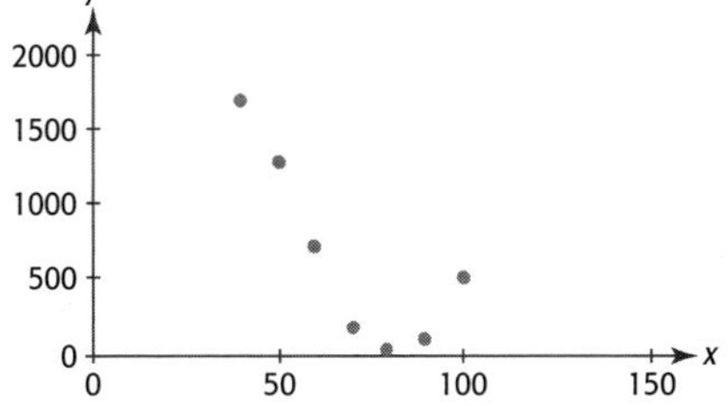

(b)

FIGURE 1.19

*Many calculators have a STAT PLOT command, and Excel has an XY (Scatter) command.

1.2 SKILLS CHECK

1. a. Complete the table of values for the function $y = x^3$, plot the values of x and y as points in a coordinate plane, and draw a smooth curve through the points.
   b. Graph the function $y = x^3$ with a graphing calculator using the viewing window $x_{min} = -4$, $x_{max} = 4$, $y_{min} = -30$, $y_{max} = 30$.
   c. Compare the graphs in parts (a) and (b).

x	-3	-2	-1	0	1	2	3
y							

2. a. Complete the table of values for the function $y = 2x^2 + 1$, plot the values of x and y as points in a coordinate plane, and draw a smooth curve through the points.
   b. Graph the function $y = 2x^2 + 1$ with a graphing calculator using the viewing window $x_{min} = -5$, $x_{max} = 5$, $y_{min} = -1$, $y_{max} = 10$.
   c. Compare the graphs in parts (a) and (b).

x	-3	-2	-1	0	1	2	3
y							

In Exercises 3–8, graph the functions with a graphing calculator using a standard viewing window.

3. $y = x^2 - 5$
4. $y = 4 - x^2$
5. $y = x^3 - 3x^2$
6. $y = x^3 - 3x^2 + 4$
7. $y = \dfrac{9}{x^2 + 1}$
8. $y = \dfrac{40}{x^2 + 4}$

For Exercises 9–12, graph the given function with a graphing calculator using (a) the standard viewing window and (b) the specified window. Which window gives a better view of the graph of the function?

9. $y = x + 20$ with $x_{min} = -10$, $x_{max} = 10$, $y_{min} = -10$, $y_{max} = 30$
10. $y = x^3 - 3x + 13$ with $x_{min} = -5$, $x_{max} = 5$, $y_{min} = -10$, $y_{max} = 30$
11. $y = \dfrac{0.4(x - 0.1)}{x^2 + 300}$ on $[-20, 20]$ by $[-0.002, 0.002]$
12. $y = -x^2 + 20x - 20$ on $[-10, 20]$ by $[-20, 90]$

For Exercises 13–16, find an appropriate viewing window for the function, using the given x-values. Then graph the function.

13. $y = x^2 + 50$, for x-values between -3 and 3.
14. $y = x^2 + 60x + 30$, for values of x between -60 and 0.
15. $y = x^3 + 3x^2 - 45x$, for values of x between -10 and 10.
16. $y = (x - 28)^3$, for x-values between 25 and 31.
17. Find a complete graph of $y = 10x^2 - 90x + 300$. (A complete graph of this function shows one turning point.)
18. Find a complete graph of $y = -x^2 + 34x - 120$. (A complete graph of this function shows one turning point.)
19. Use a calculator or a spreadsheet to find $S(t)$ for the values of t given in the following table.

t	$S(t) = 5.2t - 10.5$
12	
16	
28	
43	

20. Use a calculator or a spreadsheet to find $f(q)$ for the values of q given in the following table.

q	$f(q) = 3q^2 - 5q + 8$
-8	
-5	
24	
43	

21. Enter the data into lists and graph the scatter plot of the data, using the window [0, 110] by [0, 600].

x	20	30	40	50	60	70	80	90	100
y	500	320	276	80	350	270	120	225	250

22. Enter the data below into lists on a graphing utility and graph the scatter plot of the data, using the window $[-20, 120]$ by $[0, 80]$.

x	-15	-1	25	40	71	110	116
y	12	16	10	32	43	62	74

23. Use the table below in parts (a)–(c).

x	-3	-1	1	5	7	9	11
y	-42	-18	6	54	78	102	126

a. Use a graphing utility to graph the points from the table.
b. Use a graphing utility to graph the equation $f(x) = 12x - 6$ on the same set of axes as the data in part (a).
c. Do the points fit on the graph of the equation? Do the respective values of x and y in the table satisfy the equation in part (b)?

24. Use the table below in parts (a)–(c).

x	1	3	6	8	10
y	-10	0	15	25	35

a. Use a graphing utility to graph the points from the table.
b. Use a graphing utility to graph the equation $f(x) = 5x - 15$ on the same set of axes as the data in part (a).
c. Do the points fit on the graph of the equation? Do the respective values of x and y in the table satisfy the equation in part (b)?

25. Suppose $f(x) = x^2 - 5x$ million dollars are earned, where x is the number of years after 2000.
a. What is $f(20)$?
b. The answer to part (a) gives the number of millions of dollars earned for what year?

26. Suppose $f(x) = 100x^2 - 5x$ thousand units are produced, where x is the number of years after 2000.
a. What is $f(10)$?
b. How many units are produced in 2010, according to this function?

1.2 EXERCISES

27. *Welfare Cases* The average number of welfare cases during 1993–1998 in Niagara, Canada, is given by the model $y = -112x^2 - 107x + 15{,}056$, where x is the number of years after 1990.
a. What are the values of x that correspond to the years 1994 and 1998?
b. Find the value of y when $x = 8$. Explain what this means.
c. How many welfare cases were there in 1995, according to this model?
(Source: Regional Niagara Social Services Dept, April 25, 2000)

28. *Women in the Workforce* The number y (in thousands) of women in the workforce is given by the function

$$y = 7x^2 - 19.88x + 409.29$$

where x is the number of years after 1900.
a. Find the value of y when $x = 44$. Explain what this means.
b. Use the model to find the number of women in the workforce in 1980.
(Source: Bureau of the Census, U.S. Dept of Commerce)

29. *Internet Access* The function $P = 6.02t + 3.53$ gives the percent of households with Internet access as a function of t, the number of years after 1995.
a. What are the values of t that correspond to the years 1996 and 2004?
b. $P = f(8)$ gives the value of P for what year? What is $f(8)$?
c. What x_{min} and x_{max} should be used to set the viewing window so that t represents 1995–2005?

30. *State Lotteries* The cost of prizes and expenses of state lotteries is given by $L = 35t^2 + 740t + 1207$ million dollars, with t equal to the number of years after 1980.

a. What are the values of t that correspond to the years 1982, 1988, and 2000?
b. $P = f(4)$ gives the value of P for what year? What is $f(4)$?
c. What x_{min} and x_{max} should be used to set a viewing window so that t represents 1980–1997?

31. *Height of a Ball* If a ball is thrown into the air at 64 feet per second from the top of a 100-foot-tall building, its height can be modeled by the function $S = 100 + 64t - 16t^2$, where S is in feet and t is in seconds.
a. Graph this function on a viewing window [0, 6] by [0, 200].
b. Find the height of the ball 1 second after it is thrown and 3 seconds after it is thrown. How can these values be equal?
c. Find the maximum height the ball will reach.

32. *Depreciation* A business property valued at $600,000 is depreciated over 30 years by the straight-line method, so that its value x years after the depreciation began is

$$V = 600{,}000 - 20{,}000x$$

a. Graph this function on a viewing window [0, 30] by [0, 600,000].
b. What is the value 10 years after the depreciation is started?

33. *Earnings and Gender* A model that relates the median annual salary (in thousands of dollars) of females F, and males M, in the United States is given by $F = 0.52M + 2.78$.
a. Use a graphing utility to graph this function on the viewing window [0, 85] by [0, 65].
b. Use the graphing utility to find the median female salary that corresponds to a male salary of $63,000.
(Source: U.S. Equal Opportunity Commission)

34. *Education Spending* Federal spending (in billions of dollars) for public education during the years 1996–2001 can be modeled by the function $S = 3.32x + 23.16$, where x is the number of years after 1990.
a. Use a graphing utility to graph this function on the viewing window [0, 11] by [0, 60].
b. Find the federal spending for public education in 2001.
(Source: U.S. Dept of Education)

35. *Medical School* Based on data collected in the spring of each year between 1988 and 1997, the number of students (in thousands) of osteopathic medicine in the United States can be described by

$$S = 0.027t^2 - 4.85t + 218.93$$

where t is the number of years after 1980.
a. Graph this function on the viewing window [0, 17] by [0, 300].
b. Use technology to find S when t is 15.
c. Use the model to estimate the number of osteopathic students in 1995.
(Source: *Statistical Abstract of the United States*, 1998)

36. *State Lotteries* The cost (in millions of dollars) of prizes and expenses for state lotteries between 1980 and 1997 can be described by $L = 35.3t^2 + 740.2t + 1207.2$, where t is the number of years after 1980.
a. Graph this function on the viewing window [0, 17] by [1200, 12,000].
b. Use technology to find L when t is 16.
c. What was the cost of prizes and expenses for state lotteries in 1996?
(Source: *Statistical Abstract of the United States*, 1998)

37. *Tax Burden* Using data from the Internal Revenue Service, the per capita tax burden B (in hundreds of dollars) can be described by $B(t) = 20.37 + 1.83t$, where t is the number of years after 1980.
a. Graph this function with technology using a viewing window with $t \geq 0$ and $B(t) \geq 0$.
b. Did the tax burden increase or decrease?
(Source: Internal Revenue Service)

38. *Crime* The rate (number per 100,000 people) of juvenile arrests for crimes is given by

$$f(x) = -0.027x^2 + 5.69x + 51.15$$

where x is the number of years after 1950.
(Source: Federal Bureau of Investigation)
a. Use technology to graph this model on the viewing window [0, 50] by [0, 300].
b. This viewing window shows the graph for what time period?
c. Did the rate of juvenile arrests increase or decrease over this period?
d. Use this model to find the rate of juvenile arrests in 1960 and in 1998.
(Source: Federal Bureau of Investigation)

39. *Cost* Suppose the cost of the production and sale of x Electra dishwashers is $C(x) = 15{,}000 + 100x + 0.1x^2$ dollars. Graph this function on a viewing window with x between 0 and 50.

40. *Revenue* Suppose the revenue from the sale of x coffee makers is given by $R(x) = 52x - 0.1x^2$. Graph this function on a viewing window with x between 0 and 100.

41. *Profit* The profit from the production and sale of x laser printers is given by the function $P(x) = 200x - 0.01x^2 - 5000$, where x is the number of units produced and sold. Graph this function on a viewing window with x between 0 and 1000.

42. *Profit* The profit from the production and sale of x digital cameras is given by the function $P(x) = 1500x - 8000 - 0.01x^2$, where x is the number of units produced and sold. Graph this function on a viewing window with x between 0 and 500.

43. *Teacher Salaries* The average U.S. classroom teacher salary is given by $f(t) = 982.06t + 32{,}903.77$, where t is the number of years from 1990.

a. What inputs correspond to the years 1990–2005?

b. What outputs correspond to the inputs determined for 1990 and 2005?

c. Use the answers to parts (a) and (b) and the fact that the function increases to find an appropriate viewing window and graph this function.

(Source: www.ors2.state.sc.us/abstract)

44. *Cocaine Use* The percent of high school seniors during the years 1975–1996 who have ever used cocaine can be described by

$$y = 0.0094x^3 - 0.36x^2 + 3.35x + 8.53$$

where x is the number of years after 1975.

a. What inputs correspond to the years 1975 through 1996?

b. What outputs for y could be used to estimate the percent of seniors who have ever used cocaine?

c. Based on your answers to parts (a) and (b), choose an appropriate window and graph the equation on a graphing utility.

d. Graph the function again with a new window that gives a graph nearer the center of the screen.

e. Use this function to estimate the percent in 1990.

(Source: 1998 *World Almanac*)

45. *Runway Near-Hits* The number of runway near-hits by airplanes from 1990 to 2000 is shown in the following table. Use a graphing utility to graph the data, with x representing the number of years after 1990 and y representing the number of near-hits.

Years	Runway Near-Hits	Years	Runway Near-Hits
1990	281	1996	275
1991	242	1997	292
1992	219	1998	325
1993	186	1999	321
1994	200	2000	421
1995	240		

(Source: Federal Aviation Administration)

46. *U.S. Population* The projected population of the United States for selected years from 2000 to 2060 is shown in the table below, with the population given in millions.

a. According to this table, what will be the U.S. population in 2010?

b. Create a new table with x representing the number of years after 2000 and y representing the number of millions.

c. Use a graphing utility to graph the data from the new table as a scatter plot.

Year	Population (millions)
2000	275.3
2010	299.9
2020	324.9
2030	351.1
2040	377.4
2050	403.7
2060	432

(Source: U.S. Census Bureau)

47. *Hotel Values* The table on page 40 gives the annual rental value of an average hotel room for the years 2000–2005.

a. Let x represent the number of years from 2000 and y represent the value per room in thousands of dollars and sketch the scatter plot of the data.

b. Graph the equation $y = 0.973x^2 - 4.667x + 73.950$ on the same axes as the scatter plot.

Index of Hotel Values	
Year	Value per Room
2000	$73,978
2001	$70,358
2002	$68,377
2003	$68,192
2004	$71,691
2005	$74,584

(Source: Pennsylvania State University)

48. *Health Insurance* The table below gives the national monthly health insurance premiums paid by employers and employees for the years 2000–2004.
  a. Let x represent the number of years from 2000 and y represent the monthly premium dollars and sketch the scatter plot of the data.
  b. Graph the equation $y = 5.00x^2 + 21.91x + 378.60$ on the same axes as the scatter plot.

Years	Monthly Premiums ($)
2000	379.72
2001	401.20
2002	448.66
2003	485.42
2004	547.29

(Source: www.eip.sc.gov)

49. *Unemployment Rate* The U.S. civilian unemployment rate (as a percent) is given by the table below.
  a. According to this table, what was the unemployment rate in 2003?
  b. Graph the data from this table as a scatter plot, using the number of years after 1995 as x.
  c. Graph the equation $y = 0.0583x^4 - 1.315x^3 + 10.636x^2 - 36.622x + 50.019$ on the same axes as the scatter plot.

Year	Unemployment (%)
1998	5.1
1999	4.5
2000	4.9
2001	4.8
2002	3.9
2003	3.5

50. *Dropout Rates* The table below gives the dropout rates (as percents) for students in grades 9–12 in Vermont schools during given years.
  a. What is the dropout rate in 2004, according to the data?
  b. Graph the data from this table as a scatter plot, using the number of years after 1960 as x.
  c. On the same axes as the scatter plot, graph the equation $y = 0.0000388x^4 - 0.00336x^3 + 0.0369x^2 - 0.629x + 5.495$.

Year	Dropout (%)
1960	5.5
1970	4.9
1980	7.1
1990	5.6
2000	4
2004	5.6

1.3 Linear Functions

Key Concepts

- Linear functions
- Constant functions
- Identity functions
- Slope of a line
- Slope and y-intercept of a line
- Parallel and perpendicular line
- Constant rate of change

Section Preview Ritalin Use

Health care professionals are concerned that too many children are now taking Ritalin and that some young people are using Ritalin as a recreational drug. The increase in the use of Ritalin is evident by looking at the graph in Figure 1.20, which shows Ritalin consumption in grams per 100,000 persons as a function of the number of years after 1990. Because the graph of the function that models the consumption is a line, the function is called a **linear function**. The function that models the consumption of Ritalin over the years 1990–1996 is

$$y = 225.304x + 493.432$$

where x is the number of years after 1990 and y is measured in grams per 100,000 people. The average rate at which the use of Ritalin was increasing over the years 1990–1996 is constant and is the same as the slope of the line in Figure 1.20.

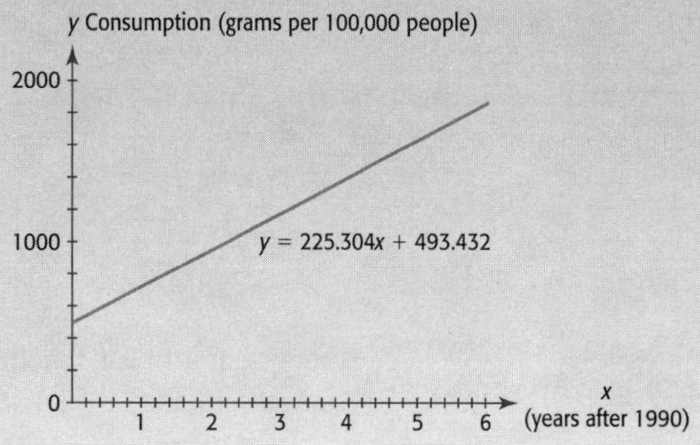

FIGURE 1.20

(Source: Index of Leading Cultural Indicators)

In this section, we investigate linear functions, discuss slope, intercepts, and average rate of change, and learn how to write equations of lines.

Linear Functions

A function whose graph is a line is a **linear function**.

LINEAR FUNCTION

A linear function is a function that can be written in the form $f(x) = ax + b$ where a and b are constants.

A function with equation of the form $y = 0x + b$, or $y = b$, is a special linear function called a **constant function**, and the function with equation $y = 1x + 0$, or $y = x$, is called the **identity function**.

If x and y are in separate terms in an equation and are each to the first power (and not in a denominator), we can rewrite the equation relating them in the form

$$y = ax + b$$

for some constants a and b. So the original equation is a linear equation that represents a linear function. If no restrictions are stated or implied by the context of the problem situation, and if its graph is not a horizontal line, both the domain and range of a linear function consist of the set of all real numbers. Note that an equation of the form $x = d$, where d is a constant, is not a function; its graph is a vertical line.

Recall that in the form $y = ax + b$, a and b represent constants and the variables x and y can represent any variables, with x representing the independent (input) variable and y representing the dependent (output) variable. For example, the function $5q + p = 400$ can be written in the form $p = -5q + 400$, so we can say that p is a linear function of q.

EXAMPLE 1 Linear Functions

Determine whether each equation represents a linear function. If so, give the domain and range.

a. $0 = 2t - s + 1$ **b.** $y = 5$ **c.** $xy = 2$

Solution

a. The equation

$$0 = 2t - s + 1$$

does represent a linear function because each of the variables t and s appear to the first power and each is in a separate term. We can solve this equation for s, getting

$$s = 2t + 1,$$

so s is a linear function of t. Because any real number can be multiplied by 2 and increased by 1, and the result is a real number, both the domain and range consist of the set of all real numbers.

b. The equation $y = 5$ is in the form $y = ax + b$, where $a = 0$ and $b = 5$, so it represents a linear function. (It is in fact a constant function, which is a special linear function.) The domain is the set of all real numbers (because $y = 5$ regardless of what x we choose), and the range is the set containing 5.

c. The equation

$$xy = 2$$

does not represent a linear function because x and y are not in separate terms, and the equation cannot be written in the form $y = ax + b$. ■

Intercepts

The points where a graph crosses or touches the x-axis and y-axis are called the ***x*-intercepts** and ***y*-intercepts**, respectively, of the graph. For example, Figure 1.21 shows that the graph of the linear function $2x - 3y = 12$ crosses the x-axis at $(6, 0)$, so the x-intercept is (6, 0). The graph crosses the y-axis at (0, −4), so the y-intercept is (0, −4).

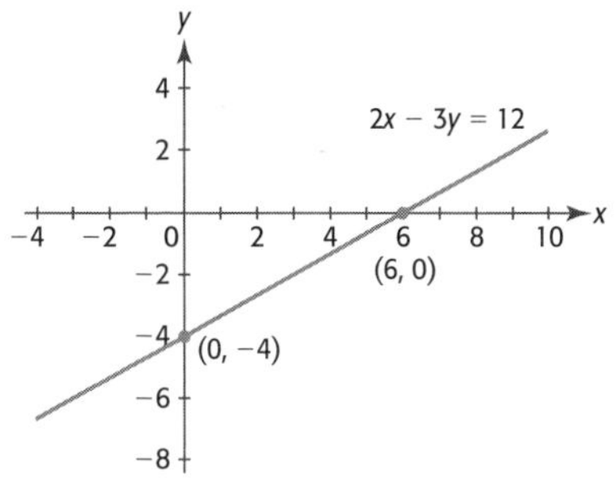

FIGURE 1.21

In this text, we will use the widely accepted convention that the x-coordinate of the x-intercept may also be called the x-intercept, and that the y-coordinate of the y-intercept may also be called the y-intercept.* The procedure for finding intercepts is a direct result of these definitions.

FINDING INTERCEPTS ALGEBRAICALLY

To find the y-intercept of a graph of $y = f(x)$, set $x = 0$ in the equation and solve for y.

To find the x-intercept(s) of the graph of $y = f(x)$, set $y = 0$ in the equation and solve for x.

For example, the y-intercept of the graph of $2x - 3y = 12$ in Figure 1.21 can be found by substituting 0 for x in the equation and solving for y.

$$\begin{aligned} 2(0) - 3y &= 12 \\ -3y &= 12 \\ y &= -4 \end{aligned}$$

Similarly, the x-intercept can be found by substituting 0 for y in the equation and solving for x.

$$\begin{aligned} 2x - 3(0) &= 12 \\ 2x &= 12 \\ x &= 6 \end{aligned}$$

*We usually call the horizontal axis intercept(s) the x-intercept(s) and the vertical axis intercept(s) the y-intercept(s), but realize that symbols other than x and y can be used to represent the input and output. For example, if $p = f(q)$, the vertical intercept is called the p-intercept, and the horizontal intercept is called the q-intercept.

The graph of a linear function has one y-intercept and has one x-intercept unless the graph is a horizontal line. The intercepts of the graph of a linear function are often easy to calculate. If the intercepts are distinct, then plotting these two points and connecting them with a line gives the graph.

> **TECHNOLOGY NOTE**
>
> When graphing with a graphing utility, finding or estimating the intercepts can help set the viewing window for the graph of a linear equation.

EXAMPLE 2 Loan Balance

A business property is purchased with a promise to pay off a \$60,000 loan plus the \$16,500 interest on this loan by making 60 monthly payments of \$1275. The amount of money, y, remaining to be paid on \$76,500 (the loan plus interest) is reduced by \$1275 each month. Although the amount of money remaining to be paid changes every month, it can be modeled by the linear function

$$y = 76{,}500 - 1275x$$

where x is the number of monthly payments made. We recognize that only integer values of x from 0 to 60 apply to this application.

a. Find the x-intercept and the y-intercept of the graph of this linear equation.

b. Interpret the intercepts in the context of this problem situation.

c. How should x and y be limited in this model so that they make sense in the application?

d. Use the intercepts and the results of part (c) to sketch the graph of the given equation.

Solution

a. To find the x-intercept, set $y = 0$ and solve for x.

$$0 = 76{,}500 - 1275x$$

$$1275x = 76{,}500$$

$$x = \frac{75{,}500}{1275} = 60$$

Thus, 60 is the x-intercept.

To find the y-intercept, set $x = 0$ and solve for y.

$$y = 76{,}500 - 1275(0)$$

$$y = 76{,}500$$

Thus, 76,500 is the y-intercept.

b. The x-intercept corresponds to the number of months that must pass before the amount owed is \$0. Therefore, a possible interpretation of the x-intercept is "The loan is paid off in 60 months." The y-intercept corresponds to the total (loan plus interest) that must be repaid 0 months after purchase—that is, when the purchase is made. Thus, the y-intercept tells us "A total of \$76,500 must be repaid."

c. We know that the total time to pay the mortgage is 60 months. A value of x larger than 60 will result in a negative value of y, which makes no sense in the application, so x varies from 0 to 60. The output, y, is the total amount owed at any time during the loan. The amount owed cannot be less than 0, and the value of the loan plus interest will be at its maximum, 76,500, when time is 0. Thus, the values of y vary from 0 to 76,500.

d. The graph intersects the horizontal axis at (60, 0) and intersects the y-axis at (0, 76,500), as indicated in Figure 1.22. Because we know that the graph of this equation is a line, we can simply connect the two points to obtain this first-quadrant graph.

Figure 1.22 shows the graph of the function on a viewing window determined by the context of the application.

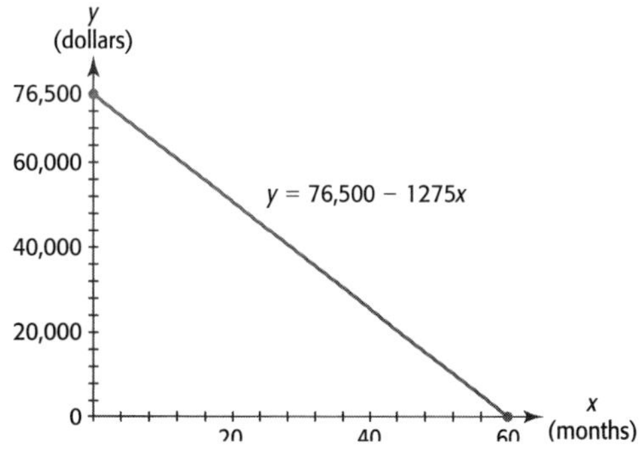

FIGURE 1.22

■

Slope of a Line

Consider two stairways: One goes from the park entrance to the shogun shrine at Nikko, Japan, and rises vertically 1 foot for every 1 foot of horizontal increase, and the second goes from the Tokyo subway to the street and rises vertically 20 centimeters for every 25 centimeters of horizontal increase. To find which set of stairs would be easier to climb, we can find the steepness, or **slope**, of each stairway. If a board is placed along the steps of each stairway, its slope is a measure of the incline of the stairway.

$$\text{shrine stairway steepness} = \frac{\text{vertical increase}}{\text{horizontal increase}} = \frac{1 \text{ foot}}{1 \text{ foot}} = 1$$

$$\text{subway stairway steepness} = \frac{\text{vertical increase}}{\text{horizontal increase}} = \frac{20 \text{ cm}}{25 \text{ cm}} = 0.8$$

The shrine stairway has a slope that is larger than that of the subway stairway, so it is steeper than the subway stairway. Thus, the subway steps would be easier to climb. In general, we define the slope of a line as follows.

SLOPE OF A LINE

The slope of a line is defined as

$$\text{slope} = \frac{\text{vertical change}}{\text{horizontal change}} = \frac{\text{rise}}{\text{run}}$$

The slope can be found by using any two points on the line (see Figure 1.23). If a nonvertical line passes through the two points, P_1 with coordinates (x_1, y_1) and P_2 with coordinates (x_2, y_2), its slope, denoted by m, is found by using

$$m = \frac{y_2 - y_1}{x_2 - x_1}$$

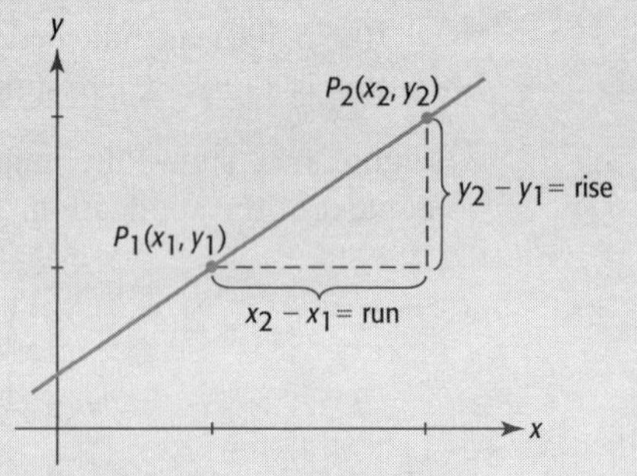

FIGURE 1.23

The slope of a vertical line is undefined because $x_2 - x_1 = 0$ and division by 0 is undefined.

The slope of any given nonvertical line is a constant. Thus, the same slope will result regardless of which two points on the line are used in its calculation.

EXAMPLE 3 Calculating the Slope of a Line

a. Find the slope of the line passing through the points $(-3, 2)$ and $(5, -4)$. What does the slope mean?

b. Find the slope of the line joining the x-intercept point and y-intercept point in the mortgage situation of Example 2.

Solution

a. We choose one point as P_1 and the other point as P_2. Although it does not matter which point is chosen to be P_1, it is important to keep the correct order of the terms in the numerator and denominator of the slope formula. Letting $P_1 = (-3, 2)$ and $P_2 = (5, -4)$ and substituting in the slope formula gives

$$m = \frac{-4 - 2}{5 - (-3)} = \frac{-6}{8} = -\frac{3}{4}$$

Note that letting $P_1 = (5, -4)$ and $P_2 = (-3, 2)$ gives the same slope:

$$m = \frac{2 - (-4)}{-3 - 5} = \frac{6}{-8} = -\frac{3}{4}$$

A slope of $-\frac{3}{4}$ means that, from a given point on the line, by moving 3 units down and 4 units to the right, or by moving 3 units up and 4 units to the left, we arrive at another point on the line.

b. Because $x = 60$ is the x-intercept in part (b) of Example 2, (60, 0) is a point on the graph. The y-intercept is $y = 76{,}500$, so (0, 76,500) is a point on the graph. Recall that x is measured in months and that y has units of dollars in the real-world setting (that is, the *context*) of Example 2. Substituting in the slope formula, we obtain

$$m = \frac{76{,}500 - 0}{0 - 60} = \frac{76{,}500}{-60} = -1275$$

This slope means that the amount owed decreases by \$1275 each month. ■

As Figure 1.24(a) to (d) indicates, the slope describes the direction of a line as well as the steepness.

THE RELATION BETWEEN ORIENTATION OF A LINE AND ITS SLOPE

1. The slope is *positive* if the line *rises* upward toward the right.

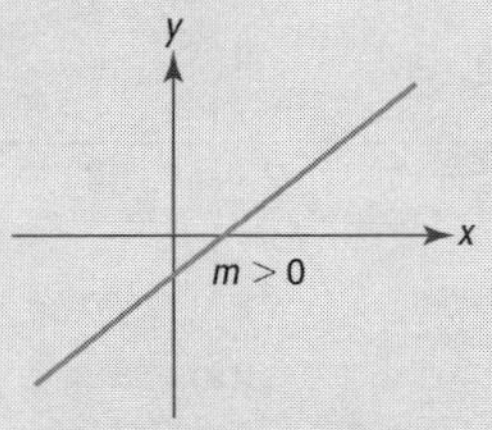

FIGURE 1.24(a)

2. The slope is *negative* if the line *falls* downward toward the right.

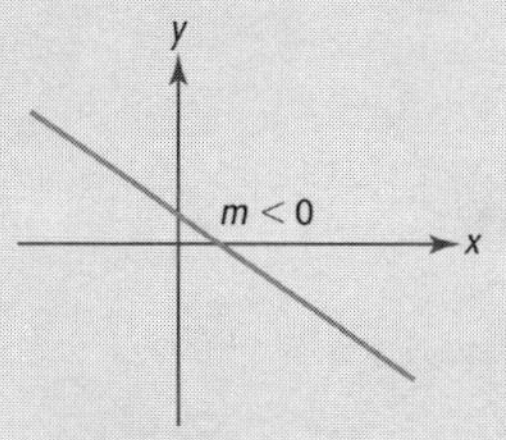

FIGURE 1.24(b)

3. The slope of a *horizontal line* is 0 because a horizontal line has a vertical change (rise) of 0 between any two points on the line.

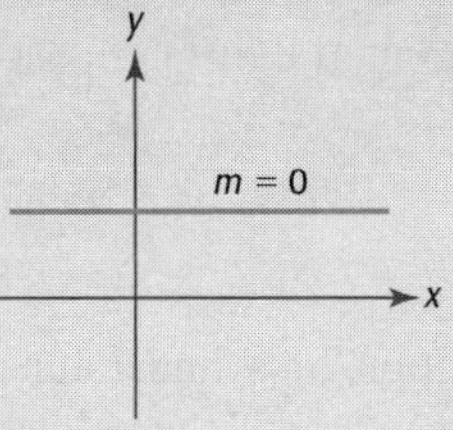

FIGURE 1.24(c)

4. The slope of a *vertical line* does not exist because a vertical line has a horizontal change (run) of 0 between any two points on the line.

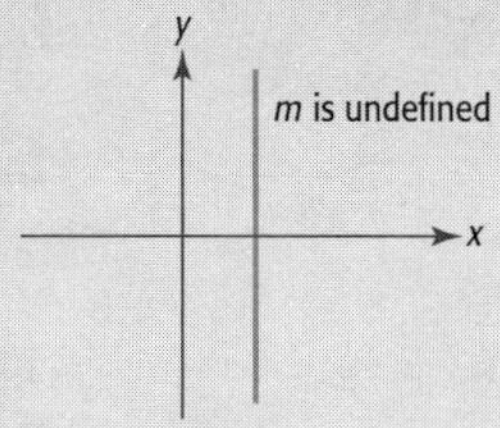

FIGURE 1.24(d)

Remembering how the orientation of a line is related to its slope can help you check that you have the correct order of the points in the slope formula. For instance, if you find that the slope of a line is positive and you see that the line falls downward from left to right when you observe its graph, you know that there is a mistake either in the slope calculation or in the graph that you are viewing.

Slope and y-Intercept of a Line

There is an important connection between the slope of the graph of a linear equation and its equation when it is written in the form $y = f(x)$. To investigate this connection, we can graph the equation

$$y = 3x + 2$$

with a graphing utility (Figure 1.25(a)). By using the TABLE feature of the graphing utility for values of x equal to 0, 1, 2, 3, and 4, we can see that each time that x increases by 1, y increases by 3 (see Figure 1.25(b)). Thus the slope is

$$\frac{\text{change in } y}{\text{change in } x} = \frac{3}{1} = 3$$

From this table, we see that the y-intercept is 2, the same value as the constant term of the equation. Observe also that the slope of the graph of $y = 3x + 2$ is the same as the coefficient of x.

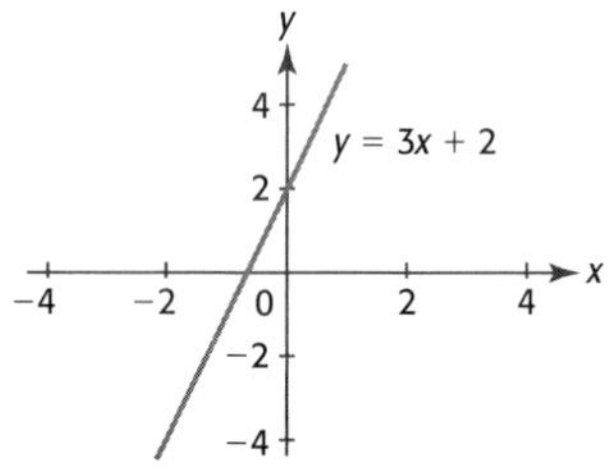

(a)

x	y_1
0	2
1	5
2	8
3	11
4	14

(b)

FIGURE 1.25

Because we denote the slope of a line by m, we have the following.

SLOPE AND Y-INTERCEPT OF A LINE

The slope of the graph of the equation $y = mx + b$ is m, and the y-intercept of the graph is b.

Thus, when the equation of a linear function is written in the form $y = mx + b$ or $f(x) = mx + b$, we can "read" the values of the slope and y-intercept of its graph.

EXAMPLE 4 Loan Balance

As we saw in Example 2, the amount of money y remaining to be paid on the loan of \$60,000 with \$16,500 interest is

$$y = 76{,}500 - 1275x$$

where x is the number of months the mortgage has been paid.

a. What is the slope and y-intercept of the graph of this function?

b. How does the amount owed on the loan change as the number of months increases?

Solution

a. Writing this equation in the form $y = mx + b$ gives $y = -1275x + 76{,}500$. The coefficient of x is -1275, so the slope is $m = -1275$; the constant term is 76,500, so the y-intercept is $b = 76{,}500$.

b. The slope of the line indicates that the amount owed decreases by \$1275 each month. This can be verified from the graph of the function and the table of sample inputs and outputs in Figure 1.26. ■

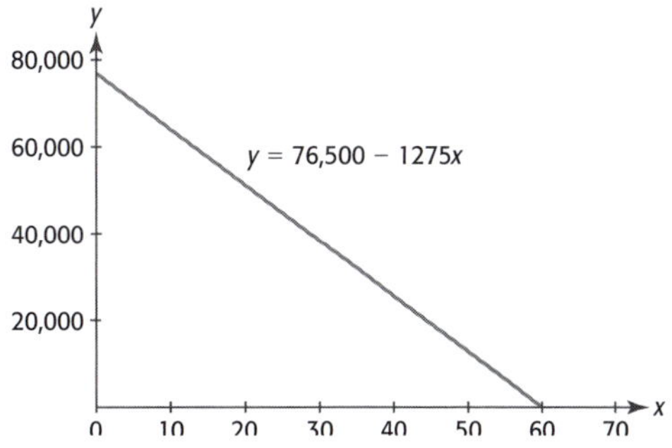

Months (x)	Amount Owed (y)
1	75,225
2	73,950
3	72,675
4	71,400
5	70,125
6	68,850
7	67,575
8	66,300
9	65,025
10	63,750

FIGURE 1.26

Constant Rate of Change

The function $y = 76{,}500 - 1275x$, whose graph is shown in Figure 1.26, gives the amount owed as a function of the months remaining on the mortgage. The coefficient of x, -1275, indicates that for each additional month, the value of y changes by -1275 (see the table in Figure 1.26). That is, the amount owed decreases at the **constant rate** -1275 each month. Note that the **constant rate of change** of this linear function is equal to the **slope** of its graph. This is true for all linear functions.

> **CONSTANT RATE OF CHANGE**
>
> The rate of change of the linear function $y = mx + b$ is the constant m, the slope of the graph of the function.

Note: The rate of change in an applied context should include appropriate units of measure.

EXAMPLE 5 Ritalin

The graph in Figure 1.27 shows Ritalin consumption in grams per 100,000 persons as a function of the number of years after 1990. The function that models the consumption of Ritalin over the years 1990–1996 is

$$y = 225.304x + 493.432$$

where x is the number of years after 1990 and y is measured in grams per 100,000 people.

a. What is the slope of the graph of this model?

b. What is the rate at which Ritalin consumption grew during this period? (Source: Index of Leading Cultural Indicators.)

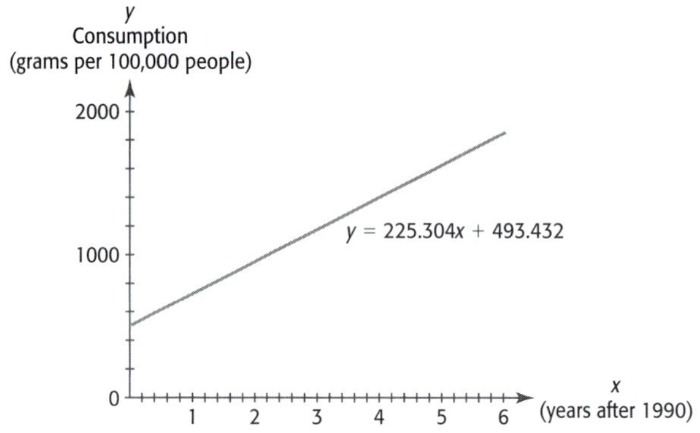

FIGURE 1.27

Solution

a. The coefficient of x in the linear function is 225.304, so the slope of the line is 225.304.

b. The slope of the line in Figure 1.27 is the rate of change of the function, so the use of Ritalin increased at a rate of 225.304 grams per 100,000 people each year. ■

For total cost, total revenue, and profit functions that are linear, their rates of change are called **marginal cost, marginal revenue**, and **marginal profit**, respectively.

Parallel and Perpendicular Lines

Two distinct nonvertical lines that have the same slope are parallel, and two nonvertical parallel lines have the same slope. In general, two lines are perpendicular if they intersect at right angles. However, the graphs of perpendicular lines may not appear to be at right angles if the axes on which they are graphed have different scales (This would happen if a standard window is used on a graphing calculator, for example.) We can determine if two lines are perpendicular by comparing the slopes of the lines. Two nonvertical and nonhorizontal lines are perpendicular if their slopes are negative reciprocals of each other.

EXAMPLE 6 Parallel and Perpendicular Lines

Determine whether the pairs of lines are parallel, perpendicular, or neither.

a. $y = -3x + 9$
$3x + y = 7$

b. $2x - 3y = 3$
$3x + 2y = -1$

c. $x - 2y = 4$
$2x - y = 3$

Solution

a. The slope of the graph of $y = -3x + 9$ is -3. Solving $3x + y = 7$ for y gives $y = -3x + 7$, so the slope is -3. Because these slopes are equal, we conclude that the lines are parallel.

b. Solving $2x - 3y = 3$ for y gives $y = \frac{2}{3}x - 1$, so the slope is $\frac{2}{3}$. Solving $3x + 2y = -1$ for y gives $y = -\frac{3}{2}x - \frac{1}{2}$, so the slope is $-\frac{3}{2}$. Because these slopes are negative reciprocals, we conclude that the lines are perpendicular.

c. Solving $x - 2y = 4$ for y gives $y = \frac{1}{2}x - 2$, so the slope is $\frac{1}{2}$. Solving $2x - y = 3$ for y gives $y = 2x - 3$, so the slope is 2. Because these slopes are not equal and not negative reciprocals, we conclude that the lines are neither parallel nor perpendicular. ■

1.3 SKILLS CHECK

1. Which of the following functions are linear?
a. $y = 3x^2 + 2$ **b.** $3x + 2y = 12$
c. $y = \frac{1}{x} + 2$

2. Find the slope of the line through (4, 6) and (28, −6).

3. Find the slope of the line through (8, −10) and (8, 4).

4. Find the slope of the line through (−6, 5) and (−2, 5).

In Exercises 5–8, (a) find the x- and y-intercepts of the graph of the given equation, if they exist, and (b) graph the equation.

5. $5x - 3y = 15$ **6.** $x + 5y = 17$

7. $3y = 9 - 6x$ **8.** $y = 9x$

9. If a line is horizontal, then its slope is ____. If a line is vertical, then its slope is _____.

10. Based on the graph of the line, determine whether the slope of the graph of the line is positive, negative, 0, or undefined in a–d below.

a.

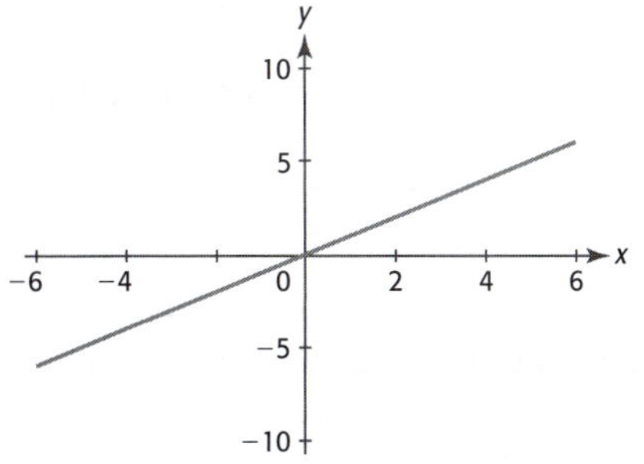

b.

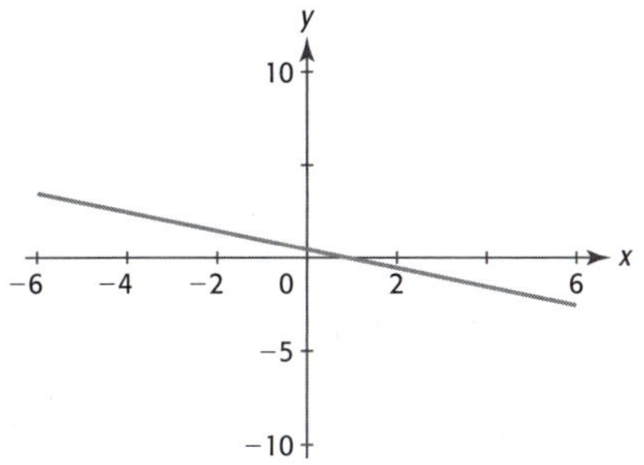

c.

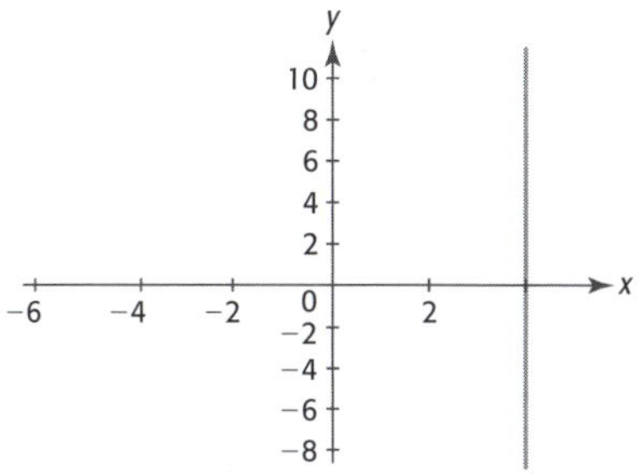

d.

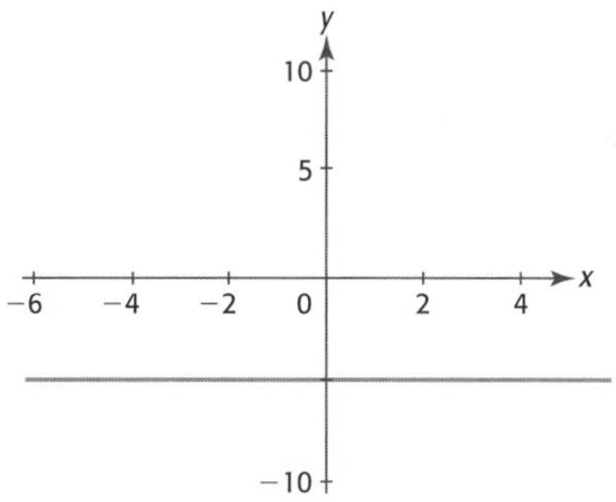

For Exercises 11–14, give the slope of the line (if it exists) and the y-intercept (if it exists).

11. $y = 4x + 8$ **12.** $3x + 2y = 7$

13. $5y = 2$ **14.** $x = 6$

For each of the functions in Exercises 15–17, do the following:

a. Find the slope and y-intercept (if possible) of the graph of the function.
b. Determine if the graph is rising or falling.
c. Graph each function on a window with the given x-range and a y-range that shows a complete graph.

15. $y = 4x + 5$; [−5, 5]

16. $y = 0.001x - 0.03$; [−100, 100]

17. $y = 50{,}000 - 100x$; [0, 500]

18. Rank the functions in Exercises 15–17 in order of increasing steepness.

For each of the functions in Exercises 19–22, find the rate of change.

19. $y = 4x - 3$

20. $y = \frac{1}{3}x + 2$

21. $y = 300 - 15x$

22. $y = 300x - 15$

23. If a linear function has the points $(-1, 3)$ and $(4, -7)$ on its graph, what is the rate of change of the function?

24. If a linear function has the points (2, 1) and (6, 3) on its graph, what is the rate of change of the function?

25. Are the lines in Exercises 23 and 24 parallel, perpendicular, or neither?

26. One line passes through the points $(-2, 3)$ and (5, 8), and a second line has equation $5x - 7y = 35$. Are the lines parallel, perpendicular, or neither?

27. **a.** Does graph (i) or graph (ii) represent the identity function?
b. Does graph (i) or graph (ii) represent a constant function?

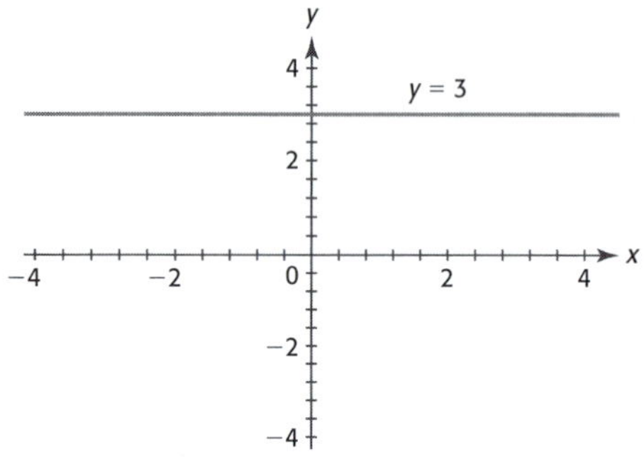

(i)

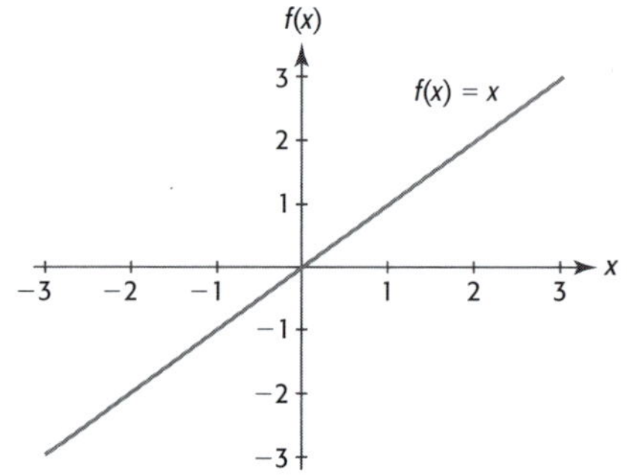

(ii)

28. What is the slope of the identity function?

29. **a.** What is the slope of the constant function $y = k$?
b. What is the rate of change of a constant function?

30. What is the rate of change of the identity function?

1.3 EXERCISES

In Exercises 31–34, determine whether or not the function is linear. If it is, tell whether the line that is the graph of the function rises or falls as the input increases.

31. *Reading Tests* The average reading score of 17-year-olds on the National Assessment of Progress tests is given by $y = 0.155x + 244.37$ points, where x is the number of years past 1970.
(Source: U.S. Dept. of Education)

32. *Women in the Workforce* The number y (in thousands) of women in the workforce is given by the function $y = 7x^2 - 19.88x + 409.29$, where x is the number of years from 1900.
(Source: Bureau of the Census, U.S. Dept of Commerce)

33. *Marriage Rate* When x is the number of years past 1950, the percent of unmarried women who get married each year is given by $M(x) = -0.762x + 85.284$.
(Source: Index of Leading Cultural Indicators)

34. *Voters* The percent of the population voting in presidential elections is given by $V(t) = -0.356t + 65.404$, where t is the number of years after 1950.
(Source: Federal Election Commission)

35. *Marijuana Use* The percent p of high school seniors using marijuana daily can be related to x, the number of years after 1990, by the equation $30p - 19x = 30$.
a. Find the x-intercept of the graph of this function.
b. Find and interpret the p-intercept of the graph of this function.
c. Graph the function, using the intercepts. What values of x on the graph represent years 1990 and after?

36. *Depreciation* An $828,000 building is depreciated for tax purposes by its owner, using the straight-line depreciation method. The value of the building after x months of use is given by $y = 828{,}000 - 2300x$ dollars.

a. Find and interpret the y-intercept of the graph of this function.
b. Find and interpret the x-intercept of the graph of this function.
c. Use the intercepts to graph the function for nonnegative x- and y-values.

37. *Life Insurance* The monthly rates for a \$100,000 life insurance policy for males aged 27–32 are shown in the table.

Age (years), x	27	28	29
Premium (dollars per month), y	11.81	11.81	11.81

Age (years), x	30	31	32
Premium (dollars per month), y	11.81	11.81	11.81

a. Could the data in this table be modeled by a constant function or the identity function?
b. Write an equation whose graph contains the data points in the table.
c. What is the slope of the graph of the function found in part (b)?
d. What is the rate of change of the data in the table?

38. *Eating Asparagus* The per capita consumption of asparagus between 1990 and 1996 is shown in the table.

Year	1990	1991	1992	1993
Asparagus Consumption (pounds per person)	0.6	0.6	0.6	0.6

Year	1994	1995	1996
Asparagus Consumption (pounds per person)	0.6	0.6	0.6

(Source: *Statistical Abstract of the United States*, 1998)

a. Sketch the data as a scatter plot, with y equal to the consumption and x equal to the year.
b. Could the data in the table be modeled by a constant function or the identity function?
c. Write the equation of a function that fits the data points.
d. Sketch a graph of the function you found in part (c) on the same axes as the scatter plot.

39. *Internet Recruiting* The percentage of Fortune Global 500 firms that actively recruited workers on the Internet from 1998 through 2000 can be modeled by $P(x) = 26.5x - 194.5$ percent, where x is the number of years after 1990.
a. What is the slope of the graph of this function?
b. Interpret the slope as a rate of change.
(Source: *Time*, August 16, 1999)

40. *Cigarette Use* For the years 1975–1991, the percent p of high school seniors who have tried cigarettes can be modeled by $p = 75.4509 - 0.7069t$, where t is the number of years after 1975.
a. Is the rate of change of the percent positive or negative?
b. What does the answer to part (a) tell us about the percent of seniors who tried cigarettes during this period?

41. *Crickets* The number of times per minute n that a cricket chirps can be modeled as a function of the Fahrenheit temperature T. The data can be approximated by the function

$$n = \frac{12T}{7} - \frac{52}{7}$$

a. Is the rate of change of the number of chirps positive or negative?
b. What does this tell us about the relationship between temperature and the number of chirps?

42. *Tax Burden* The per capita tax burden T (in hundreds of dollars) can be described by $T(t) = 20.37 + 1.834t$, where t is the number of years after 1980.
a. What is the slope of the graph of this function?
b. What is the rate of growth of the per capita tax burden per year?
(Source: Internal Revenue Service)

43. *Earnings and Minorities* According to the U.S. Equal Employment Opportunity Commission, the relation between the median annual salaries of minorities and whites can be modeled by the function $M = 0.959W - 1.226$, where M and W represent the median annual salary (in thousands of dollars) for minorities and whites, respectively.
a. Is this function a linear function?
b. What is the slope of the graph of this function?
c. Interpret the slope as a rate of change.
(Source: *Statistical Abstract of the United States*, 1993)

44. *Voting* Between 1950 and 1996, the percent of the voting population who voted in presidential elections

is given by $p = 65.4042 - 0.3552x$, where x is the number of years after 1950.

a. What is the slope of the graph of this function? Write a sentence interpreting this value.

b. What is the annual rate of change in the percent of the eligible population who voted in presidential elections between 1950 and 1996?

(Source: Federal Election Commission)

45. *Marijuana Use* The percent p of high school seniors using marijuana daily can be modeled by $30p - 19x = 30$, where x is the number of years after 1990.

a. Use this model to determine the slope of the graph of this function if x is the independent variable.

b. What is the rate of change of the percent of high school seniors using marijuana per year?

(Source: Index of Leading Cultural Indicators)

46. *Seawater Pressure* In seawater, the pressure p is related to the depth d according to the model $33p - 18d = 496$, where d is the depth in feet and p is in pounds per square inch.

a. What is the slope of the graph of this function?

b. Interpret the slope as a rate of change.

47. *Advertising Impact* An advertising agency has found that when it promotes a new product in a city the weekly rate of change R of the number of people who are aware of it x weeks after it is introduced is given by $R = 3500 - 70x$. Find the x- and R-intercepts and then graph the function on a viewing window that is meaningful in the application.

48. *ATM Transactions* The dollar volume of transactions at automatic teller machines (ATMs) is modeled by $D(x) = 0.137x - 5.09$ billion dollars, where x is the number of terminals (in thousands).

a. What was the value of the transactions when 50,000 ATMs were available, according to this model?

b. This model approximates the dollar value of transactions and does not fit the data exactly. The equation cannot model the data for values of x that give negative dollar values. Test integer values of x to find the values for which this function fails to model the data because it gives negative dollar values.

(Source: Electronic Funds Transfer Association, 1993).

49. *Internet Users* The percent of U.S. population with Internet access can be modeled by the function

$$y = 5.74x + 14.61$$

where x is the number of years after 1995.

a. Find the slope and the y-intercept of the graph of this equation.

b. What interpretation could be given to the y-intercept?

c. What interpretation could be given to the slope?

(Source: Jupiter Media Metrix)

50. *Wireless Service Spending* The total amount spent in the United States for wireless communication services S (in billions of dollars) can be modeled by the function

$$S = 6.205 + 11.23t$$

where t is the number of years after 1995.

a. Find the slope and the y-intercept of the graph of this equation.

b. What interpretation could be given to the y-intercept?

c. What interpretation could be given to the slope?

(Source: Cellular Telecommunications and Internet Association)

51. *Depreciation* Suppose the cost of a business property is \$1,920,000 and a company depreciates it with the straight-line method. Suppose v is the value of the property after x years, and the line representing the value as a function of years passes through the points (10, 1,310,000) and (20, 700,000).

a. What is the slope of the line through these points?

b. What is the annual rate of change of the value of the property?

52. *Men in the Workforce* The number of men in the workforce (in millions) for the years from 1890 to 1990 can be approximated by the linear model determined by connecting the points (1890, 18.1) and (1990, 68.5).

a. Find the annual rate of change of the model whose graph is the line connecting these points.

b. What does this tell us about men in the workforce?

53. *Profit* A company charting its profits notices that the relationship between the number of units sold x and the profit P is linear. If 300 units sold results in \$4650 profit and 375 units results in \$9000 profit, find the marginal profit, which is the rate of change of the profit.

54. *Cost* A company buys and retails baseball caps and the total cost function is lincar. The total cost for 200 caps is \$2690, and the total cost of 500 caps is \$3530. What is the marginal cost, which is the rate of change of the function?

55. *Marginal Cost* Suppose the monthly total cost for the manufacture of golf balls is $C(x) = 3450 + 0.56x$, where x is the number of balls produced each month.

a. What is the slope of the graph of the total cost function?
b. What is the marginal cost (rate of change of the cost function) for the product?
c. What is the cost of each additional ball that is produced in a month?

56. *Marginal Cost* Suppose the monthly total cost for the manufacture of 19-inch television sets is $C(x) = 2546 + 98x$, where x is the number of TVs produced each month.
a. What is the slope of the graph of the total cost function?
b. What is the marginal cost for the product?
c. Interpret the marginal cost for this product.

57. *Marginal Revenue* Suppose the monthly total revenue for the manufacture of golf balls is $R(x) = 1.60x$, where x is the number of balls sold each month.
a. What is the slope of the graph of the total revenue function?
b. What is the marginal revenue for the product?
c. Interpret the marginal revenue for this product.

58. *Marginal Revenue* Suppose the monthly total revenue for the manufacture of television sets is $R(x) = 198x$, where x is the number of TVs sold each month.
a. What is the slope of the graph of the total revenue function?
b. What is the marginal revenue for the product?
c. Interpret the marginal revenue for this product.

59. *Profit* The profit for a product is given by $P(x) = 19x - 5060$, where x is the number of units produced and sold. Find the marginal profit for the product.

60. *Profit* The profit for a product is given by the function $P(x) = 939x - 12{,}207$, where x is the number of units produced and sold. Find the marginal profit for the product.

1.4 Equations of Lines

Key Concepts

- Writing equations of lines
- General form
- Slope-intercept form
- Point-slope form
- Horizontal line
- Vertical line
- Average rate of change
- Difference quotient
- Secant line
- Approximately linear data

Section Preview Depreciation

If a \$750,000 building were depreciated by the straight-line method over a period of 25 years, then the value of the property would be reduced by 1/25 of the original value, or by the **constant rate** of \$30,000, each year. We can use this information to write the linear equation that models the value y of the property as a function of the number of years x that it is depreciated because the rate of change is constant. In this section, we learn how to write the equation of a linear function from information about the line, such as the slope and a point on the line or two points on the line. We also discuss **average rates of change** for data that can be modeled by nonlinear functions, slopes of secant lines, and how to create linear models that approximate data that are "nearly" linear.

Writing Equations of Lines

Creating a linear model from data involves writing a linear equation that describes the mathematical situation. If we know two points on a line or the slope of the line and one point, we can write the equation of the line.

Recall that if a linear equation has the form $y = mx + b$, then the coefficient of x is the slope of the line that is the graph of the equation, and the constant b is the y-intercept of the line. Thus if we know the slope and the y-intercept of a line, we can write the equation of this line.

SLOPE-INTERCEPT FORM OF THE EQUATION OF A LINE

The slope-intercept form of the equation of a line with slope m and y-intercept b is

$$y = mx + b$$

EXAMPLE 1 Appliance Repair

An appliance repairman charges \$60 for a service call plus \$25 per hour for each hour spent on the repair. If a linear function models his service call charges, write the equation of the function.

Solution

Let x represent the number of hours spent on the appliance repair and let y be the service call charge in dollars. The slope of the line is the amount that the charge increases for every hour of work done, so the slope is 25. Because \$60 is the basic charge before any time is spent on the repair, 60 is the y-intercept of the line. Substituting these values in the slope-intercept form of the equation gives the equation of the linear function modeling this situation. When the repairman works x hours on the service call, the charge is

$$y = 25x + 60 \text{ dollars}$$ ■

We next consider a form of an equation of a line that can be written if we know the slope and a point. If the slope of a line is m, then the slope between a fixed point (x_1, y_1) and any other point (x, y) on the line is also m. That is,

$$m = \frac{y - y_1}{x - x_1}$$

Solving for $y - y_1$ (that is, multiplying both sides of this equation by $x - x_1$) gives the **point-slope form** of the equation of a line.

POINT-SLOPE FORM OF THE EQUATION OF A LINE

The equation of the line with slope m that passes through a known point (x_1, y_1) is

$$y - y_1 = m(x - x_1)$$

EXAMPLE 2 Using a Point and Slope to Write an Equation of a Line

Write equations for the lines that pass through the point $(-1, 5)$ and have

a. slope $\frac{3}{4}$ **b.** slope 0 **c.** undefined slope

Solution

a. We are given $m = \frac{3}{4}$, $x_1 = -1$, and $y_1 = 5$. Substituting in the point-slope form, we obtain

$$y - 5 = \frac{3}{4}(x - (-1))$$

$$y - 5 = \frac{3}{4}(x + 1)$$

$$y - 5 = \frac{3}{4}x + \frac{3}{4}$$

$$y = \frac{3}{4}x + \frac{23}{4}$$

Figure 1.28 shows the graph of this line with rise 3 and run 4.

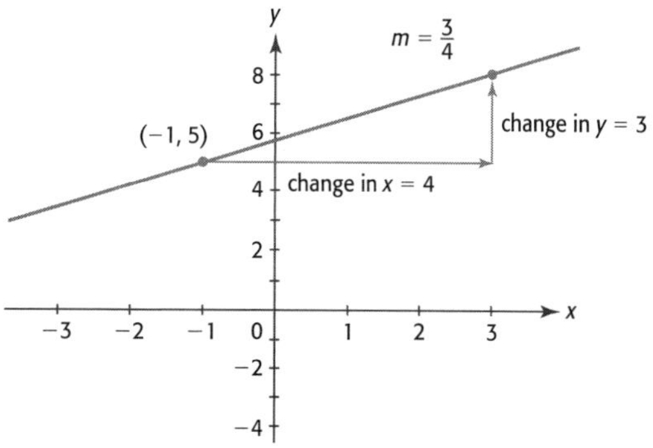

FIGURE 1.28

b. If $m = 0$, the point-slope form gives us the equation

$$y - 5 = 0(x - (-1))$$

$$y = 5$$

Because the output is always the same value, the graph of this linear function is a horizontal line (Figure 1.29(a)).

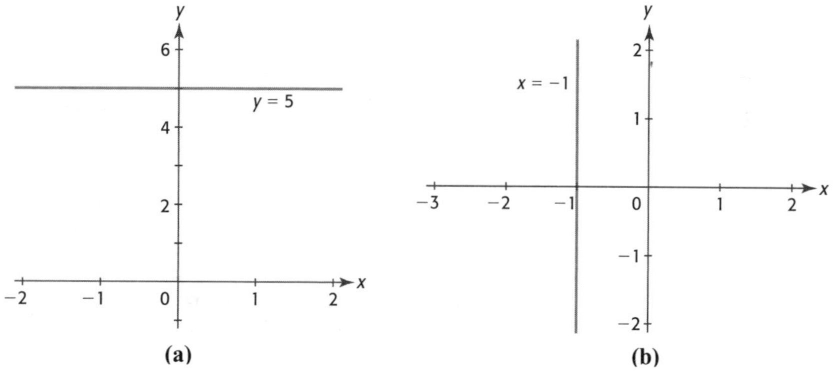

FIGURE 1.29

c. Because m is undefined, we cannot use the point-slope form to write the equation of this line. Lines with undefined slope are vertical lines. Every point on the vertical line through $(-1, 5)$ has an x-coordinate of -1. Thus, the equation of the line is $x = -1$. Note that this equation does not represent a function (Figure 1.29(b)). ■

As we saw in Example 2, parts (b) and (c), there are special forms when the lines are horizontal or vertical.

VERTICAL AND HORIZONTAL LINES

A vertical line has the form $x = a$, where a is a constant and a is the x-coordinate of any point on the line.

A horizontal line has the form $y = b$, where b is a constant and b is the y-coordinate of any point on the line.

In part (a) of Example 2, we found that the equation of the line was

$$y = \frac{3}{4}x + \frac{23}{4}$$

By multiplying both sides of this equation by 4 and writing the new equation with x and y on the same side of the equation, we have a new form of the equation:

$$4y = 3x + 23, \quad \text{or} \quad 3x - 4y = -23$$

This equation is called the **general form** of the equation of the line in Figure 1.28.

GENERAL FORM OF THE EQUATION OF A LINE

$ax + by = c$ where a, b, and c are real numbers, with a and b not both equal to 0.

If we know two points on a line, we can find the slope of the line and use either of the points and this slope to write the equation of the line.

EXAMPLE 3 Using Two Points to Write an Equation of a Line

Write the equation of the line that passes through the points $(-1, 5)$ and $(2, 4)$.

Solution

Because we know two points on the line, we can find the slope of the line.

$$m = \frac{4 - 5}{2 - (-1)} = \frac{-1}{3} = -\frac{1}{3}$$

We can now substitute one of the points and the slope in the point-slope form to write the equation. Using the point $(-1, 5)$ gives

$$y - 5 = -\frac{1}{3}[x - (-1)], \quad \text{or} \quad y - 5 = -\frac{1}{3}x - \frac{1}{3}, \quad \text{so} \quad y = -\frac{1}{3}x + \frac{14}{3}$$

Using the point $(2, 4)$ gives

$$y - 4 = -\frac{1}{3}(x - 2)$$

$$y - 4 = -\frac{1}{3}x + \frac{2}{3}$$

$$y = -\frac{1}{3}x + \frac{14}{3}$$

Notice that both equations are the same regardless of which of the two given points is used in the point-slope form. ■

In summary, these are the forms we have discussed for the equation of a line.

FORMS OF LINEAR EQUATIONS

General form:	$ax + by = c$	where a, b, and c are real numbers, with a and b not both equal to 0.
Point-slope form:	$y - y_1 = m(x - x_1)$	where m is the slope of the line and (x_1, y_1) is a point on the line.
Slope-intercept form:	$y = mx + b$	where m is the slope of the line and b is the y-intercept.
Vertical line:	$x = a$	where a is a constant, and a is the x-coordinate of any point on the line. The slope is undefined.
Horizontal line:	$y = b$	where b is a constant, and b is the y-coordinate of any point on the line. The slope is 0.

If the rate of change of the outputs with respect to the inputs is a constant, and we know two points that satisfy the conditions of the application, then we can write the equation of the linear function that models the application.

EXAMPLE 4 Inmate Population

The number of people (in millions) in U.S. prisons or jails grew at a constant rate from 1990 to 2000, with 1.15 million people incarcerated in 1990 and 1.91 million incarcerated in 2000.

a. Write the equation that models the number N of prisoners as a function of the year x.

b. The Bureau of Justice Statistics projected that 2.29 million people will be incarcerated in 2005. Does your model agree with this projection? (Source: Bureau of Justice Statistics)

Solution

a. The rate of growth is constant, so the points fit on a line, and a linear function can be used to find the model for the number of incarcerated people as a function of the years. The rate of change (and slope of the line) is given by

$$m = \frac{1.91 - 1.15}{2000 - 1990} = \frac{0.76}{10} = 0.076$$

Substituting in the point-slope form of the equation of a line (with either point) gives the equation of the line that contains the two points and thus models the application.

$$N - 1.91 = 0.076(x - 2000) \quad \text{or} \quad N = 0.076x - 150.09$$

b. Substituting 2005 for x in the function gives

$$N = N(2005) = 0.076(2005) - 150.09 = 2.29$$

Thus the projection that 2.29 million people will be incarcerated in 2005 agrees with this model. ■

Average Rate of Change

For a function whose graph is not a line, the slope may vary from point to point on the curve. The calculation and interpretation of the slope of a curve are topics you will study if you take a calculus course. However, there is a quantity called the **average rate of change** that can be applied to any function relating two variables.

In general, we can find the average rate of change of a function between two input values if we know how much the function outputs change between the two input values.

AVERAGE RATE OF CHANGE

The average rate of change of $f(x)$ with respect to x over the interval from $x = a$ to $x = b$ (where $a < b$) is calculated as

$$\text{average rate of change} = \frac{\text{change in } f(x) \text{ values}}{\text{corresponding change in } x \text{ values}} = \frac{f(b) - f(a)}{b - a}$$

How is average rate of change over some interval of points related to the slope of a line connecting the points? For any function, the average rate of change between two points is the slope of the line joining the two points. Such a line is called a **secant line**.

EXAMPLE 5 Toyota Sales

The total Toyota hybrid vehicle units sold for the years between 1997 and 2001 can be approximated by the model

$$S(x) = 2821x^3 - 75{,}653x^2 + 674{,}025x - 1{,}978{,}335$$

where x is the number of years after 1990. A graph of $S(x)$ is shown in Figure 1.30. Because $S(7) = 446$, there were 446 units sold in 1997; $S(9) = 16{,}506$, so the number sold in 1999 was 16,506. (Source: www.toyota.com)

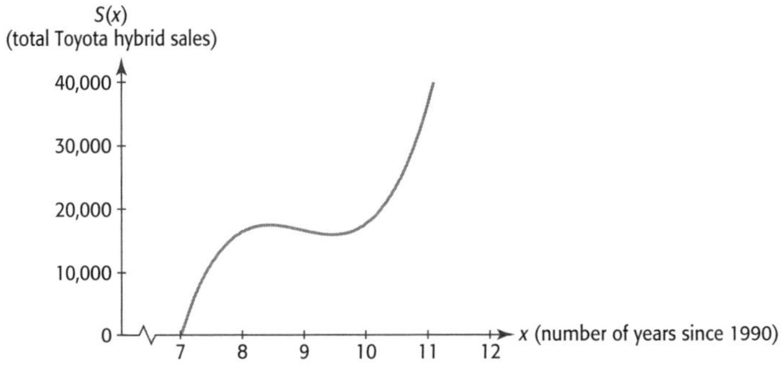

FIGURE 1.30

a. Find the average rate of change of Toyota hybrid sales between 1997 and 1999.

b. Interpret your answer to part (a).

c. What is the relationship between the slope of the secant line joining the points (7, 446) and (9, 16,506) and the answer to part (a)?

Solution

a. We find the average rate of change between these two points to be

$$\frac{16{,}506 - 446}{9 - 7} = \frac{16{,}060 \text{ units}}{2 \text{ years}} = 8030 \text{ units per year}$$

b. A possible interpretation is this: On average between 1997 and 1999, 8030 Toyota hybrid units were sold per year.

c. The slope of this line, shown in Figure 1.31, is the same as the average rate of change of the sales between 1997 and 1999 found in part (a). ■

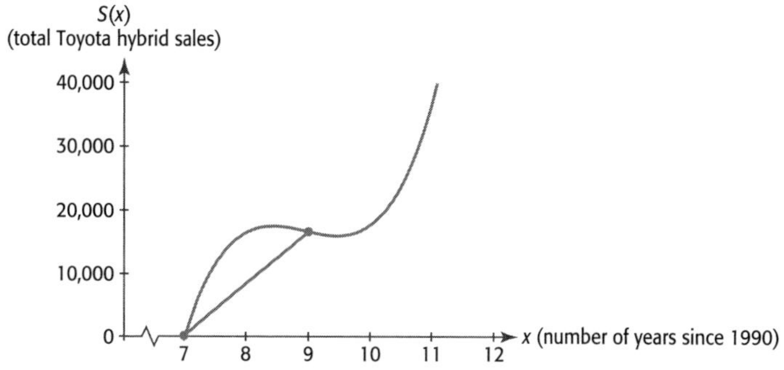

FIGURE 1.31 ■

An important skill in mathematics is finding the average rate of change of a function from a point $(x, f(x))$ to the point $(x + h, f(x + h))$ on the graph of $y = f(x)$. This average rate of change is given by the **difference quotient**

$$\frac{f(x + h) - f(x)}{h}$$

EXAMPLE 6 Average Rate of Change

For the function $f(x) = x^2 + 1$, whose graph is shown in Figure 1.32, find:

a. $f(x + h)$ **b.** $f(x + h) - f(x)$

c. The average rate of change $\dfrac{f(x + h) - f(x)}{h}$

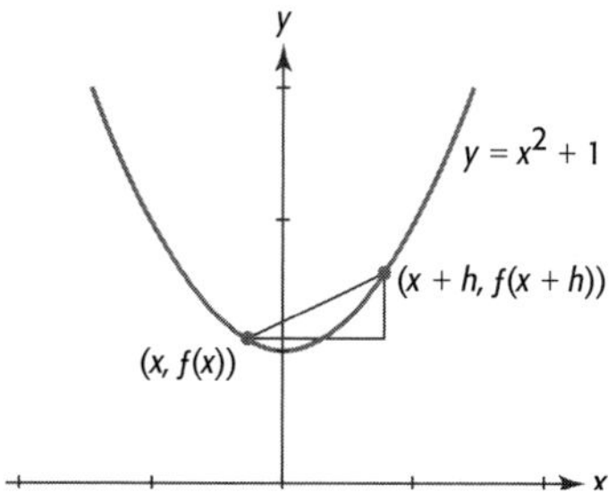

FIGURE 1.32

Solution

a. $f(x + h)$ is found by substituting $x + h$ for x in $f(x) = x^2 + 1$:

$$f(x + h) = (x + h)^2 + 1 = (x^2 + 2xh + h^2) + 1$$

b. $f(x + h) - f(x) = (x^2 + 2xh + h^2) + 1 - (x^2 + 1)$

$$= x^2 + 2xh + h^2 + 1 - x^2 - 1 = 2xh + h^2$$

c. $\dfrac{f(x + h) - f(x)}{h} = \dfrac{2xh + h^2}{h} = 2x + h$ ■

Approximately Linear Data

Real-life data are rarely perfectly linear, but some sets of real data points lie sufficiently close to a line so that two points can be used to create a linear function that models the data. Consider the following example.

EXAMPLE 7 High School Enrollment

Table 1.11 and Figure 1.33(a) show the enrollment (in thousands) in grades 9–12 of U.S. public and private schools for the years 1990–2002.

a. Create a scatter plot of the data. Does a line fit the data points exactly?

b. Find the average rate of change of the high school enrollment between 1990 and 2002.

c. Write the equation of the line determined by this rate and one of the two points.

TABLE 1.11

Year, x	Enrollment, y (thousands)
1990	12,488
1991	12,703
1992	12,882
1993	13,093
1994	13,376
1995	13,697
1996	14,060
1997	14,272
1998	14,428
1999	14,623
2000	14,802
2001	15,058
2002	15,332

(Source: U.S. Dept of Education)

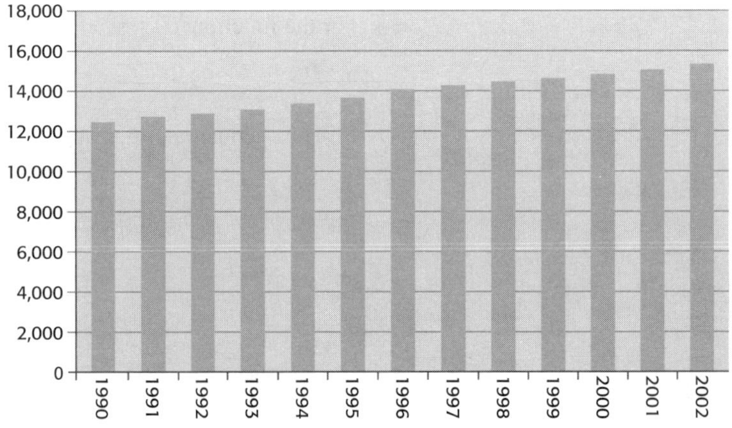

FIGURE 1.33(a)

Solution

a. A scatter plot is shown in Figure 1.33(b). The graph shows that the data do not fit a linear function exactly, but that a linear function could be used as an approximate model for the data.

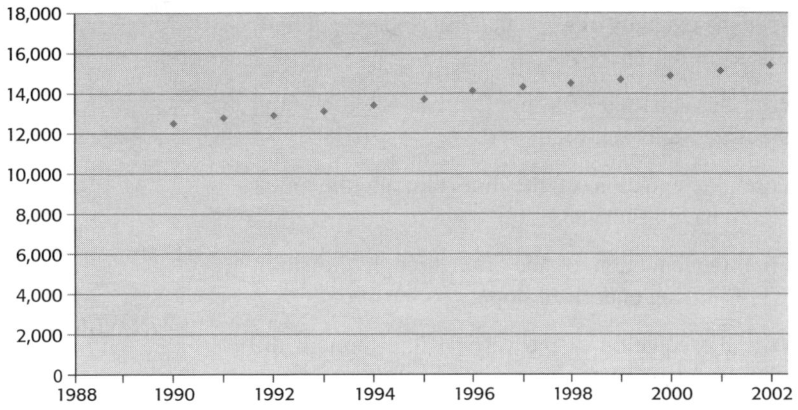

FIGURE 1.33(b)

b. To find the average rate of change between 1990 and 2002, we use the points (1990, 12,488) and (2002, 15,332).

$$\frac{15{,}332 - 12{,}488}{2002 - 1990} = \frac{2844}{12} = 237$$

This means that the high school enrollments have grown at an average rate of 237 thousand per year (because the outputs are in thousands).

c. Because the enrollment growth is approximately linear, the average rate of change can be used as the slope of the graph of a linear function describing the enrollments from 1990 through 2002. The graph connects the points (1990, 12,488) and (2002, 15,332), so we can use either of these two points to write the equation of the line. Using the point (1990, 12,488) and $m = 237$ gives the equation.

$$y - 12{,}488 = 237(x - 1990)$$
$$y - 12{,}488 = 237x - 471{,}630$$
$$y = 237x - 459{,}142$$

This equation is a model of the enrollment, in thousands, as a function of the year. ■

The equation in Example 7 approximates a model for the data, but it is not the best possible model for the data. In Section 1.6 we will see how technology can be used to find the linear function that is the best fit for a set of data of this type.

1.4 SKILLS CHECK

1. Write the equation of the line that has slope 4 and y-intercept $\frac{1}{2}$.

2. Write the equation of the line that has slope 5 and y-intercept $\frac{1}{3}$.

3. Write the equation of the line that has slope $\frac{1}{3}$ and y-intercept 3.

4. Write the equation of the line that has slope $-\frac{1}{2}$ and y-intercept -8.

5. Write the equation of the line that has slope $-\frac{3}{4}$ and y-intercept 2.

6. Write the equation of the line that has slope 3 and y-intercept $\frac{2}{5}$.

7. Write the equation of the line through the point $(-1, 4)$ having slope $m = 5$.

8. Write the equation of the line through the point $(-4, 3)$ having slope $m = -\frac{1}{2}$.

9. Write the equation of a line that has slope $-\frac{3}{4}$ and passes through $(4, -6)$.

10. Write the equation of a line that has slope $-\frac{2}{3}$ and passes through $(-3, 6)$.

11. Write the equation of the line through the point $(-1, 4)$ having slope $m = 0$.

12. Write the equation of the line through the point $(-1, 4)$ having undefined slope.

13. Write the equation of the vertical line through the point $(9, -10)$.

14. Write the equation of the horizontal line through the point $(9, -10)$.

15. Write the equation of the line that passes through $(-2, 1)$ and $(4, 7)$.

16. Write the equation of the line that passes through $(3, 2)$ and $(5, 8)$.

17. Write the equation of the line that passes through $(-1, 3)$ and $(2, 6)$.

18. Write the equation of the line that passes through $(-3, 4)$ and $(4, -3)$.

19. Write the equation of the line that passes through $(-4, 2)$ and $(-3, 5)$.

20. Write the equation of the line that passes through $(2, -5)$ and $(5, -6)$.

21. Write the equation of the line that passes through $(9, 2)$ and $(9, 5)$.

22. Write the equation of the line that passes through $(5, 2)$ and $(-3, 2)$.

23. Write the equation of the line having x-intercept -5 and y-intercept 4.

24. Write the equation of the line having x-intercept 4 and y-intercept -5.

For Exercises 25 and 26, write the equation of the line whose graph is shown.

25.

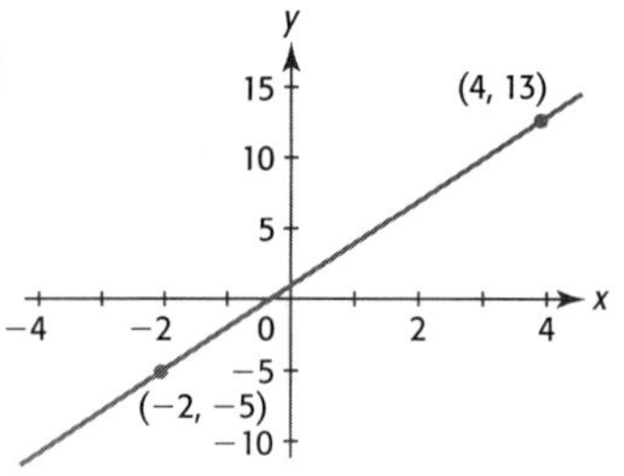

26.

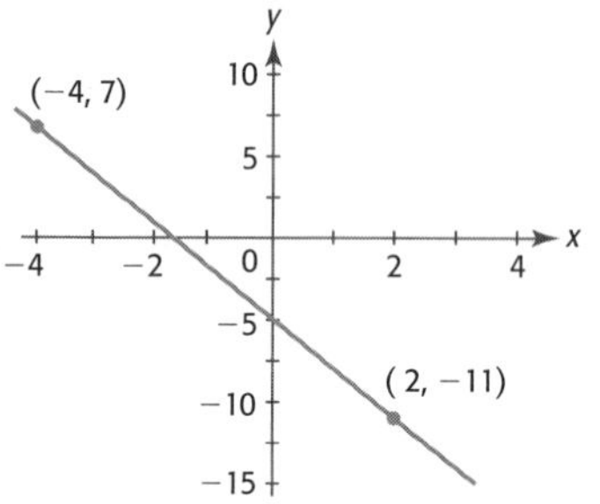

27. Write the equation of a function if its rate of change is -15 and $y = 12$ when $x = 0$.

28. Write the equation of a function if its rate of change is -8 and $y = -7$ when $x = 0$.

29. For the function $y = x^2$, compute the average rate of change between $x = -1$ and $x = 2$.

30. For the function $y = x^3$, compute the average rate of change between $x = -1$ and $x = 2$.

For the functions given in Exercises 31–34, find

$$\frac{f(x + h) - f(x)}{h}$$

31. $f(x) = 45 - 15x$

32. $f(x) = 32x + 12$

33. $f(x) = 2x^2 + 4$

34. $f(x) = 3x^2 + 1$

In Exercises 35–36, use the table, which gives a set of input values x and the corresponding outputs y that satisfy a function.

35. **a.** Would you say that a linear function could be used to model the data? Explain.
b. Write the equation of the linear function that fits the data.

x	10	20	30	40	50
y	585	615	645	675	705

36. **a.** Verify that the values satisfy a linear function.
b. Write the equation of the linear function that fits the data.

x	1	7	13	19
y	−5	8.5	17.5	26.5

1.4 EXERCISES

37. *Utility Charges* Palmetto Electric determines its monthly bills for residential customers by charging a base price of \$8.95 plus an energy charge of 9.35 cents for each kilowatt-hour (KWh) used. Write an equation for the monthly charge y (in dollars) as a function of x, the number of KWh used.

38. *Phone Bills* For interstate calls, AT&T charges 7 cents per minute plus a base charge of \$4.95 each month. Write an equation for the monthly charge y as a function of the number of minutes of use.

39. *Depreciation* A business uses straight-line depreciation to determine the value y of a piece of machinery over a 10-year period. Suppose the original value (when $t = 0$) is equal to \$36,000 and its value is reduced by \$3600 each year. Write the linear equation that models the value y of this machinery at the end of year t.

40. *Sleep* Each day, a young person should sleep 8 hours plus $\frac{1}{4}$ hour for each year the person is under 18 years of age. Write the equation relating hours of sleep y to age x, for $6 \leq x \leq 18$.

41. *Population Growth* Assume the growth of the population of Del Webb's Sun City Hilton Head community was linear from 1996 to 2000, with a population of 198 in 1996 and a rate of growth of 705 per year.
    a. Write an equation for the population P of this community as a function of the number of years x after 1996.
    b. Use the function to estimate the population in 2002.
    c. For what years in the population more than 4,428?
    (Source: Island Packet, March 2000)

42. *SAT Scores* The composite SAT score for the Beaufort County School District was 952 points in 1994, and the average rate of increase was 0.51 point per year. Write a linear function that models the SAT scores y as a function of x, the number of years after 1994.
(Source: Island Packet, October 1999)

43. *Depreciation* A business uses straight-line depreciation to determine the value y of an automobile over a 5-year period. Suppose the original value (when $t = 0$) is equal to \$26,000 and the salvage value (when $t = 5$) is equal to \$1000.
    a. By how much has the automobile depreciated over the 5 years?
    b. By how much is the value of the automobile reduced at the end of each of the 5 years?
    c. Write the linear equation that models the value s of this automobile at the end of year t.

44. *Retirement* For Pennsylvania state employees for whom the average of the three best yearly salaries is \$75,000, the retirement plan gives an annual pension of 2.5% times 75,000, multiplied by the number of years of service. Write the linear function that models the pension P in terms of the number of years of service, y.
(Source: Pennsylvania State Retirement Fund)

45. *Patrol Cars* The Beaufort County Sheriff's office assigns a patrol car to each of its deputy sheriffs, who keeps the car 24 hours a day. As the population of the county grew, the number of deputies and the number of patrol cars also grew. The data in the following table could describe how many patrol cars were necessary as the number of deputies increased. Write the function that models the relationship between the number of deputies and patrol cars.

Number of Deputies	10	50	75	125	180	200	250
Number of Patrol Cars in Use	10	50	75	125	180	200	250

46. *Life Insurance* The monthly premiums for a \$100,000 life insurance policy for males aged 27–32 are shown in the table below. Write the linear model for the data in the table.

Age (years)	27	28	29
Premium (dollars per month)	11.81	11.81	11.81

Age (years)	30	31	32
Premium (dollars per month)	11.81	11.81	11.81

47. *Profit* A company charting its profits notices that the relationship between the number of units sold, x, and the profit, P, is linear. If 300 units sold results in \$4650 profit and 375 units results in \$9000 profit, write the equation that models its profit.

48. *Cost* A company buys and retails baseball caps. The total cost function is linear, the total cost for 200 caps is \$2680, and the total cost of 500 caps is \$3530. Write the equation that models this cost function.

49. *Depreciation* Suppose the cost of a business property is \$1,920,000 and a company depreciates it with the straight-line method If v is the value of the property after x years and the line representing the value as a function of years passes through the points (10, 1,310,000) and (20, 700,000), write the equation that gives the annual value of the property.

50. *Depreciation* Suppose the cost of a business property is \$860,000 and a company depreciates it with the straight-line method. Suppose y is the value of the property after t years.
   a. What is the value at the beginning of the depreciation (when $t = 0$)?
   b. If the property is completely depreciated ($y = 0$) in 25 years, write the equation of the line representing the value as a function of years.

51. *Cigarette Use* The percent of high school seniors who smoke cigarettes can be modeled by a linear function $p = f(t)$, where t is the number of years after 1975. If two points on the graph of this function are (6, 32.7) and (26, 32.1), write the equation of this function.

52. *Earnings and Race* Data from 2003 for various age groups show that for each \$100 increase in median weekly income for whites, the median weekly income for blacks increases by \$61.90. Also, for these workers, the median weekly income for whites was \$676 and for blacks was \$527. Write the equation that gives the median weekly income for blacks as a function of the median weekly income for whites.
(Source: U.S. Department of Labor)

53. *Blood Alcohol Percent* The table below gives the number of drinks and the resulting blood alcohol percent for a 90-lb woman. ("One drink" is equal to 1.25 oz of 80 proof liquor, 12 oz of regular beer, or 5 oz of table wine, and many states have set .08% as the legal limit for driving under the influence.)
   a. The rate of change in blood alcohol percent per drink for a 90-lb woman is a constant. What is it?
   b. Write the equation of the function that models the blood alcohol percent as a function of the number of drinks.

Number of Drinks	0	1	2	3	4
Blood Alcohol Percent	0	.05	.10	.15	.20

Number of Drinks	5	6	7	8	9
Blood Alcohol Percent	.25	.30	.35	.40	.45

(Source: Pennsylvania Liquor Control Board)

54. *Drinking and Driving* The following table gives the number of drinks and the resulting blood alcohol percent for a 180-lb man legally considered driving under the influence (DUI).
   a. The average rate of change of the blood alcohol percent with respect to the number of drinks is a constant. What is it?
   b. Use the rate of change and one point determined by a number of drinks and the resulting blood alcohol percent to write the equation of a linear model for this data.

Number of Drinks	5	6	7	8	9	10
Blood Alcohol Percent	.11	.13	.15	.17	.19	.21

(Source: Pennsylvania Liquor Control Board)

55. *Men in the Workforce* The number of men in the workforce (in millions) for selected decades from 1890 to 1990 is shown in the figure on page 67. The decade is defined by the year at the beginning of the decade and $g(t)$ is defined by the average number of men (in millions) in the workforce during the decade (indicated by the point on the graph within the decade). This data can be approximated by the linear model determined by the line connecting (1890, 18.1) and (1990, 68.5).
   a. Write the equation of the line connecting these two points to find a linear model for this data.
   b. Does this line appear to be a reasonable fit to the data points?
   c. How does the slope of this line compare with the average rate of change in the function during this period?

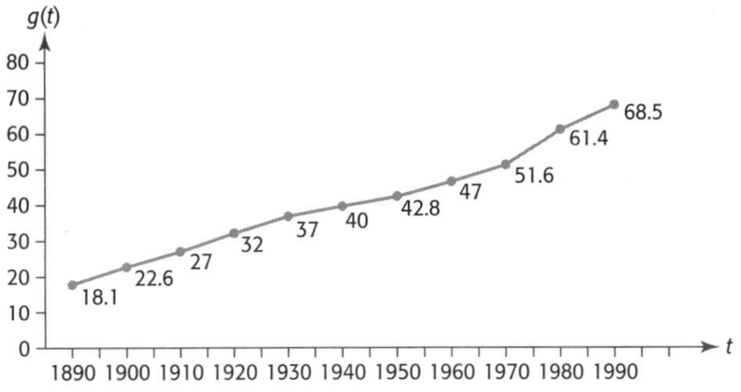

(Source: 1998 *World Almanac*)

56. *Farm Workers* The figure below shows the percent p of U.S. workers in farm occupations for selected years t.

a. Write the equation of a line connecting the points (1820, 71.8) and (1994, 2.6), with values rounded to two decimal places.

b. Does this line appear to be a reasonable fit to the data points?

c. Interpret the slope of this line as a rate of change of the percent of farm workers.

d. Can the percent p of U.S. workers in farm occupations continue to fall at this rate for 20 more years? Why or why not?

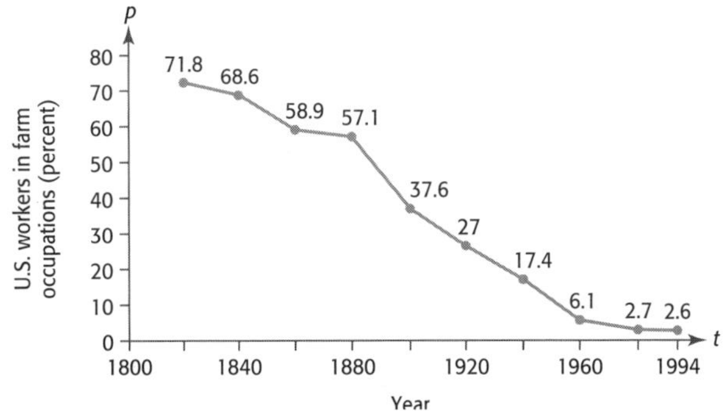

(Source: U.S. Department of Agriculture)

57. *Education Spending* The figure below gives the amount (in billions of dollars) spent on education during the years 1996–2001.

a. What is the average annual rate of change in spending between 1996 and 2001?

b. If a line were drawn connecting the data points (1996, 23) and (2001, 40.1), what is the slope of the line?

c. Is the rate of change in spending for education the same each year?

d. Can this data be modeled exactly by a linear equation? Explain.

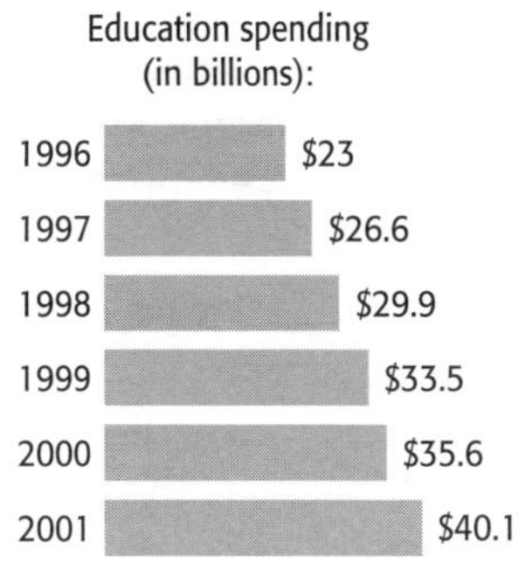

(Source: U.S. Department of Education)

58. *Enrollment Projection* The figure below shows enrollment projections for three schools. (Outputs are measured in students.)

a. What is the slope of the line joining the two points in the figure that show the Beaufort enrollment projections?

b. Find the average rate of change in projected enrollment for students in Beaufort schools between 2000 and 2005.

c. How should this information be used when planning future school building in Beaufort?

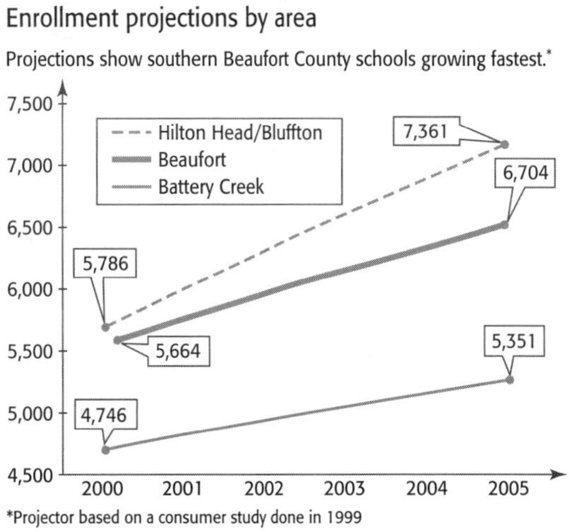

(Source: Beaufort County School District)

59. *Teenage Mothers* The percent of births to teenage mothers that are out-of-wedlock can be approximated by a linear function of the number of years after 1950. The percent was 15 in 1960 and 76 in 1996.

a. What is the slope of the line joining the points (10, 15) and (46, 76)? Round the answer to two decimal places.

b. What is the average rate of change in the percent of teenage out-of-wedlock births over this period?
c. What does this average rate of change tell about the out-of-wedlock births to teenage mothers?
d. Write the equation of the line joining the two given points. Round the values to two decimal places.
(Source: Father Facts)

60. *Voting* The percent of eligible people voting in presidential elections can be expressed as a linear function of the years 1960–1996. The percent was 63.1 in 1960 and 55.1 in 1992.
a. What is the slope of the line joining the given points?
b. What is the average rate of change in the percent voting in these elections? Interpret this value.
c. Write the equation of the line joining these two given points.
(Source: Federal Election Commission)

61. *Prison Population* The graph showing the total number of prisoners in state and federal prisons for the years 1960 through 1997 is shown in the figure. There were 212,953 prisoners in 1960 and 1,197,590 in 1997. (Source: U.S. Bureau of Justice Statistics)
a. What is the average rate of growth of the prison population from 1960 to 1997?
b. What is the slope of the line connecting the points satisfying the conditions above?
c. Write the equation of the secant line joining these two points on the curve.
d. Can this secant line be used to make a good estimate of total number of prisoners in 1998?
e. Points associated with what two years could be used to write a linear equation that will give a better estimate of total number of prisoners in 1998?

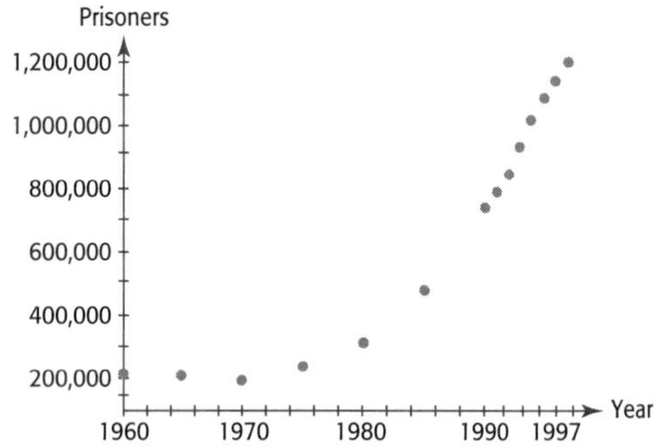

62. *Investment* The graph of the future value of an investment of \$1000 for x years earning interest at a rate of 8% compounded continuously is shown in the following figure. The \$1000 investment is worth approximately \$1083 after 1 year and about \$1492 after 5 years.
a. What is the average rate of change of the future value over the 4-year period?
b. Interpret this average rate of change.
c. What is the slope of the line connecting the points satisfying the conditions above?
d. Write the equation of the secant line joining the two given points.

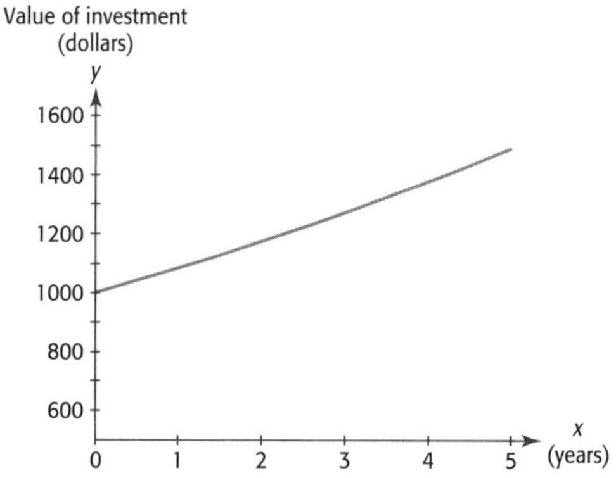

63. *Working Age* The scatter plot below projects the ratio of the working-age population to the elderly.
a. Does the data appear to fit a linear function?
b. The data points shown in the scatter plot from 2010 to 2030 are projections made from a mathematical model. Do those projections appear to be made with a linear model? Explain.
c. If the ratio is projected to be 3.9 in 2010 and 2.2 in 2030, what is the average annual rate of change of the data over this period of time?
d. Write the equation of the line joining these two points.

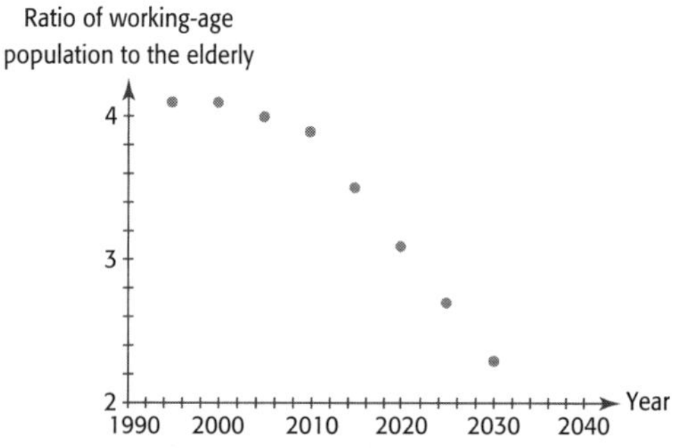

(Source: *Newsweek*, December 1999)

64. *Women in the Workforce* The number of women in the workforce (in millions) for selected years from 1890 to 1990 is shown in the following figure.
a. Would the data in the scatter plot be modeled well by a linear function? Why or why not?
b. The number of women in the workforce was 3.704 million in 1890 and 16.443 million in 1950. What

is the average rate of change in the number of women in the workforce during this period? Round the answer to four decimal places.

c. If the number of women in the workforce was 16.443 million in 1950 and 59.531 million in 1990, what is the average rate of change in the number of women in the workforce during this period?

d. Is it reasonable that these two average rates of change are different? How can you tell this from the graph?

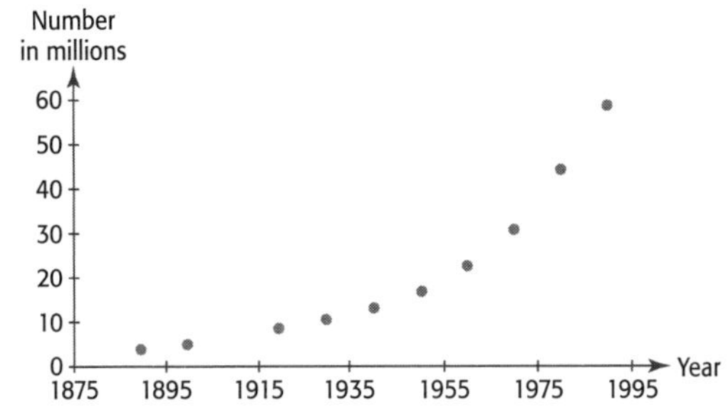

(Source: *Newsweek*, December 1999)

65. *U.S. Population* The total U.S. population for selected years from 1950 to 1995 is shown in the table below, with the population given in thousands. Using x to represent the number of years from 1950 and y to represent the number of thousands of people:

a. Find the average annual rate of change in population during these years, with the appropriate units.

b. Write the equation of the line connecting the points associated with 1950 and 1995.

c. Use the equation to estimate the population in 1975. Does it agree with the population shown in the table?

d. Why might the values be different?

Year	Population (thousands)	Year	Population (thousands)
1950	152,271	1975	215,973
1955	165,931	1980	227,726
1960	180,671	1985	238,466
1965	194,303	1990	249,948
1970	205,052	1995	263,044

(Source: U.S. Census Bureau)

66. *Smoking Cessation* The table below gives the percent of people over age 20 who once smoked and quit smoking between 1965 and 1990.

a. Create a new table with the data aligned so that the years are represented by the number of years past 1965.

b. An equation that models the data in the new table is $y = 0.723x + 29.246$. Show the graph of the data and the equation on the same set of axes.

c. Do the graphs indicate that a linear model fits the data reasonably well?

Year	People Who Have Quit Smoking (percent)	Year	People Who Have Quit Smoking (percent)
1965	29.8	1979	39
1966	29.3	1980	38.9
1970	35.2	1983	41.8
1975	36	1985	45
1976	37	1987	44.9
1977	36.6	1990	48.4
1978	38.3		

(Source: Indiana Tobacco Control Center, U.S. Centers for Disease Control)

67. *Taxes* The table on the next page shows some sample incomes and the income tax due for each income, and the figure below shows a graph of the data points.

a. Is the rate of change of tax with respect to income bracket constant for the incomes shown in the table?

b. What is the rate of change in tax per $1 in income?

c. Write a linear equation to represent income tax due as a function of taxable income for the incomes in the table.

d. Verify that the model fits the data by evaluating the function at $x = 30{,}100$ and at $x = 30{,}300$ and comparing the resulting y-values with the income tax due for these taxable incomes.

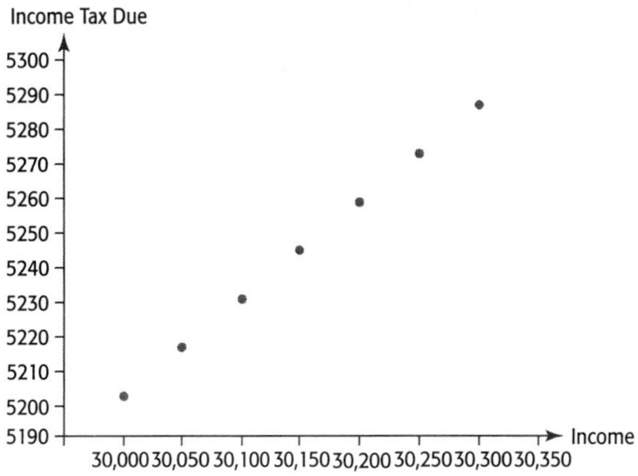

Taxable Income	Income Tax Due
$30,000	$5203
30,050	5217
30,100	5231
30,150	5245
30,200	5259
30,250	5273
30,300	5287

(Source: U.S. Federal tax table for 1997)

68. *Social Agency* A social agency provides emergency food and shelter to two groups of clients. The first group has x clients who need an average of $300 for emergencies, and the second group has y clients who need an average of $200 for emergencies. The agency has $100,000 to spend for these two groups.
   a. Write the equation that gives the number of clients who can be served in each group.
   b. Find the y-intercept and the slope of the graph of this equation. Interpret each value.
   c. If 10 clients are added from the first group, what happens to the number served in the second group?

1.5 Algebraic and Graphical Solutions of Linear Equations

Key Concepts

- Algebraic solutions of linear equations
- Solving real-world application problems
- Zero of a function
- Solutions, zeros, and x-intercepts
- Graphical solution of linear equations
- x-intercept method
- Intersection method
- Formulas
- Solving an equation for a specified variable

Section Preview Prison Sentences

The average sentence length and the average time served in state prisons for various crimes in 1996 is shown in Table 1.12. This data can be used to create a linear function which is an approximate model for the data:

$$y = 0.55x - 2.886$$

This function describes the 1996 mean time y served in prison for a crime as a function of the mean sentence length x, where x and y are each measured in months. Assuming that this model remains valid after 1996, it predicts that the mean time served on a 5-year (60 months) sentence is $0.55(60) - 2.886 \approx 30$ months; that is, approximately $2\frac{1}{2}$ years.

TABLE 1.12

Average Sentence (months)	Average Time Served (months)
62	30
85	45
180	95
116	66
92	46
61	33
56	26

(Source: U.S. Department of Justice, "Truth in Sentencing in State Prisons," January 1999)

To find the sentence that would give an expected time served of 10 years (120 months), we use algebraic or graphical methods to solve the equation

$$120 = 0.55x - 2.866$$

for x. In this section, we use additional algebraic methods and graphical methods to solve linear equations in one variable.

Algebraic Solution of Linear Equations

We can use algebraic, graphical, or a combination of algebraic and graphical methods to solve linear equations. Sometimes it is more convenient to use algebraic solution methods rather than graphical solution methods, especially if an exact solution is desired. The steps used to solve a linear equation in one variable algebraically follow.

STEPS FOR SOLVING A LINEAR EQUATION IN ONE VARIABLE

1. If a linear equation contains fractions, multiply both sides of the equation by a number that will remove all denominators from the equation. If there are two or more fractions, use the least common denominator (LCD) of the fractions.
2. Remove any parentheses or other symbols of grouping.
3. Perform any additions or subtractions to get all terms containing the variable on one side and all other terms on the other side of the equation. Combine like terms.
4. Divide both sides of the equation by the coefficient of the variable.
5. Check the solution by substitution in the original equation. If a real-world solution is desired, check the algebraic solution for reasonableness in the real-world situation.

EXAMPLE 1 Algebraic Solutions

a. Solve for x: $\dfrac{2x - 3}{4} = \dfrac{x}{3} + 1$

b. Solve for y: $y - \dfrac{1}{2}\left(\dfrac{y}{2}\right) = -6$

Solution

a.

$$\frac{2x - 3}{4} = \frac{x}{3} + 1$$

$$12\left(\frac{2x - 3}{4}\right) = 12\left(\frac{x}{3} + 1\right)$$ Multiply both sides by the LCD, 12.

$$3(2x - 3) = 12\left(\frac{x}{3} + 1\right)$$ Simplify the fraction on the left.

$$6x - 9 = 4x + 12$$ Remove parentheses using the distributive property.

$$2x = 21$$ Subtract $4x$ from both sides and add 9 to both sides.

$$x = \frac{21}{2}$$ Divide both sides by 2.

Check the result: $\dfrac{2\left(\dfrac{21}{2}\right) - 3}{4} \stackrel{?}{=} \dfrac{\dfrac{21}{2}}{3} + 1 \Rightarrow \dfrac{9}{2} = \dfrac{9}{2}$

b. $y - \dfrac{1}{2}\left(\dfrac{y}{2}\right) = -6$

$y - \dfrac{y}{4} = -6$ Remove parentheses first to find the LCD.

$4\left(y - \dfrac{y}{4}\right) = 4(-6)$ Multiply both sides by the LCD, 4.

$4y - y = -24$ Remove parentheses using the distributive property.

$3y = -24$ Combine like terms.

$y = -8$ Divide both sides by 3.

Check the result: $-8 - \dfrac{1}{2}\left(\dfrac{-8}{2}\right) \stackrel{?}{=} -6 \Rightarrow -6 = -6$ ■

As the next example illustrates, we solve an application problem that is set in a real-world context by using the same solution methods. However, you must remember to include units of measure with your answer and check that your answer makes sense in the problem situation.

EXAMPLE 2 Credit Card Debt

It is hard for people to pay off credit card debts in a reasonable period of time because of high interest rates. The interest paid on a \$10,000 debt over 3 years is approximated by

$$y = 175.393x - 116.287 \text{ dollars}$$

when the interest rate is x%. What is the interest rate if the interest is \$1637.60? (Source: Consumer Federation of America)

Solution

To answer this question, we solve the linear equation

$$1637.60 = 175.393x - 116.287$$
$$1753.887 = 175.393x$$
$$x = 9.9998$$

Thus, the interest is \$1637.60 if the interest rate is approximately 10%.
(Note that if you check the approximate answer, you are checking only for the reasonableness of the estimate.) ■

EXAMPLE 3 Stock Market

For a period of time, a man is very successful speculating on an Internet stock, with its value growing to \$100,000. However, the stock value drops rapidly until its value is reduced by 40%. What percent increase will have to occur before the latest value returns to \$100,000?

Solution

The value of the stock after the 40% loss is $100{,}000 - .40(100{,}000) = 60{,}000$. To find the percent p of increase that is necessary to return the value to 100,000, we solve

$$60{,}000 + 60{,}000p = 100{,}000$$
$$60{,}000p = 40{,}000$$
$$p = \frac{40{,}000}{60{,}000} = \frac{2}{3} = 66\frac{2}{3}\%$$

Thus, the stock must increase by $66\frac{2}{3}\%$ for it to return to a value of \$100,000. ■

Solutions, Zeros, and x-Intercepts

Because x-intercepts are x values that make the output of the function equal to 0, these intercepts are also called **zeros** of the function.

ZERO OF A FUNCTION

Any number a for which $f(a) = 0$ is called a **zero** of the function $f(x)$. If a is real, a is an x-intercept of the graph of the function.

The zeros of the function are values that make the function equal to 0, so they are also solutions to the equation $f(x) = 0$.

The following three concepts are numerically the same:

The x-intercepts of the graph of $y = f(x)$
The real zeros of the function f
The real solutions to the equation $f(x) = 0$

The following example illustrates the relationships that exist among x-intercepts of the graph of a function, zeros of the function, and solutions to associated equations.

EXAMPLE 4 Relationships Among x-Intercepts, Zeros, and Solutions

For the function $f(x) = 13x - 39$, find:

a. $f(3)$ **b.** The zero of $f(x) = 13x - 39$

c. The x-intercept of the graph of $y = 13x - 39$

d. The solution to the equation $13x - 39 = 0$

Solution

a. Evaluate the output $f(3)$ by substituting the input 3 for x in the function:

$$f(3) = 13(3) - 39 = 0$$

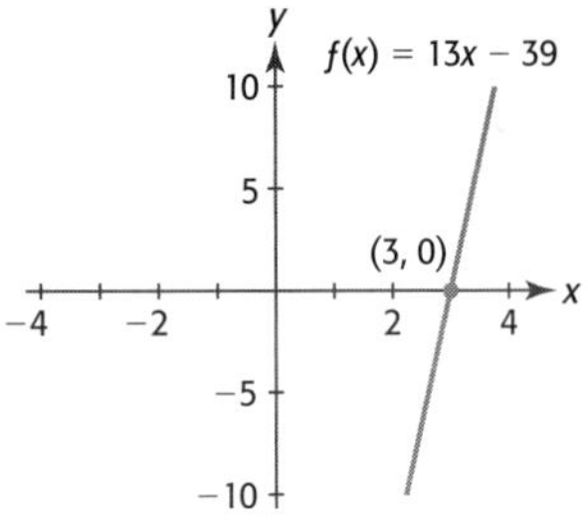

(a)

	A	B
1	x	y
2	3	0

(b)

FIGURE 1.34

b. Because $f(x) = 0$ when $x = 3$, we say that 3 is a zero of the function.

c. The x-intercept of a graph occurs at the value of x where $y = 0$, so the x-intercept is $x = 3$.

The only x-intercept is $x = 3$ because the graph of $f(x) = 13x - 39$ is a line that crosses the x-axis at only one point (Figure 1.34(a)).

d. The solution to the equation $13x - 39 = 0$ is $x = 3$ because $13x - 39 = 0$ gives $13x = 39$ or $x = 3$. ■

We can use TRACE, CALC VALUE, or TABLE on a graphing utility to check the reasonableness of a solution. A graph of $y = 13x - 39$ with x between -4 and 4 and y between -10 and 10 is given in Figure 1.34(a). We can use this graph to confirm that $y = 0$ when $x = 3$. A spreadsheet for $y_1 = 13x - 39$, shown in Figure 1.34(b), also confirms that $f(3) = 0$.

Graphical Solution of Linear Equations

We can also view the graph or use TRACE on a graphing utility to obtain a quick estimate of an answer. TRACE may or may not provide the exact solution to an equation, but it will provide an estimate of the solution. If your calculator or computer has a graphical or numerical solver, the solver can be used to obtain the solution to the equation $f(x) = 0$. Recall that an x-intercept of the graph of $y = f(x)$, a zero of $f(x)$, and the solution to the equation $f(x) = 0$ are all different names for the same input value. If the graph of the linear function $y = f(x)$ is not a horizontal line, we can find the one solution to the nonconstant linear equation $f(x) = 0$ as described below. We call this solution method the ***x*-intercept method**.

Zero of a function
Any number a for which f(a)=0 is called a zero of a function f(x).
If a is real, a is the x intercept of the graph
a is also the solution of the equation f(x)=0

SOLVING A LINEAR EQUATION USING THE *X*-INTERCEPT METHOD WITH GRAPHING UTILITIES

1. Rewrite the equation to be solved with 0 (and nothing else) on one side of the equation.
2. Enter the nonzero side of the equation found in the previous step in the equation editor of your graphing utility and graph the line in an appropriate viewing window. Be certain that you can see the line cross the horizontal axis on your graph.
3. Find the x-intercept by inspection with TRACE or by using the graphical solver that is often called ZERO. The x-intercept is the value of x that makes the equation equal to zero, so it is the solution to the equation. The value of x found by this method is sometimes a decimal approximation of the exact solution rather than the exact solution.

EXAMPLE 5 Graphical Solution

Solve $\frac{2x - 3}{4} = \frac{x}{3} + 1$ for x using the x-intercept method.

Solution

To solve the equation using the x-intercept method, we first rewrite the equation with 0 on one side.

$$0 = \frac{x}{3} + 1 - \frac{2x - 3}{4}$$

Next, enter the right-hand side of the equation,

$$\frac{x}{3} + 1 - \frac{2x - 3}{4}$$

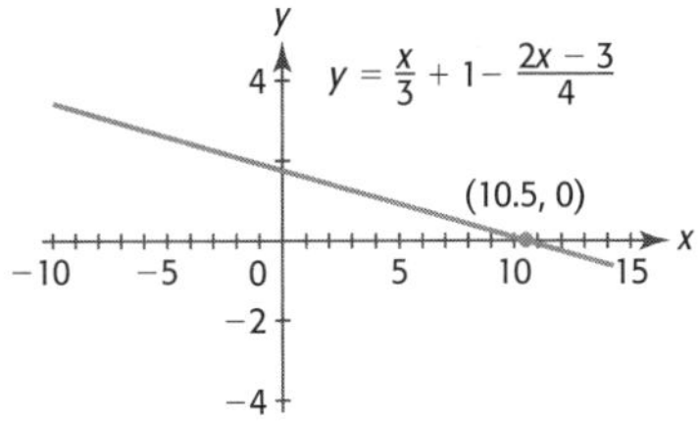

FIGURE 1.35

in the equation editor of your graphing utility as

$$y_1 = (x/3) + 1 - (2x - 3)/4$$

and graph this function. Use parentheses as needed to preserve the order of operations. You should obtain a graph similar to the one seen in Figure 1.35, which was obtained using the viewing window $[-10, 15]$ by $[-4, 4]$. However, any graph in which you can see the x-intercept will do. Using ZERO gives the x-intercept of the graph, 10.5. (See Figure 1.35.) This is the value of x that makes $y = 0$, and thus the original equation true, so it is the solution to this linear equation.

You can determine if $x = 10.5 = \frac{21}{2}$, found graphically, is the exact answer by substituting this value into the original equation. Both sides of the original equation are equal when $x = 10.5$, so it is the exact solution. ■

We can also use the intersection method described below to solve linear equations:

SOLVING A LINEAR EQUATION USING THE INTERSECTION METHOD

1. Enter the left side of the equation as y_1 and the right side as y_2. Graph both of these equations on a window that shows their point of intersection.
2. Find the point of intersection of the two graphs. This is the point where $y_1 = y_2$. The x-value of this point is the value of x that makes the two sides of the equation equal, so it is the solution to the original equation.

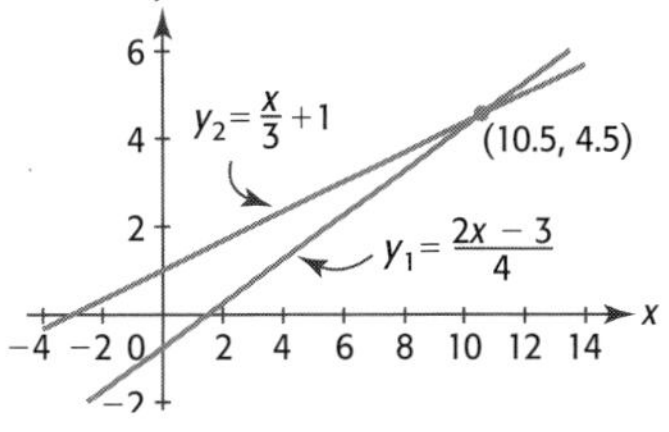

FIGURE 1.36

The equation in Example 5,

$$\frac{2x - 3}{4} = \frac{x}{3} + 1$$

can be solved using the intersection method. Enter the left side as

$$y_1 = (2x - 3)/4 \text{ and the right side as } y_2 = (x/3) + 1$$

Graphing these two functions gives the graphs in Figure 1.36. The point of intersection is found to be $(10.5, 4.5)$. So, the solution is $x = 10.5$ (as we found in Example 5).

EXAMPLE 6 Criminal Sentences

Solving $120 = 0.55x - 2.886$, where x is measured in months, gives the sentence for a crime that would give an expected time served of 10 years. Solve this equation by using the

a. x-intercept method.

b. intersection method.

Solution

a. We solve the linear equation $120 = 0.55x - 2.886$ by using the x-intercept method as follows:

Rewrite the equation in a form with 0 on one side:

$$0 = 0.55x - 2.886 - 120$$

Combine the constant terms to obtain

$$0 = 0.55x - 122.886$$

Enter $y_1 = 0.55x - 122.886$ and graph this equation (Figure 1.37). Using ZERO gives an x-intercept of approximately 223 (Figure 1.37). Thus, if a prisoner receives a sentence of 223 months, we would expect him or her to serve 120 months, or 10 years.

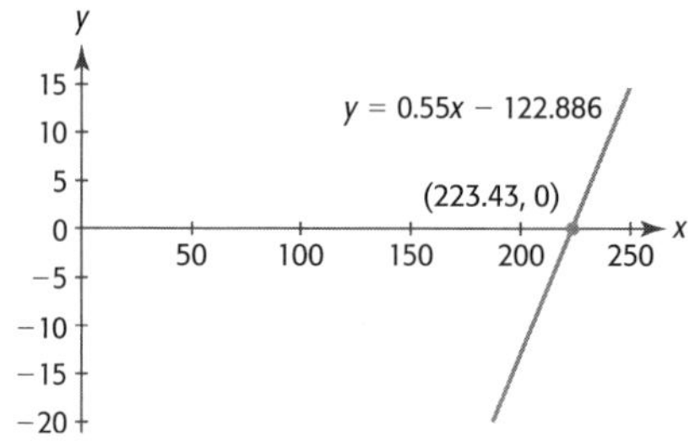

FIGURE 1.37

b. To use the intersection method, enter $y_1 = 0.55x - 2.886$ and $y_2 = 120$, graph these equations, and find the intersection of the lines. Figure 1.38 shows the point of intersection, and again we see that x is approximately 223.

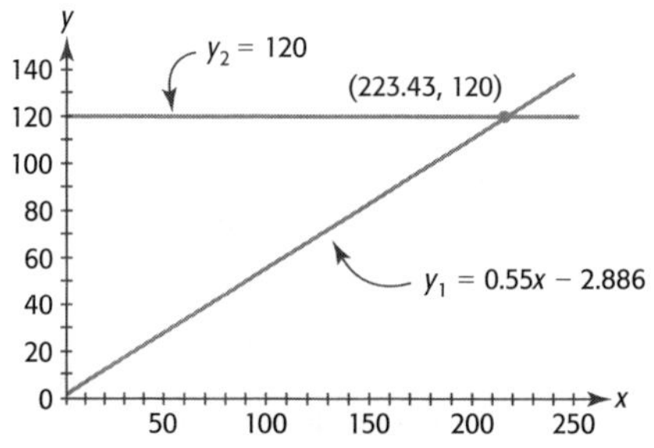

FIGURE 1.38

■

Spreadsheet Solution

We can use the **Goal Seek** feature of Excel to find the x-intercept of the graph of a function; that is, the zeros of the function. To solve

$$120 = 0.55x - 2.886$$

we enter the nonzero side of $0 = 0.55x - 122.886$ as shown in Table 1.13. Using **Goal Seek** with cell B2 set to value 0 gives the value of x that solves the equation as shown in Table 1.14.

TABLE 1.13

	A	B
1	x	y
2		= 0.55*A2 − 122.886

TABLE 1.14

	A	B
1	x	y
2	223.43	0

Formulas; Solving an Equation for a Specified Linear Variable

An equation that contains two or more letters that represent constants or variables is called a **literal equation**. Formulas are examples of literal equations. If one of two or more of the variables in an equation is present only to the first degree, we can solve for that variable by treating the other variables as constants and using the same steps that we used to solve a linear equation in one variable. This is sometimes useful because it is necessary to get equations in the proper form to enter them in a graphing utility.

EXAMPLE 7 Simple Interest

The formula for the future value of an investment of $\$P$ at simple interest rate r for t years is $A = P(1 + rt)$. Solve the formula for r, the interest rate.

Solution

$$A = P(1 + rt)$$

$$A = P + Prt \quad \text{Remove parentheses.}$$

$$A - P = Prt \quad \text{Get the term containing } r \text{ by itself on one side of the equation.}$$

$$r = \frac{A - P}{Pt} \quad \text{Divide both sides by } Pt.$$

■

EXAMPLE 8 Writing an Equation in Functional Form and Graphing Equations

Solve each of the following equations for y so that y is expressed as a function of x. Then graph the equation on a graphing utility with a standard viewing window.

a. $2x - 3y = 12$

b. $x^2 + 4y = 4$

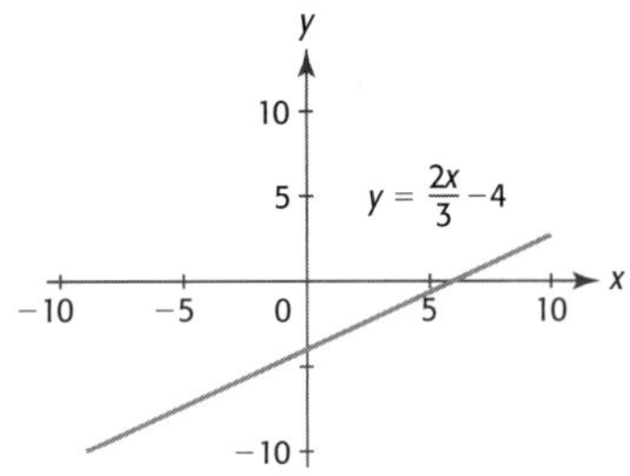

FIGURE 1.39

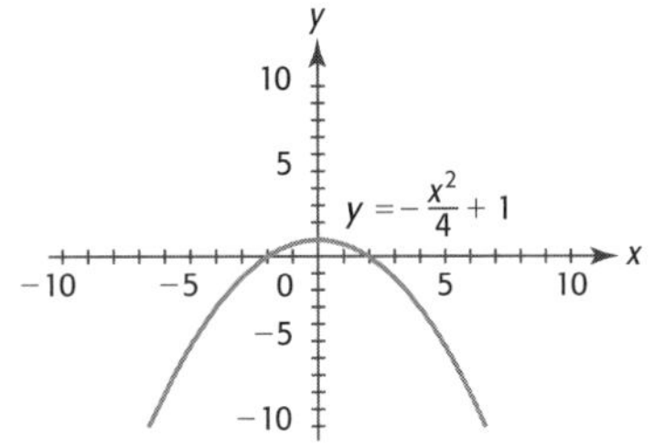

FIGURE 1.40

Solution

a. Because y is to the first power in the equation, we solve the equation for y using linear equation solution methods:

$$2x - 3y = 12$$

$$-3y = -2x + 12 \qquad \text{Isolate the term involving } y.$$

$$y = \frac{2x}{3} - 4 \qquad \text{Divide both sides by } -3\text{, the coefficient of } y.$$

The graph of this equation is shown in Figure 1.39.

b. The variable y is to the first power in the equation

$$x^2 + 4y = 4$$

so we solve for y by using linear solution methods. (Note that to solve this equation for x would be more difficult; this method will be discussed in Chapter 2.)

$$4y = -x^2 + 4 \qquad \text{Isolate the term containing } y.$$

$$y = \frac{-x^2}{4} + 1 \qquad \text{Divide both sides by 4, the coefficient of } y.$$

The graph of this equation is shown in Figure 1.40. ■

1.5 SKILLS CHECK

In Exercises 1–12, solve the following equations.

1. $5x - 14 = 23 + 7x$
2. $3x - 2 = 7x - 24$
3. $3(x - 7) = 19 - x$
4. $5(y - 6) = 18 - 2y$
5. $x - \frac{5}{6} = 3x + \frac{1}{4}$
6. $3x - \frac{1}{3} = 5x + \frac{3}{4}$
7. $\frac{5(x - 3)}{6} - x = 1 - \frac{x}{9}$
8. $\frac{4(y - 2)}{5} - y = 6 - \frac{y}{3}$
9. $5.92t = 1.78t - 4.14$
10. $0.023x + 0.8 = 0.36x - 5.266$
11. $\frac{3}{4} + \frac{1}{5}x - \frac{1}{3} = \frac{4}{5}x$
12. $\frac{2}{3}x - \frac{6}{5} = \frac{1}{2} + \frac{5}{6}x$

For Exercises 13–16, find: (a) the solution to the equation $f(x) = 0$, (b) the x-intercept of the graph of $y = f(x)$, and (c) the zero of $f(x)$.

13. $f(x) = 32 + 1.6x$
14. $f(x) = 15x - 60$
15. $f(x) = \frac{3}{2}x - 6$
16. $f(x) = \frac{x - 5}{4}$

In Exercises 17 and 18, you are given a table showing input and output values for a given function $y_1 = f(x)$. Using this table, find (if possible) (a) *the x-intercept of the graph of $y = f(x)$,* (b) *the y-intercept of the graph of $y = f(x)$, and,* (c) *the solution to the equation $f(x) = 0$.*

17. 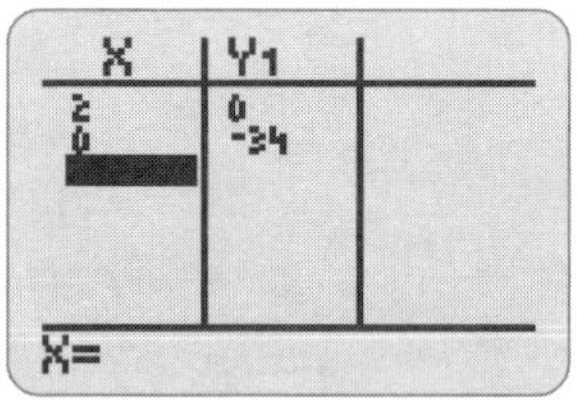

18. 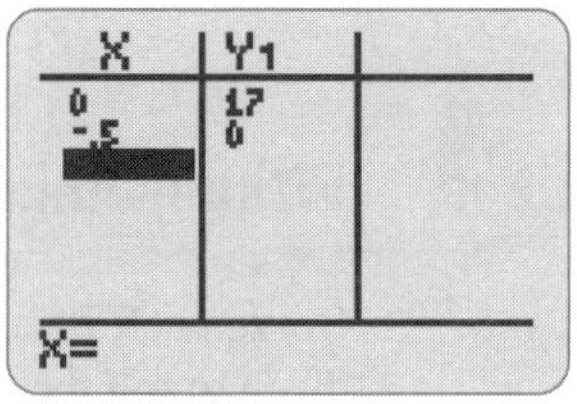

In Exercises 19 and 20, you are given the graph of a certain function $y = f(x)$ and the zero of that function. Using this graph, find (a) the x-intercept of the graph of $y = f(x)$ and, (b) the solution to the equation $f(x) = 0$.

19.

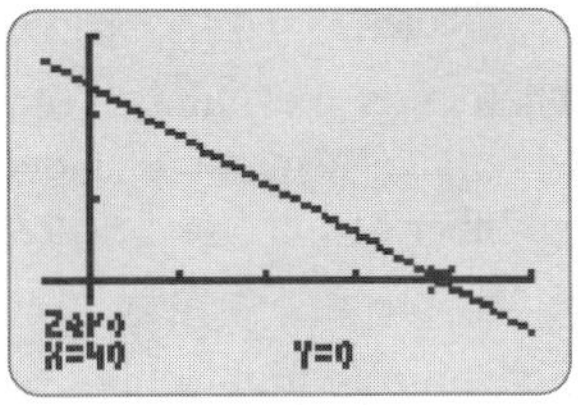

20.

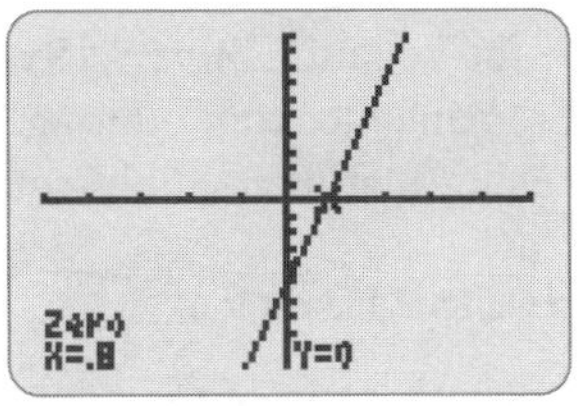

In Exercises 21–28, solve the equations using graphical methods.

21. $4x - 24 = 27 - 3x$ **22.** $3x - 8 = 15x + 4$

23. $3(s - 8) = 5(s - 4) + 6$

24. $5(2x + 1) + 5 = 5(x - 2)$

25. $\frac{3t}{4} - 2 = \frac{5t - 1}{3} + 2$

26. $4 - \frac{x}{6} = \frac{3(x - 2)}{4}$

27. $\frac{t}{3} - \frac{1}{2} = \frac{t + 4}{9}$ **28.** $\frac{s - 5}{4} + x = \frac{s}{2} + \frac{1}{3}$

29. Solve $A = P(1 + rt)$ (a) for t and (b) for P.

30. Solve $V = \frac{1}{3}\pi r^2 h$ for h.

31. Solve $5F - 9C = 160$ for F.

32. Solve $4(a - 2x) = 5x + \frac{c}{3}$ for x.

33. Solve $\frac{P}{2} + A = 5m - 2n$ for n.

34. Solve $y - y_1 = m(x - x_1)$ for x.

In Exercises 35–38, solve the following equations for y and graph them with a standard window on a graphing utility.

35. $5x - 3y = 5$ **36.** $3x + 2y = 6$

37. $x^2 + 2y = 6$ **38.** $4x^2 + 2y = 8$

1.5 EXERCISES

39. *Depreciation* An \$828,000 building is depreciated for tax purposes by its owner using the straight-line depreciation method. The value of the building, y, after x months of use is given by $y = 828{,}000 - 2300x$ dollars.

a. After how many years will the value of the building be \$690,000?

b. For what months is the value of the building greater than 690,000?

40. *Temperature Conversion* The equation $5F - 9C = 160$ gives the relationship between Fahrenheit and Celsius temperature measurements. What Fahrenheit measure is equivalent to a Celsius measurement of 20°?

41. *Investments* The future value of a simple interest investment is given by $S = P(1 + rt)$. What principal P must be invested for $t = 5$ years at the simple interest rate $r = 10\%$ so that the future value grows to \$9000?

42. *Temperature–Humidity Index* The temperature–humidity index I is $I = t - 0.55(1 - h)(t - 58)$, where t is the air temperature in degrees Fahrenheit and h is the relative humidity expressed as a decimal. If the air temperature is 80°, find the humidity h (as a percent) that gives an index value of 79.

43. *Earnings and Minorities* The median annual salary (in thousands of dollars) of minorities M is related to the median salaries of whites W by $M = 0.959W - 1.226$. What is the median annual salary for whites when the median annual salary for minorities is \$50,560?
(Source: *Statistical Abstract of the United States*, 1993)

44. *Internet Brokerage Accounts* The number of brokerage accounts on the Internet in year t can be modeled by $B(t) = 3.303t - 6591.560$ million accounts. If this model remains valid, in what year were there 14.44 million accounts?
(Source: Gomez Advisors)

45. *Game Show Question* The following question was worth $32,000 on the game show *Who Wants to Be a Millionaire?*: "At what temperature are the Fahrenheit and Celsius temperature scales the same?" Answer this question. Recall that Fahrenheit and Celsius temperatures are related by $5F - 9C = 160$.

46. *Out-of-Wedlock Births* The percent of all teen mothers who are unmarried can be given by the equation $y = 1.78x - 3.998$, where x is the number of years from 1950. If this model remains accurate, in what year will the percent have reached 80?
(Source: Father Facts)

47. *Reading Score* The average reading score on the National Assessment of Progress tests is given by $y = 0.155x + 255.37$, where x is the number of years past 1970. In what year would the average reading score be 259.4 if this model is accurate?

48. *Tax Burden* The per capita tax burden B can be described by $B(t) = 20.37 + 1.834t$ hundred dollars, where t is the number of years past 1980. In what year does this model indicate that the per capita tax burden was $4788?
(Source: Internal Revenue Service)

49. *Mobile-Phone Bills* The average mobile-phone local monthly bill between 1995 and 1998 can be modeled by $B(x) = -3.963x + 51.172$ dollars, where x is the number of years after 1995. If the model remains accurate beyond 1998, in what year does it indicate that the monthly bill is $35.32?
(Source: *Newsweek*, March 20, 2000)

50. *Marriage Rate* The percent of unmarried women who get married in any particular year is given by $y = -0.0762x + 8.5284$, where x is the number of years after 1950. During what year does this model predict the percent will be 6.09?
(Source: Index of Leading Cultural Indicators)

51. *U.S. Civilian Population* The U.S. population, excluding armed forces, can be modeled for the years 1970–1998 by the function $p = 2351x + 201{,}817$, where p is in thousands of people and x is in years from 1970. During what year does the model estimate the population to be 258,241,000?
(Source: www.census.gov/statab)

52. *Internet Workers* The percent of Fortune Global 500 firms that actively recruited workers on the Internet between 1998 and 2000 can be modeled by $P(x) = 26.5x - 62$ percent, where x is the number of years after 1995. In what year does this model indicate that 44% of the firms recruited on the Internet?
(Source: iLogos.com)

53. *Inmates* The total number of inmates in custody between 1990 and 1998 in state and federal prisons is given approximately by $y = 68.476x + 728.654$ thousand prisoners, where x is the number of years after 1990. If the model remains accurate long enough, in what year will the number of inmates be 797,130?
(Source: Bureau of Justice Statistics)

54. *Marijuana Use* The percent p of high school seniors using marijuana daily can be related to x, the number of years after 1990, by the equation $30p - 19x = 1$. Assuming the model remains accurate, during what year was the percent using marijuana daily equal to 3.2%?
(Source: Index of Leading Cultural Indicators)

55. *Presidential Elections* The percent of the population voting in presidential elections has been estimated to be $p = 65.4042 - 0.3552x$, where x is the number of years after 1950.
(Source: Federal Election Commission)

a. During what year does this model predict that the percent voting in a presidential election will be 49%?

b. Of those persons eligible to vote, 53% cast a ballot in the 2000 presidential election. Did this model accurately describe the percent voting in the 2000 election? What does this answer tell you about using models for predictions?

56. *Personal Income* U.S. personal income increased between 1993 and 1997 according to the model $I(x) = 341.28x + 4459.78$ billion dollars, where x is the number of years after 1990. In what year does this model predict that U.S. personal income would reach $7190 billion if this model is accurate beyond 1997?
(Source: ECONOMAGIC)

57. *Cigarette Advertising* The total U.S. cigarette advertising and promotional expenditures from 1975 to 1996 can be modeled by the equation $y = 277.318x - 1424.766$, where y is measured in millions of dollars and x is the number of years from 1970. If this model remains accurate, in what year did the spending exceed $6 billion?
(Source: Federal Trade Commission)

58. *Mobile-Phone Subscribers* The number of mobile-phone subscribers (in millions) between 1995 and 1999 can be modeled by $S(x) = 11.75x + 32.95$, where x is the number of years after 1995. In what year does this model indicate that there were 68,200,000 subscribers? (Source: *Newsweek*, March 20, 2000)

59. *Grades* To earn an A in a course, a student must get an average score of at least 90 on five tests. If her first four test scores are 92, 86, 79, and 96, what score does she need on the last test to obtain a 90 average?

60. *Grades* To earn an A in a course, a student must get an average score of at least 90 on three tests and a final exam. His final exam score counts double, and his lowest score is removed if the final score is higher than the lowest score. If his first three test scores are 86, 79, and 96, what is the lowest score he needs on the last test to obtain a 90 average?

61. *Tobacco Judgment* As part of the largest-to-date damage award in a jury trial in history, the July 2000 penalty handed down against the tobacco industry included a $74 billion judgment against Philip Morris. If this amount was 94% of this company's 1999 revenue, how much was Philip Morris's 1999 revenue? (Source: *Newsweek*, July 24, 2000)

62. *Tobacco Judgment* As part of the largest-to-date damage award in a jury trial in history, the July 2000 penalty handed down against the tobacco industry included a $36 billion judgment against R. J. Reynolds. If this amount was 479% of this company's 1999 revenue, how much was R. J. Reynolds 1999 revenue? (Source: *Newsweek*, July 24, 2000)

63. *Sales Commission* A salesman earns $50,000 in commission in 1 year and then has his commission reduced by 20% the next year. What percent increase in commission over the second year will give him $50,000 in the third year?

64. *Salaries* A man earning $100,000 per year has his salary reduced by 5% because of a reduction in his company's market. A year later, he receives $104,500 in salary. What percent raise from the reduced salary does this represent?

65. *Sales Tax* The total cost of a new automobile, including a 6% sales tax on the price of the automobile, is $29,998. How much of the total cost of this new automobile is sales tax?

66. *Wildlife Management* In wildlife management, the capture–mark–recapture technique is used to estimate the population size of certain types of fish or animals. To estimate the population, we enter information in the equation

$$\frac{\text{total in population}}{\text{total number marked}} = \frac{\text{total number in second capture}}{\text{number found marked in second capture}}$$

Suppose that 50 sharks are caught along a certain shoreline, marked, and then released. If a second capture of 50 sharks gives 20 sharks that have been marked, what is the resulting population estimate?

1.6 Fitting Lines to Data Points: Modeling Linear Functions

Key Concepts

- Fitting lines to data points
- Linear regression
- Modeling data
- Discrete function
- Continuous function
- Applying models
- Goodness of fit
- Constant first differences

Section Preview Earnings and Gender

Table 1.15 shows the earnings of year-round full-time workers by gender and educational attainment. Figure 1.41 shows the scatter plot of the data, and it appears that a line would approximately fit along these data points. We can determine the relationship between the two sets of earnings by creating a linear equation that gives the female annual earnings as a function of the male annual earnings. This model can be used to estimate an annual female salary from a given annual male salary. In this section, we will learn how to create linear models from data points, how to use graphing utilities to create linear models, and how to use the model to answer questions about the data.

TABLE 1.15

Educational Attainment	Average Annual Earnings for Males ($ thousand)	Average Annual Earnings for Females ($ thousand)
Less than ninth grade	18.743	12.392
Some high school	18.908	12.057
High school graduate	30.414	18.092
Some college	33.614	20.241
Associate's degree	40.047	25.079
Bachelor's degree or more	66.810	36.755

(Source: U.S. Bureau of the Census)

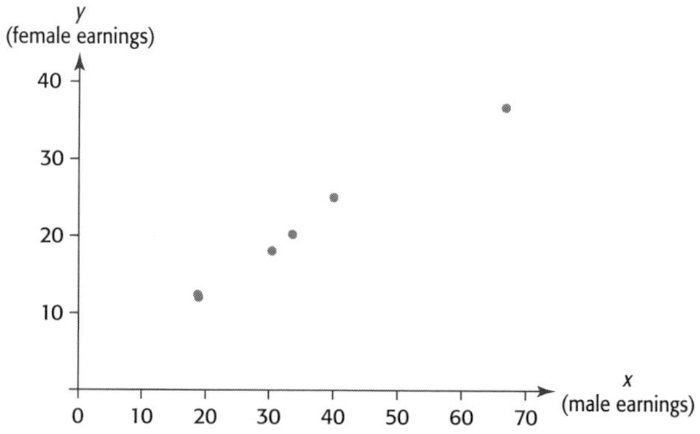

FIGURE 1.41

Fitting Lines to Data Points

When real-world information is collected as numerical information called *data*, technology can be used to determine the pattern exhibited by the data (provided that a recognizable pattern exists). These patterns can often be described by mathematical functions.

Before discussing methods of creating linear models for real-world data, we compare the graphs of several functions and a data set to see which of the graphs visually gives the best fit for the data. For now, we informally define the "best fit" line as the one that appears to come the closest to all the data points.

EXAMPLE 1 Health Service Employment

Table 1.16 gives the number of full- and part-time employees in offices and clinics of dentists for selected years between 1970 and 1998. (Source: U.S. Bureau of the Census)

TABLE 1.16

Year	1970	1975	1980	1985	1990	1995	1998
Employees (in thousands)	222	331	415	480	580	644	666

a. Draw a scatter plot of the data with the x-value of each point representing the number of years after 1970 and the y-value representing the number of dental employees (in thousands) corresponding to that year.

b. Individually graph each of the following three equations on the same graph as the scatter plot, and determine which of the lines appears to give the best fit to the data.

$$y_1 = -12x + 660$$
$$y_2 = 13x + 220$$
$$y_3 = 16x + 242$$

Solution

a. By using x as the number of years after 1970, we have aligned the data with $x = 0$ representing 1970, $x = 1$ representing 1971, and so forth. Enter into the lists of a graphing utility (or an Excel spreadsheet) the aligned input data representing the years in Table 1.16 and the output data representing the number of employees shown in the second row of Table 1.16. Figure 1.42(a) shows the lists containing the data. The graph of these data points is shown in Figure 1.42(b). The window for this scatter plot can be set manually or with a command on the graphing utility such as ZOOMSTAT, which is used to automatically set the window and display the graph.

L1	L2
0	222
5	331
10	415
15	480
20	580
25	644
28	666

(a)

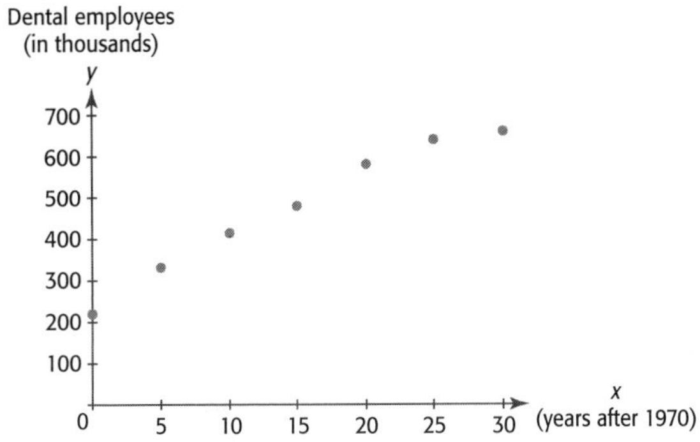

(b)

FIGURE 1.42

b. The graphs of the data and

$$y_1 = -12x + 660$$

shown in Figure 1.43(a) show that the line is falling (its slope is negative), but the data points follow a rising pattern. Thus, an equation with a positive slope would be a better fit. Figure 1.43(b) shows that the graph of

$$y_2 = 13x + 220$$

is a better fit to the data than the graph in Figure 1.43(a), but a line with a steeper (larger) slope would be better. The graph of

$$y_3 = 16x + 242$$

is the line that appears to fit better than either of the other two lines (Figure 1.43(c)). Thus, of the three linear functions given to model the number of full and part-time employees in dentists' offices and clinics, the best is $y_3 = 16x + 242$ thousand employees, where x is the number of years after 1970. ■

In Example 1, it was not difficult to choose which of the three given lines fits the data the best. However, is the line in Figure 1.43(c) the best-fitting line of all possible lines that could be drawn on the scatter plot? How do we determine the equation of the best-fit line? We now discuss answers to these questions.

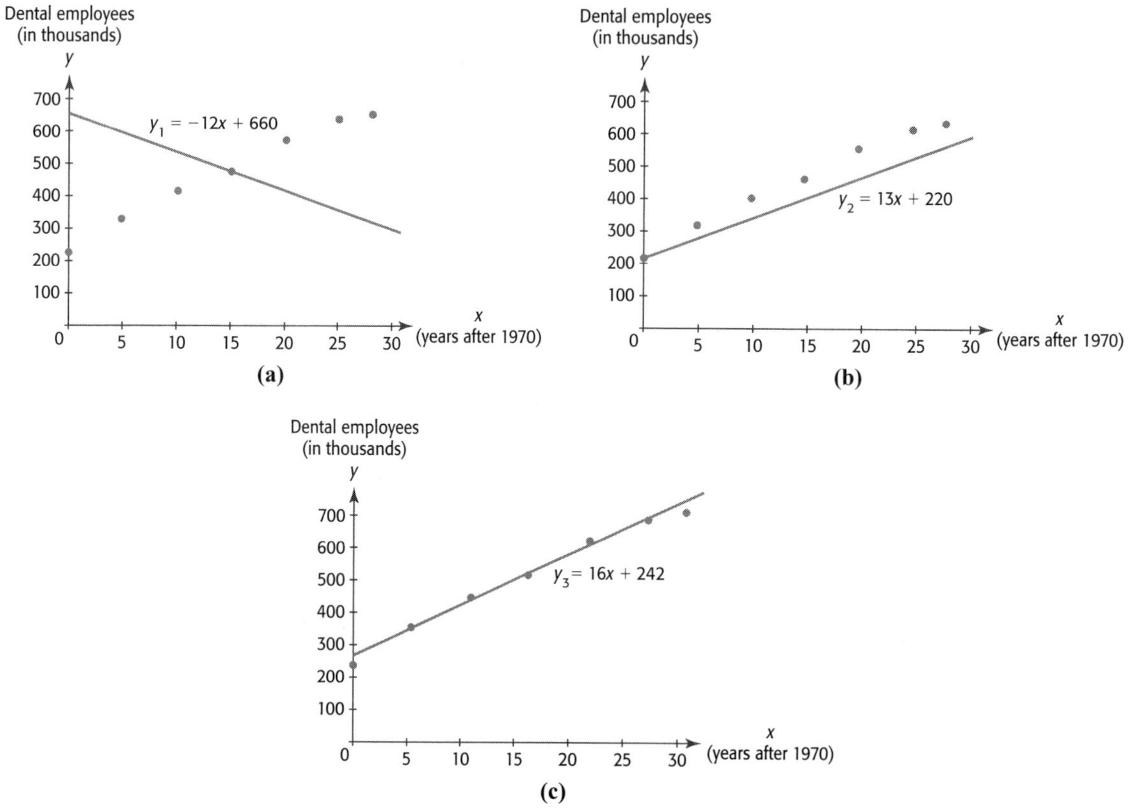

FIGURE 1.43

Linear Regression

The points determined by the data in Table 1.16 do not all lie on a line, but we can determine the equation of the line that is the "best fit" for these points by using a procedure called **linear regression**. This procedure defines the *best-fit line* as the line for which the sum of the squares of the vertical distances from the data points to the line is a minimum. For this reason, the linear regression procedure is also called the **least-squares method**.

The vertical distance between a data point and the corresponding point on a line is simply how much the line misses going through the point—that is, the difference in the y-values of the data point and the point on the line. If we call this difference in outputs d_i (where i takes on the values from 1 to n for n data points), the least-squares method* requires that

$$d_1^2 + d_2^2 + d_3^2 + \cdots + d_n^2$$

be as small as possible. The line for which this sum of squared differences is as small as possible is called the *linear regression line* or *least-squares line* and is the one that we consider to be the **best-fit line** for the data.

To illustrate the linear regression process, consider again the two lines in Figures 1.43(b) and 1.43(c) and the data points in Table 1.16. Figures 1.44 and 1.45 indicate the vertical distances for each of these lines. Below is the calculation of the sum of squared differences for each line.

Figure 1.44:
$$\begin{aligned} d_1^2 + d_2^2 + d_3^2 + \cdots + d_7^2 &= (222 - 220)^2 + (331 - 285)^2 + (415 - 350)^2 + (480 - 415)^2 \\ &\quad + (580 - 480)^2 + (644 - 545)^2 + (666 - 584)^2 \\ &= 4 + 2116 + 4225 + 4225 + 10{,}000 + 9801 + 6724 \\ &= 37{,}095 \end{aligned}$$

Figure 1.45:
$$\begin{aligned} d_1^2 + d_2^2 + d_3^2 + \cdots + d_7^2 &= (222 - 242)^2 + (331 - 322)^2 + (415 - 402)^2 + (480 - 482)^2 \\ &\quad + (580 - 562)^2 + (644 - 642)^2 + (666 - 690)^2 \\ &= 400 + 81 + 169 + 4 + 324 + 4 + 576 \\ &= 1558 \end{aligned}$$

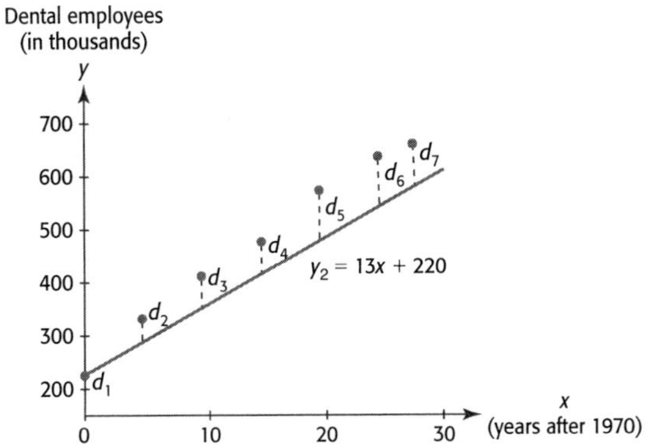

FIGURE 1.44

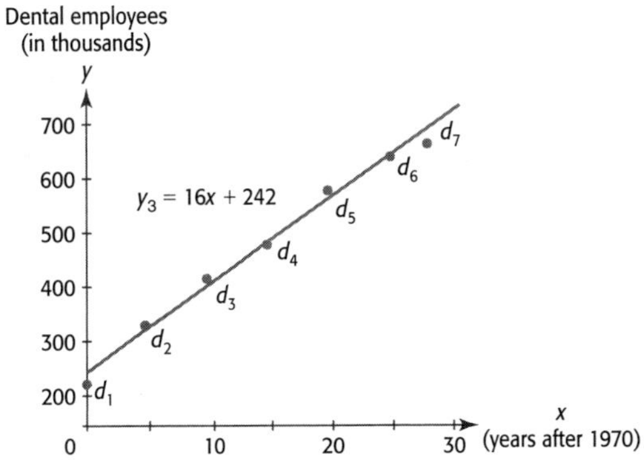

FIGURE 1.45

*The sum of the squared differences is often called SSE, the *sum of squared errors*. The development of the equations that lead to the minimum SSE is a calculus topic.

It is not difficult to see that the vertical distances are smaller for the graph of

$$y_3 = 16x + 242$$

We see numerically from the sum of squared differences calculations, as well as visually, from Figures 1.44 and 1.45, that the line

$$y_3 = 16x + 242$$

is the line that seems to better model the data when given a choice of the lines y_2 and y_3. However, is this line the best-fit line for these data? Remember that the best-fit line must have the smallest sum of squared differences of *all* lines that can be drawn through the data! We will see that the line which is the best fit for these data points, with coefficients rounded to three decimal places, is

$$y = 15.910x + 242.758$$

The sum of squared differences for the regression line is approximately 1550.46, which is slightly smaller than 1558.

Development of the formulas that give the best-fit line for a set of data is beyond the scope of this text, but graphing calculators, computer programs, and spreadsheets have built-in formulas or programs that give the equation of the best-fit line. That is, we can use technology to find the best linear model for the data. The calculation of the sum of squared differences for the dental employees data was given only to illustrate what the best-fit line means. You will not be asked to calculate, nor do we use from this point on, the value of the sum of squared differences in this text.

We illustrate the use of technology to find the regression line by returning to the data of Example 1. To find the equation that gives the number of employees in dental offices as a function of the years past 1970, we use the following steps:

(a)

(b)

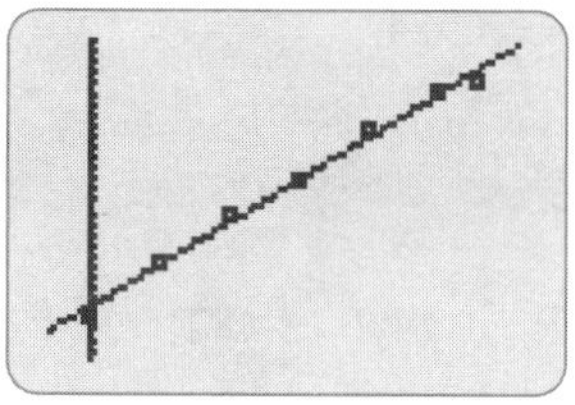

[0, 30] by [200, 700]

(c)

FIGURE 1.46

MODELING DATA

Step 1: Enter the data into lists of a graphing utility.

Step 2: Create a scatter plot of the data to see if a linear model is reasonable. The data should appear to follow a linear pattern with no distinct curvature.

Step 3: Use the graphing utility to obtain the linear equation that is the best fit for the data. (Figure 1.46(a) shows the equation for Example 1, which can be approximated by $y = 15.910x + 242.758$.)

Step 4: Graph the linear function (unrounded) and the data points on the same graph to see how well the function fits the data. (The equation entered into an equation editor and the graph of the data and the best-fit line for Example 1 are shown in Figures 1.46(b) and 1.46(c), respectively.)

Step 5: Report the equation and/or numerical results in a way that makes sense in the context of the problem, with the appropriate units, and with the variables identified.

Figure 1.46(c) shows that

$$y = 15.910x + 242.758$$

is a good model of the number of full and part-time employees in dentists' offices and clinics, where y is in thousands of employees and x is the number of years after 1970.

The screens for the graphing utility that you use may vary slightly from those given in this text. Also, the regression line you obtain is dependent on your particular technology and may have some decimal places slightly different from those shown here.

When modeling a set of data, it is important to be careful when rounding coefficients in equations and when rounding during calculations. We will use the following guidelines in this text.

TECHNOLOGY NOTE

After a model for a data set has been found, it can be rounded for reporting purposes. However, do not use a rounded model in calculations, and do not round answers during the calculation process unless otherwise specified. When the model is used to find numerical answers, the answers should be rounded in a way that agrees with the context of the problem, and with no more accuracy than the original output data.

EXAMPLE 2 Earnings and Gender

Table 1.17 shows the earnings of year-round full-time workers by gender and educational attainment.

a. Let x represent earnings for males, let y represent earnings for females, and create a scatter plot of the data.

b. Create a linear model that expresses females' annual earnings as a function of males' earnings.

c. Graph the linear function and the data points on the same graph, and discuss how well the function models the data.

TABLE 1.17

Educational Attainment	Average Annual Earnings for Males ($ thousand)	Average Annual Earnings for Females ($ thousand)
Less than ninth grade	18.743	12.392
Some high school	18.908	12.057
High school graduate	30.414	18.092
Some college	33.614	20.241
Associate's degree	40.047	25.079
Bachelor's degree or more	66.810	36.755

(Source: U.S. Bureau of the Census)

Solution

a. From Table 1.17 enter the data in the lists of a graphing utility. Figure 1.47(a) on the next page shows a list of the data points. The scatter plot of all the data points from Table 1.17 is shown in Figure 1.47 (b).

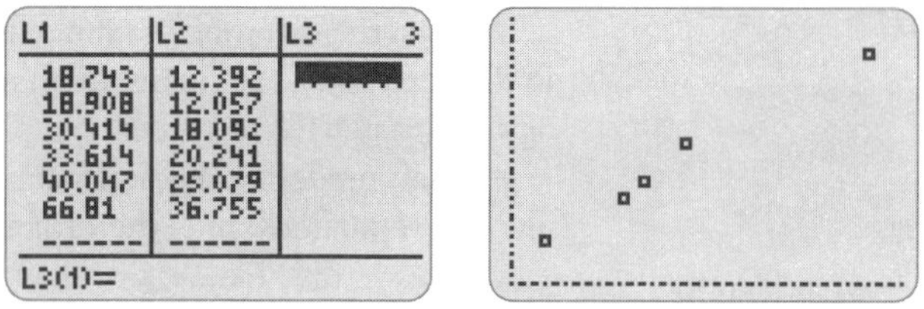

[14, 72] by [7, 41]

(a) (b)

FIGURE 1.47

FIGURE 1.48

b. The points in the scatter plot in Figure 1.47 (b) exhibit a nearly linear pattern, so a line could possibly be used to model these data. The equation of the line that is the best fit for the data points can be found with a graphing calculator (Figure 1.48). The equation, rounded to four decimal places, is

$$y = 0.5177x + 2.775$$

Remember that a *model* gives not only the equation but also a description of the variables and their units of measure. The model

$$y = 0.5177x + 2.775 \text{ thousand dollars}$$

where x is in thousands of dollars, expresses women's annual earnings y as a function of men's annual earnings x.

c. Enter the unrounded equation into the equation editor and graph the model along with the data points.* We observe that the line follows the general trend indicated by the data (Figure 1.49 (a) and 1.49 (b)). Note that not all points lie on the graph of the equation, even though this is the line that is the best fit for the data.

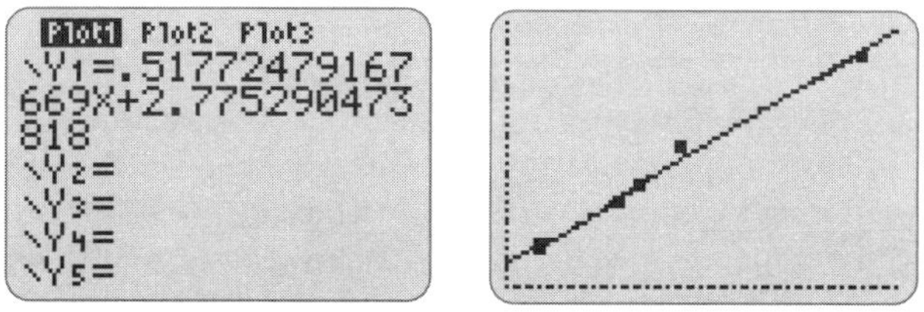

[14, 72] by [7, 41]

(a) (b)

FIGURE 1.49 ■

Modeling Discrete Data with Linear Functions

As we have seen, linear equations are used to model applications even though some points on the line do not fit the application. For example, suppose a pizza shop charges $2 for a small pizza and $0.75 for each topping, with up to five toppings available. The total price of the pizza, y, is determined by the number of toppings, x. The relationship between these variables can be expressed by a table with integer x-values from 0 to 5

*Even though we write rounded models found from data given in this text, all calculations are done using the unrounded model found by the graphing utility.

or the graph (scatter plot) of this **discrete function** (with a finite number of inputs) shown in Figure 1.50 (a). For ease of discussion, we can also model this application with the linear function $y = 2 + 0.75x$, with no restrictions on the domain because the points from the discrete function fit on the graph of this **continuous function**. We use the term **continuous function** to describe a function whose inputs can be any real number or any real number between two specified values (informally, a continuous function is a function whose graph can be drawn without lifting pen from paper). Figure 1.50 (b) shows the graph of $y = 2 + 0.75x$, which is the continuous model of this discrete application. In the context of this application, x and y only make sense when they are both nonnegative, so we draw the graph in the first quadrant.

If a continuous function is used to model data in an application that only has a discrete interpretation, then we cannot use the model to estimate outputs for every real number input. For example, we found a continuous model giving the price of the pizza as a function of the number of toppings, but a noninteger number input does not make sense in this situation. The only inputs that make sense in this case are integers representing the numbers of toppings. Nevertheless, in situations like this, we can use continuous models to solve problems, but then give a discrete interpretation to the solution.

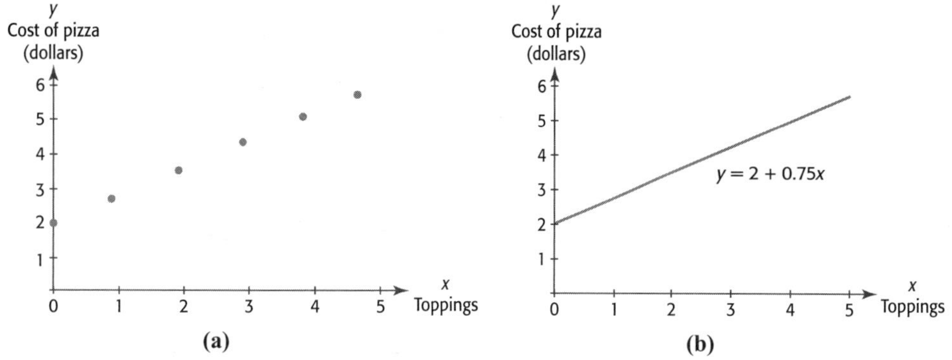

FIGURE 1.50

EXAMPLE 3 U.S. Population

The total U.S. population for selected years beginning in 1960 and projected to 2050 is shown in Table 1.18 on the next page, with the population given in millions.

a. Align the data to represent the number of years from 1960, and draw a scatter plot of the data.

b. Create a linear equation that is the best fit for these data, where y is in millions and x is the number of years from 1960.

c. Graph the equation of the linear model on the same graph with the scatter plot and discuss how well the model fits the data.

d. Use the model to estimate the population in 1997 and compare it with the population given in the table.

TABLE 1.18

Year	Population (millions)	Year	Population (millions)
1960	180.671	1995	263.044
1965	194.303	1998	270.561
1970	205.052	2000	281.422
1975	215.973	2003	294.043
1980	227.726	2025	358.030*
1985	238.466	2050	408.695*
1990	249.948		

*Estimates
(Source: U.S. Census Bureau)

Solution

a. The aligned data has $x = 0$ representing 1960, $x = 5$ representing 1965, and so forth. Figure 1.51(a) shows the first seven entries using the aligned data. The scatter plot of these data is shown in Figure 1.51(b). The coordinates of the first two points in the scatter plot are (0, 180.671) and (5, 194.303).

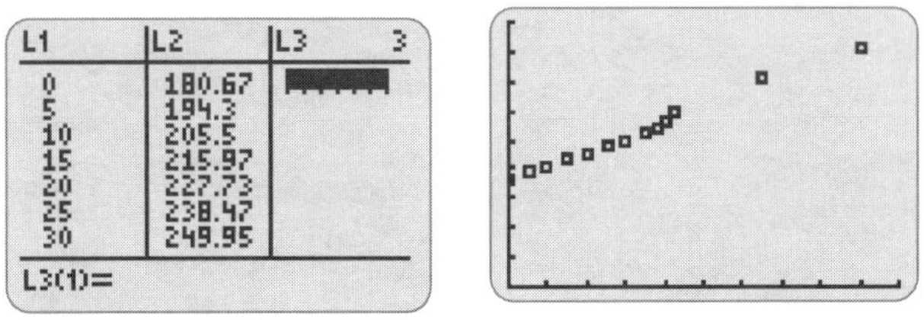

[0, 100] by [0, 450]

(a) (b)

FIGURE 1.51

b. The equation of the best-fit line is found by using linear regression in the graphing calculator. Rounding the decimals to three places, the linear model for the U.S. population is

$$y = 2.607x + 177.195 \text{ million}$$

where x is the number of years from 1960.

c. Enter the complete (unrounded) equation in the graphing calculator (Figure 1.52(a)) and graph it along with the scatter plot. As shown in Figure 1.52(b), the graph of the best-fit line is very close to the data points. However, the points do not all fall on the line because the U.S. population did not increase by exactly the same amount each year.

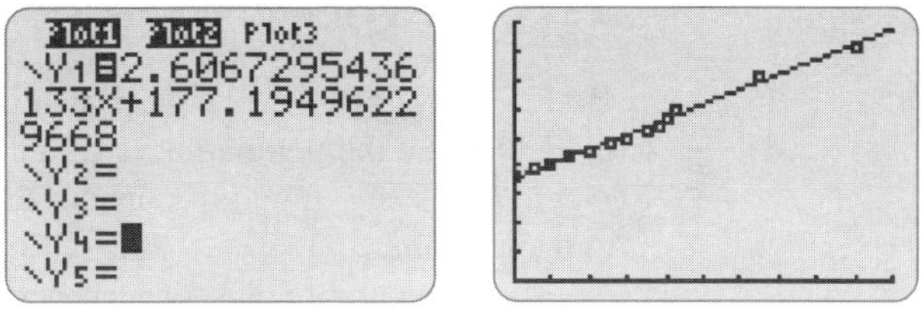

[0, 100] by [0, 450]

(a) (b)

FIGURE 1.52

d. Evaluating the unrounded model

$$y = 2.6067295436133x + 177.19496229668$$

at $x = 37$ gives an estimated population of 273.644 million, or 273,644,000, in 1997. The actual U.S. population was 281,421,996 in the 2000 census, so the estimate from the model is not unreasonable. ■

Spreadsheet Solution

We can also use spreadsheets to fit the linear function that is the best fit for the data. Table 1.19 shows the Excel spreadsheet for the data of Example 3. Selecting the cells containing the data, using **Chart Wizard** to get the scatter plot of the data, and selecting **Add Trendline** gives the equation of the linear function that is the best fit for the data, along with the scatter plot and the graph of the best-fitting line. The equation and graph of the line are shown in Figure 1.53.

TABLE 1.19

	A	B
1	Year x	Population (millions) y
2	0	180.671
3	5	194.303
4	10	205.052
5	15	215.973
6	20	227.726
7	25	238.466
8	30	249.948
9	35	263.044
10	38	270.561
11	40	281.422
12	43	294.043
13	65	358.03
14	90	408.695

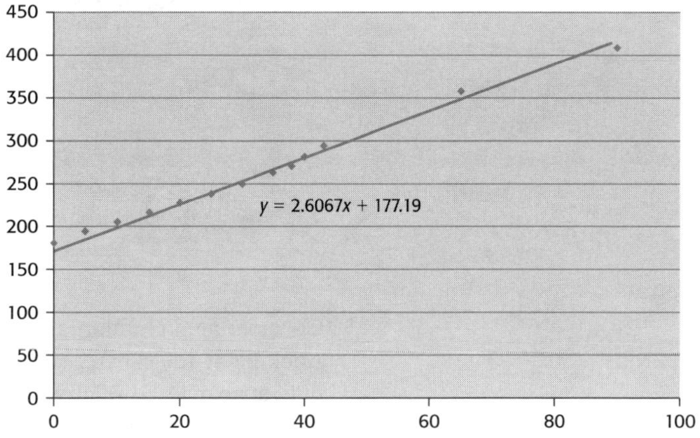

FIGURE 1.53

Applying Models

Because 1997 is a year between two given values in the table of Example 3, using the model to find the population is called **interpolation**. When a model is evaluated for a *prediction* using input(s) outside the given data points, the output is called an **extrapolation**.

Another question arises whenever data that involves time is used. If we label the x-coordinate of a point on the input axis as 1999, what time during 1999 do we mean? Does 1999 refer to the beginning of the year, the middle of the year, its end, or some other time? Because most data from which the functions in applications are derived represent end-of-year totals, we make the following convention when modeling, unless otherwise specified.

A point on the input axis indicating a time refers to the end of the time period.

For instance, the point representing the year 1999 means "at the end of 1999." Notice that this instant in time also represents the beginning of the year 2000. If, for example, data is aligned with x equal to the number of years after 1990, then any x-value greater than 9 and less than or equal to 10 represents some time in the year 2000. If a point represents anything other than the end of the period, this information will be clearly indicated. Also, if a and b are points in time, we use the phrases "from a to b" and "between a and b" to both indicate the same-time interval.

Goodness of Fit

If output values for a set of data change by the same amount when the input values change by the same amount, then we can find a linear function that models the data exactly. In this case, we say that the inputs are **uniform** and the **first differences** are constant. Let's consider the data in Table 1.20, which gives the conversion from selected Celsius temperatures to Fahrenheit temperatures. Because the Celsius temperatures (inputs) differ by a constant amount (10), we can compare the (first) differences of the outputs to see if they are constant. If we subtract successive Fahrenheit temperatures (Table 1.21), we see that all the differences are the same constant (18), which indicates that the data can be modeled exactly by a linear function. (We know that this linear model exists; its equation is $F = 9/5C + 32$.)

TABLE 1.20

Celsius Degrees (°C)	−20	−10	0	10	20	30	40
Fahrenheit Degrees (°F)	−4	14	32	50	68	86	104

TABLE 1.21

Outputs	−4		14		32		50		68		86		104
First Differences of Outputs		**18**		**18**		**18**		**18**		**18**		**18**	

Consider again Example 3 where we modeled the U.S. population for selected years. Looking at Table 1.22, notice that for uniform inputs the first differences of the outputs appear to be relatively close to the same constant, especially compared to the size of the population. (If these differences were closer to a constant, the fit would be better.)

TABLE 1.22

Uniform Inputs (years)	Outputs Population (millions)	First Differences in Output (difference in population)
1960	180.671	
1965	194.303	**13.632**
1970	205.052	**10.749**
1975	215.973	**10.921**
1980	227.726	**11.753**
1985	238.466	**10.740**
1990	249.948	**11.482**
1995	263.044	**13.096**
2000	281.422	**18.378**

So how "good" is the fit of the linear model $y = 2.6067295436133x + 177.19496229668$ to the data in Example 3? Based on observation of the graph of the line and the data points on the same set of axes (Figure 1.53), it is reasonable to say that the regression line provides a very good fit, but not an exact fit, to the data.

CONSTANT FIRST DIFFERENCES

If the first differences of data outputs are constant (for uniform inputs), a linear model can be found that fits the data exactly. If the first differences are "nearly constant," a linear model can be found that is an approximate fit for the data.

The goodness of fit of a line to a set of data points can be observed from the graph of the line and the data points on the same set of axes, and it can be measured if your graphing utility computes the **correlation coefficient**. The correlation coefficient is a number $\boldsymbol{r}$, $-1 \leq r \leq 1$, that measures the strength of the linear relationship that exists between the two variables. The closer that $|r|$ is to 1, the more closely the data points fit on the linear regression line. (There is no linear relationship between the two variables if $r = 0$.) Positive values of r indicate that the output variable increases as the input variable increases, and negative values of r indicate that the output variable decreases as the input variable increases. But the *strength* of the relationship is indicated by how close $|r|$ is to 1. For the data of Example 3, computing the correlation coefficient gives $r = .997$ (Figure 1.54), which means that the linear relationship is strong and that the linear model is an excellent fit for the data. When a calculator feature or computer program is used to fit a linear model to data, the resulting equation can be considered a best fit for the data. As we will see later in this text, other mathematical models may be better fits to some sets of data, especially if the first differences of the outputs are not close to being constant.

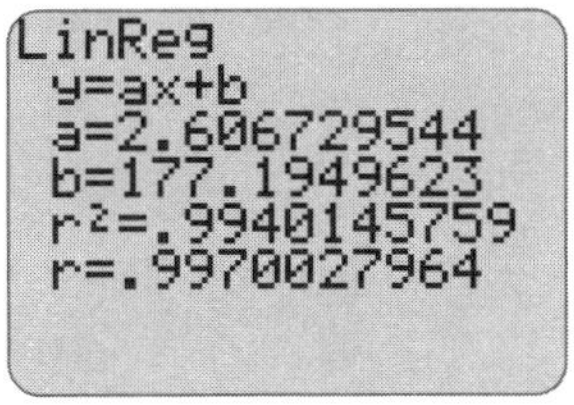

FIGURE 1.54

1.6 SKILLS CHECK

Report models to three decimal places unless otherwise specified. Use unrounded models to graph and calculate unless otherwise specified.

Discuss whether the data shown in the scatter plots in the figures for Exercises 1 and 2 should be modeled by a linear function.

1.

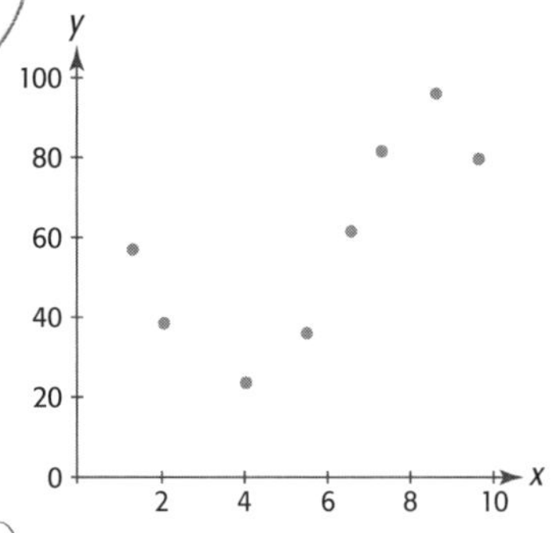

2.

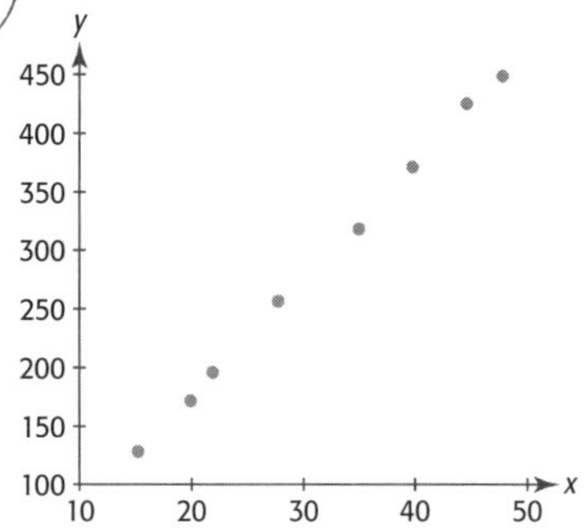

Create a scatter plot for each of the sets of data in Exercises 3 and 4.

3.

x	1	3	5	7	9
y	4	7	10	13	16

4.

x	1	2	3	5	7	9	12
y	1	3	6	1	9	2	6

5. Can the scatter plot in Exercise 3 be fit exactly or only approximately by a linear function? How do you know?

6. Can the scatter plot in Exercise 4 be fit exactly or only approximately by a linear function? How do you know?

7. Find the linear function that is the best fit for the data in Exercise 3.

8. Find the linear function that is the best fit for the data in Exercise 4.

Use the data in the following table for Exercises 9–12.

x	5	8	11	14	17	20
y	7	14	20	28	36	43

9. Construct a scatter plot of the data in the table.

10. Determine if the points plotted in Exercise 9 appear to lie near some line.

11. Create a linear model for the data in the table.

12. Use the function $y = f(x)$ created in Exercise 11 to evaluate $f(3)$ and $f(5)$.

Use the data in the table for Exercises 13–16.

x	2	5	8	9	10	12	16
y	5	10	14	16	18	21	27

13. Construct a scatter plot of the data in the table.

14. Determine if the points plotted in Exercise 13 appear to lie near some line.

15. Create a linear model for the data in the table.

16. Use the rounded function $y = f(x)$ that was reported in Exercise 15 to evaluate $f(3)$ and $f(5)$.

17. Determine which of the equations, $y = -2x + 8$ or $y = -1.5x + 8$, is the better fit for the data points (0, 8), (1, 6), (2, 5), (3, 3).

18. Determine which of the equations, $y = 2.3x + 4$ or $y = 2.1x + 6$, is the better fit for the data points (20, 50), (30, 73), (40, 96), (50, 119), (60, 142).

19. Without graphing, determine which of the following data sets are exactly linear, approximately linear, or nonlinear.

a.

x	y
1	5
2	8
3	11
4	14
5	17

b.

x	y
1	2
2	5
3	10
4	17
5	26

c.

x	y
1	6
2	7
3	12
4	14
5	18

20. Why can't first differences be used to tell if the following data is linear? (1, 3), (4, 5), (5, 7), (7, 9)

1.6 EXERCISES

Report models to three decimal places unless otherwise specified. Use unrounded models to graph and calculate unless otherwise specified.

21. *Working Age* The scatter plot in the figure below projects the ratio of the working-age population to the elderly.

a. Does the scatter plot in the figure represent discrete or continuous data?

b. Would a linear function be a good model for the data points shown in the figure? Explain.

c. Could the data shown for the years beyond 2010 be modeled by a linear function?

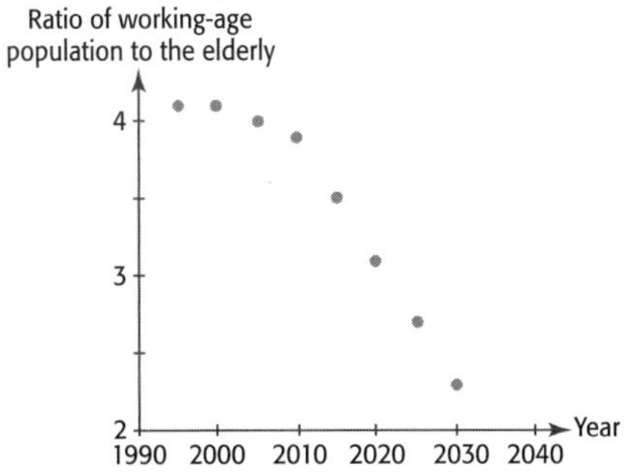

(Source: *Newsweek*, December 1999)

22. *Women in the Workforce* The number of women in the workforce for selected years from 1890 to 1990 is shown in the following figures.

a. Does the scatter plot in the figure define the number of working women as a discrete or continuous function of the year?

b. Does the graph of $y = W(x)$ shown in the figure define the number of working women as a discrete or continuous function of the year?

c. Would the data in the scatter plot be better modeled by a linear function rather than by the nonlinear function $y = W(x)$? Why or why not?

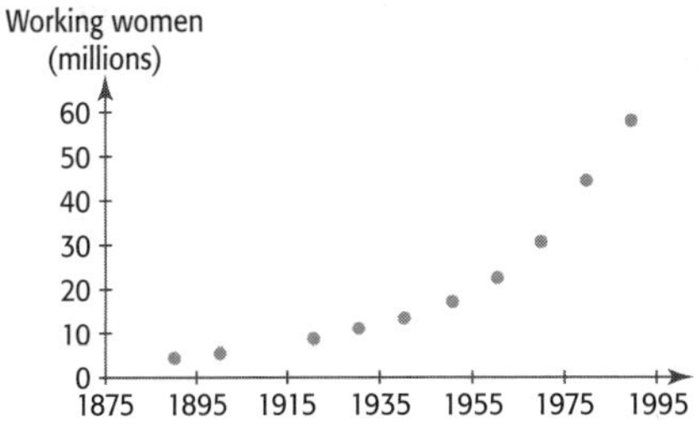

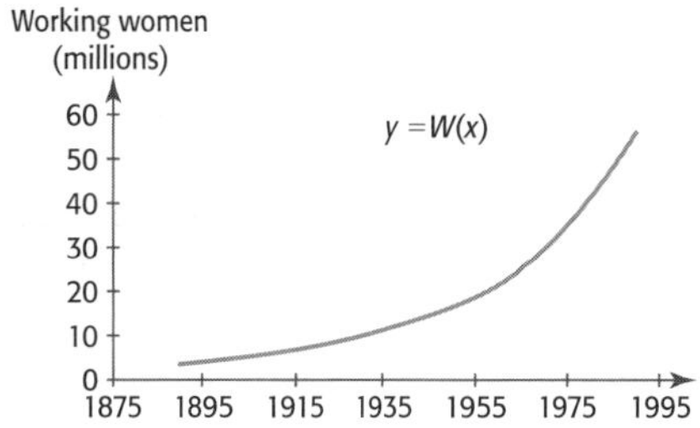

(Source: *Newsweek*, December 1999)

23. *Future Value of an Investment* If \$1000 is invested at 6% simple interest, the future value S in t years is given in the table below.

a. Is the rate of change of the future value constant for uniform inputs?

b. Can the future value be modeled by a linear function?

c. Write the equation that gives the future value as a function of the time in years.

Year (t)	0	1	2	3	4	5
Future value (S)	1000	1060	1120	1180	1240	1300

24. *Lettuce Prices* The average price $p(x)$ of iceberg lettuce, in cents per pound, in U.S. cities between February and June 1999 is a function of the month x in which the average price was determined. Does the point $(5.75, p(5.75))$ have a valid meaning in this situation? If so, interpret the coordinates of the point. If not, explain why not.

25. *Employees* The function $y = 15.910x + 242.758$ models the number of full- and part-time employees in dentists' offices and clinics, where y is in thousands of employees and x is the number of years after 1970. This model was found using data for the years between 1970 and 1998.
(Source: U. S. Bureau of the Census)

a. What does this model estimate as the number of employees in 1993? Is this interpolation or extrapolation?

b. What does this model estimate as the number of employees in 2000? Is this interpolation or extrapolation?

26. *Internet Recruiting* The percentage of Fortune Global 500 firms that actively recruited workers on the Internet between 1998 and 2000 can be modeled by $P(x) = 26.5x - 194.5$ percent, where x is the number of years after 1990.
(Source: *Time*, August 16, 1999)
   a. Is this model discrete or continuous?
   b. Does the input $x = 9$ have a valid meaning in this situation? If so, interpret $P(9)$. If not, explain why not.
   c. Does the input $x = 9.4$ have a valid meaning in this application? If so, interpret $P(9.4)$. If not, explain why not.

27. *Smoking Cessation* The data in the table below gives the percent of adults over age 20 who once smoked and quit smoking between 1965 and 1990.
   a. Using an input equal to the number of years after 1960, write a linear model for the data.
   b. In what year does the model indicate that the output is 39%?
   c. Discuss the reliability of the answer to part (b).

Number of Years After 1960	Adults Who Have Quit Smoking (%)
5	29.8
6	29.3
10	35.2
15	36
16	37
17	36.6
18	38.3
19	39
20	38.9
23	41.8
25	45
27	44.9
30	48.4

(Source: Indiana Tobacco Control Center, U.S. Centers for Disease Control)

r = 0.9653
r² = 0.9825

28. *Smoking Cessation* The following table gives the percent of people over age 20 who once smoked and quit smoking between 1965 and 1990.
   a. Align the data to the number of years after 1965, and fit a linear model to these data.
   b. Show the graph of the data and the equation on the same set of axes.
   c. Compare the model in part (a) with the one developed in Exercise 27. What parts of the equations are the same and what parts are different?

Year	Adults Who Have Quit Smoking (%)
1965	29.8
1966	29.3
1970	35.2
1975	36
1976	37
1977	36.6
1978	38.3
1979	39
1980	38.9
1983	41.8
1985	45
1987	44.9
1990	48.4

(Source: Indiana Tobacco Control Center, U.S. Centers for Disease Control)

29. *Postal Rates* The table on the next page gives the local postal rates as a function of the weight of the printed matter that is mailed.
   a. Write the linear equation that models the data.
   b. Compare the outputs from the equation with the data outputs for several different weights. How well does the model fit the data?

Weight (lb)	Postal Rate ($)	Weight (lb)	Postal Rate ($)
1.5	1.14	9	1.43
2	1.16	10	1.47
3	1.20	11	1.51
4	1.24	12	1.55
5	1.28	13	1.59
6	1.31	14	1.63
7	1.36	15	1.67
8	1.39		

(Source: 1999 *World Almanac*)

30. *SAT Scores* The average math SAT scores in Beaufort County, South Carolina, are given in the table below. Write the equation of the line that is the best fit for these data, aligning the data so that $x = 0$ in 1990. Compare the outputs from the equation with the original data for each of the years 1994 to 1999 and determine the year in which the model output is closest to the data value.

Year	Average Math SAT Score
1994	472
1995	464
1996	470
1997	471
1998	473
1999	470

(Source: Beaufort County School Report 1998–1999, *The Island Packet*)

31. *Postal Rates* Consider the data and the rounded function that models the data in Exercise 29.

a. What does the model predict the rate to be for 16 lbs. of bound printed matter?

b. If the postal rate is $1.55, what is the estimated weight from the model?

c. What does the model indicate that the increase in postal rate will be for each increase of 1 lb in weight?

d. What does the data give as the increase in postal rate for each pound increase in weight for weights from 9 lb to 14 lb?

32. *Jail Population* The figure below gives the average daily number of inmates in the Beaufort County, South Carolina, Detention Center.

a. Align the data to the number of years after 1990, and create a linear equation that models the data.

b. Assuming that the model is still appropriate, use the linear function to estimate the population in the Beaufort County jail in 2005.

c. In what year does the model indicate that the average daily number of inmates is 115?

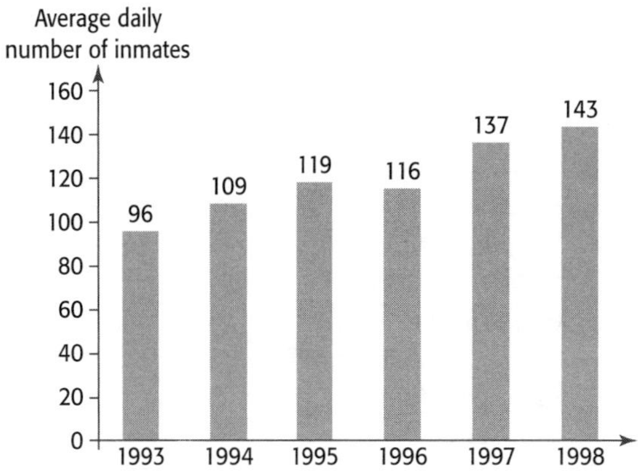

(Source: Beaufort County Detention Center, *The Island Packet*, October 1999)

33. *Marriage Rate* The marriage rate (per 1000 unmarried women) is given by the table on the next page for the years 1960 to 1997.

a. Write the linear equation that models the marriage rate as a function of years after 1950.

b. What does the model indicate the marriage rate to be in 1985?

c. For what year does the model indicate that the rate fell below 50 per 1000?

d. Discuss the reliability of your answer to part (c).

Year	Marriage Rate (marriages per 1000 unmarried women)
1960	73.5
1970	76.5
1980	61.4
1990	54.5
1991	54.2
1992	53.3
1993	52.3
1994	51.5
1995	50.8
1996	49.7
1997	49.4

(Source: Index of Leading Cultural Indicators)

34. *Volunteers in Medicine* Retired physicians and nurses donate services to treat patients in the Volunteers in Medicine Clinic on Hilton Head Island. The numbers treated free of charge during each of the years 1995 to 1999 are given in the table below.

a. Write the equation of the line that is the best fit for the data, aligning the input to be the number of years after 1990.

b. What does the model indicate as the average rate of increase of patients per year from 1995 to 1999?

Year	Patients Treated
1995	3106
1996	3417
1997	4133
1998	4246
1999	4400

(Source: "Volunteers in Medicine," *The Island Packet*, November 21, 1999)

35. *Education* The numbers of children up to 21 years of age that are served by educational programs for the disabled for the years between 1988 and 1996 are given in the accompanying table.

a. Write a linear equation that models these data, aligning the input to be the number of years after 1980 and using as output the number of thousands of children served.

b. What does the model indicate is the average rate of increase in the number of children served per year from 1988 to 1996?

c. What does this model estimate the number of children served to be in 2005? Is this an interpretation or an extrapolation?

Year	Disabled Children Served (thousands)
1988	4446
1989	4544
1990	4641
1991	4771
1992	4949
1993	5125
1994	5309
1995	5378
1996	5573

(Source: Office of Special Education and Rehabilitative Services of the U.S. Department of Education)

36. *U.S. Households with Internet Access* The following table gives the percentage of U.S. households with Internet access in various years.

Year	1996	1997	1998	1999	2000	2001
Percent	8.5	14.3	26.2	28.6	41.5	50.5

(Source: U.S. Census Bureau)

a. Create a scatter plot of the data, with x equal to the number of years from 1995.

b. Create a linear equation that models the data.

c. Graph the function and the data on the same graph, to see how well the function models (fits) the data.

37. *Gross Domestic Product* The table below gives the gross domestic product (the value of all goods and services, in billions of dollars) of the United States for selected years from 1940 to 2003.

a. Create a scatter plot of the data, with y representing the GDP in billions of dollars and x representing the number of years after 1940.

b. Find the linear function that best fits the data, with x equal to the number of years after 1940.

c. Is a line a good fit for this data?

Year	Gross Domestic Product	Year	Gross Domestic Product
1940	837	1980	3746
1945	1559	1985	4207
1950	1328	1990	4853
1955	1700	1995	5439
1960	1934	2000	9817
1965	2373	2001	10,128
1970	2847	2002	10,487
1975	3173	2003	11,004

(Source: U.S. Bureau of Economic Analysis)

38. *Giving at IBM* The amount of individual donations to nonprofit organizations and educational institutions by employees at IBM, for the years 1994–1998, is shown in the following table.

a. Write the linear equation that is the best fit for these data, with x equal to the number of years after 1990.

b. If this model is valid after 1998, for what year does the model predict that giving will reach \$50 million?

Year	Donations (\$ million)
1994	35.6
1995	35.9
1996	36.7
1997	39.2
1998	43.9

(Source: 1998 IBM Annual Report)

39. *Salaries* The table below gives the median annual income of female workers with 4 years of college for the years 1979–1989.

Year	Salary (\$)	Year	Salary (\$)
1979	13,441	1985	21,589
1980	15,143	1986	22,412
1981	16,322	1987	23,399
1982	17,405	1988	25,187
1983	18,452	1989	26,709
1984	20,257		

(Source: U.S. Bureau of Census)

a. Align the data to represent the number of years after 1970 and draw a scatter plot of the data.

b. Write the best-fit linear model for the data.

c. Graph the linear function and the data points on the same graph and discuss how well the function models the data.

d. Classify each of the following as discrete or continuous.

i. The data in the table.

ii. The scatter plot in part (a).

iii. The model in part (b).

40. *World Cigarette Production* The table below gives the per capita world production of cigarettes for selected years from 1950 to 1992.

a. Find the equation of the line that is the best fit for this data. Align the data so that $x = 0$ in 1950.

b. Is the model with equation in part (a) discrete or continuous?

c. Does the model indicate that world cigarette production was increasing or decreasing between 1950 and 1992?

d. Do you think the model with equation from part (a) would apply for the years 1993 to 2003? Why or why not?

Year	World Production (cigarettes per person)
1950	660
1955	691
1960	708
1965	767
1970	840
1975	916
1980	985
1985	1000
1990	1017
1992	984

(Source: The National Clearinghouse on Tobacco and Health)

41. *U.S. Population* The following table gives the projections of the U.S. population from 2000 to 2100.

a. Find a linear function that models this data, with x equal to the number of years from 2000 and $f(x)$ equal to the population in millions.

b. Find $f(65)$ and state what it means.

c. What does this model predict the population to be in 2080? How does this compare with the value for 2080 in the table?

Year	Population (millions)	Year	Population (millions)
2000	275.3	2060	432.0
2010	299.9	2070	463.6
2020	324.9	2080	497.8
2030	351.1	2090	533.6
2040	377.4	2100	571.0
2050	403.7		

(Source: www.census.gov/population/projections)

42. *Earnings and Race* The following table gives the median household income for whites and blacks in various years.

a. Let x represent the median household income for whites and y represent the corresponding median household income for blacks and make a scatter plot of these data.

b. Find a linear model that expresses the median household income for blacks as a function of the median household income for whites.

c. What income for blacks corresponds to \$50,000 for whites, according to the model?

	Median Household Income (dollars)	
Year	Whites	Blacks
1981	38,954	21,859
1985	40,614	24,163
1990	42,622	25,488
1995	42,871	26,842
2000	46,910	31,690
2001	46,261	30,625
2002	46,119	29,691
2003	45,631	29,645

(Source: U.S. Bureau of the Census)

43. *Internet Brokerage Accounts* The number of Internet brokerage accounts is given in the table below.

a. Write the linear equation that models the number of Internet brokerage accounts as a function of the number of years after 1990.

b. According to the model, what is the annual increase in the number of accounts?

c. Use the model to estimate when the number of accounts will be 20 million.

d. What happened in 2001 that may make this model invalid for years after 2001?

Year	Internet Brokerage Accounts (millions)
1996	1.5
1997	4.1
1998	7.1
1999	10.5
2000	14.0
2001	18.0

(Source: *Time*, May 31, 1999)

44. *Consumer Spending* The consumer spending per person per year for television subscription video services is given in the table below.

Year	Spending ($)	Year	Spending ($)
1992	101.39	1997	155.70
1993	109.02	1998	168.75
1994	110.89	1999	178.23
1995	125.54	2000	187.60
1996	140.37	2001	196.62

(Source: *Statistical Abstract of the U.S.*, 1998)

a. Find the equation of the line that gives the per capita yearly consumer spending as a function of the number of years after 1990.

b. Use the equation to determine the year in which the per capita yearly consumer spending for these television services first exceeds $250.

c. What assumption must be made for the answer to part (b) to be considered valid?

45. *Prison Sentences* The National Center for Policy Analysis calculates expected prison time by multiplying the probabilities of being arrested, of being prosecuted, and of being convicted by the length of sentence if convicted. The expected time served (in days) for each serious crime committed is shown in the table below.

a. Use the data to create a linear equation to model expected prison time in days for a serious crime as a function of the number of years after 1960.

b. Use the model to find the expected prison time in 1975.

Year	Expected Prison Time (days)
1970	10.1
1980	10.6
1985	13.2
1990	18.0
1992	18.5
1993	17.2
1994	19.5
1995	20.2
1996	21.7

(Source: National Center for Policy Analysis)

1.7 Systems of Linear Equations in Two Variables

Key Concepts

- System of equations
- Solving a system of linear equations in two variables
- Graphing method
- Substitution method
- Elimination method
- Break-even analysis
- Supply and demand
- Market equilibrium
- Nonunique solutions to systems of linear equations in two variables
- Modeling systems of equations

Section Preview Concerta

The new drug Concerta contains a time-release version of methylphenidate, the main ingredient in Ritalin. Primarily because its effects last 12–14 hours as opposed to Ritalin's 4–5 hours, sales of this drug increased rapidly while Ritalin sales decreased slightly. Looking at the graphs of the market shares of the two drugs in Figure 1.55(a), we can see that they intersect at a point, and we interpret this to mean that Concerta market share equaled that of Ritalin in the first few months after it was released. The graphs appear to intersect near the point representing 7 weeks after August 25 (2000) and a 7% market share. (Other drugs, mostly generic, account for the remaining market share.) If we wrote the equations that represented these graphs with x representing weeks and y representing market-share percent, the point of intersection of the two graphs would represent the *simultaneous* solution of the two equations because both equations would be satisfied by the coordinates of the point.

In this section, we solve systems of linear equations in two variables graphically, by substitution, and by the elimination method.

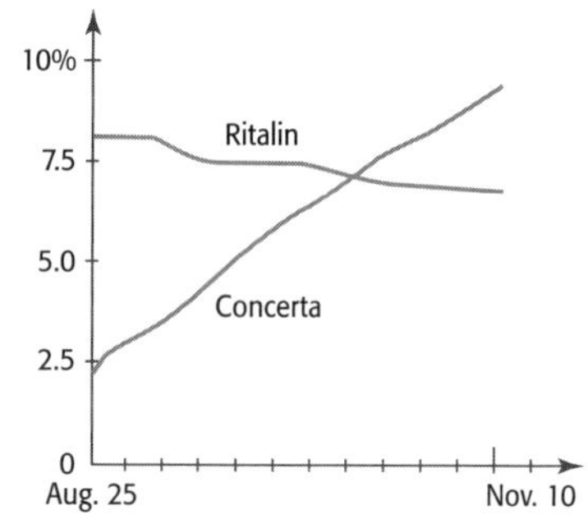

FIGURE 1.55(a)

(Source: *Newsweek*, December 4, 2000)

Graphical Solution of Systems

In Section 1.5, we used the intersect method to solve a linear equation by first graphing functions representing the expressions on each side of the equation and then finding the intersection of these graphs. For example, to solve

$$3000x - 7200 = 5800x - 8600$$

we can graph

$$y_1 = 3000x - 7200 \text{ and } y_2 = 5800x - 8600$$

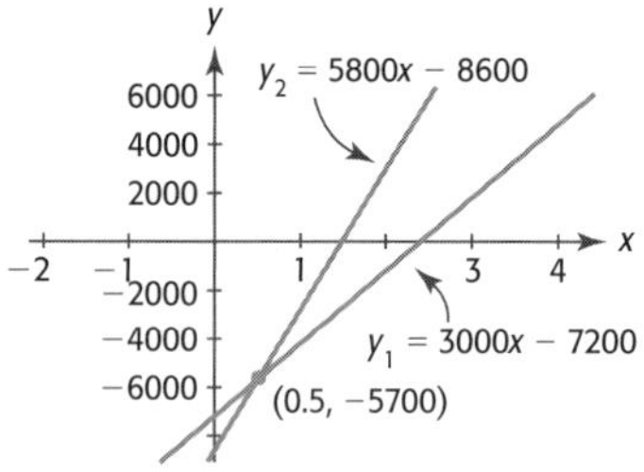

FIGURE 1.55(b)

and find the point of intersection to be $(0.5, -5700)$. The x-coordinate of the point of intersection of the lines is the value of x that satisfies the original equation, $3000x - 7200 = 5800x - 8600$. Thus, the solution to this equation is $x = 0.5$ (Figure 1.55(b)). In this example, we were actually using a graphical method to solve a **system of two equations in two variables** denoted by

$$\begin{cases} y = 3000x - 7200 \\ y = 5800x - 8600 \end{cases}$$

The coordinates of the point of intersection of the two graphs give the x and y values that satisfy both equations **simultaneously**, and these values are called the **solution** to the system. The following example uses the graphical method to solve a system of equations in two variables.

EXAMPLE 1 Break Even

A company is said to **break even** from the production and sale of a product if the total revenue equals the total cost—that is, if $R(x) = C(x)$. Because profit $P(x) = R(x) - C(x)$, we can also say that the company breaks even if the profit for the product is zero.

Suppose a company has its total revenue for a product given by $R = 5585x$ and its total cost given by $C = 61{,}740 + 440x$, where x is the number of thousands of tons of the product that is produced and sold per year. The company is said to break even when the total revenue equals the total cost—that is, when $R = C$. Find the number of thousands of tons of the product that gives break even and how much the revenue and cost are at that level of production.

Solution

We graph the revenue function as $y_1 = 5585x$ and the cost function as $y_2 = 61{,}740 + 440x$ with a graphing utility on a window that contains the point of intersection. (Figure 1.59(a)). We find the point of intersection to be (12, 67,020) (Figure 1.56(b)).

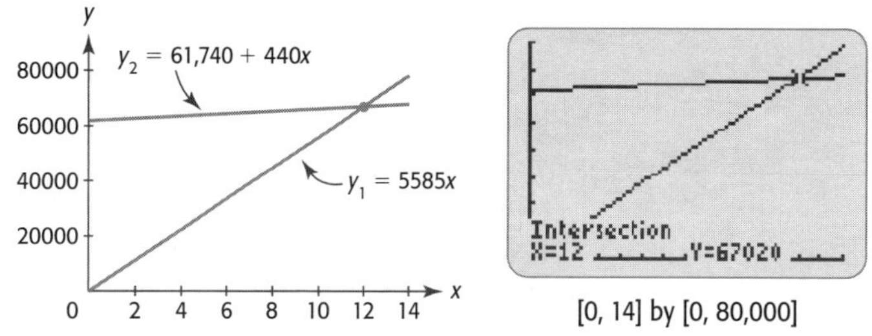

[0, 14] by [0, 80,000]

(a) (b)

FIGURE 1.56

Thus the company will break even on this product if 12 thousand tons of the product are sold, when both the cost and revenue equal \$67,020. ■

It is frequently necessary to solve each equation for a variable so that the equation can be graphed with a graphing utility. It is also necessary to find a viewing window that contains the point of intersection. Consider the following example.

EXAMPLE 2 Solving a System of Linear Equations

Solve the system

$$\begin{cases} 3x - 4y = 21 \\ 2x + 5y = -9 \end{cases}$$

Solution

To solve this system with a graphing utility, we first solve both equations for y. The solution follows:

$$\begin{aligned} 3x - 4y &= 21 \\ -4y &= 21 - 3x \\ y &= \frac{21 - 3x}{-4} = \frac{3x - 21}{4} \end{aligned} \qquad \begin{aligned} 2x + 5y &= -9 \\ 5y &= -9 - 2x \\ y &= \frac{-9 - 2x}{5} \end{aligned}$$

Graphing these equations with a window that contains the point of intersection (Figure 1.57(a)), and finding the point of intersection (Figure 1.57(b)) gives $x = 3$, $y = -3$, so the solution is $(3, -3)$.

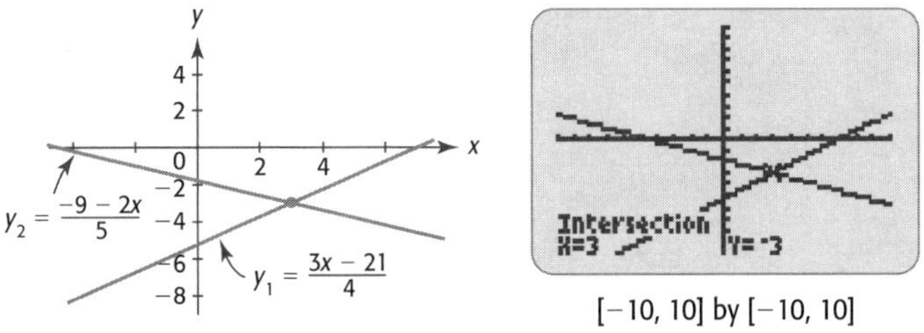

[−10, 10] by [−10, 10]

(a) (b)

FIGURE 1.57

The system of linear equations discussed in Example 2 has a unique solution, shown as the point of intersection of the graphs. It is possible that two equations in a system describe the same line. When this happens, the equations are equivalent, and the values that satisfy one equation are also solutions to the other equation, and to the system. Such a system is a **dependent system**. If a system contains two equations whose graphs are parallel lines, they have no point in common, and thus the system has no solution. Such a system of equations is **inconsistent**. Figure 1.58(a)–(c) represents these three situations: systems that have a unique solution, many solutions (dependent system), and no solution (inconsistent system), respectively. Note that the slopes of the lines are equal in Figure 1.58(b) and in Figure 1.58(c).

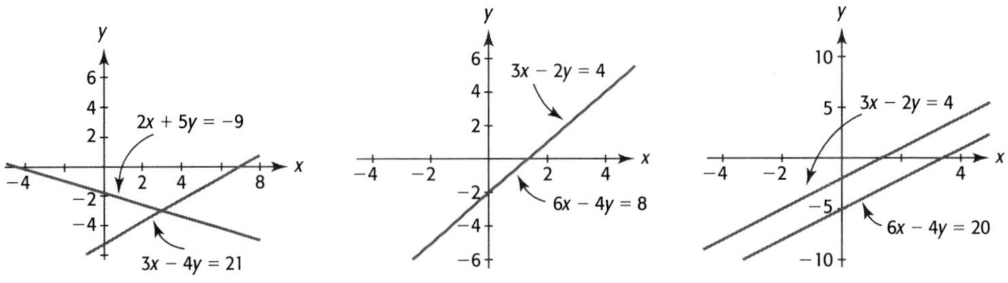

Intersecting lines — unique solution (a)

Same lines — infinite number of solutions Dependent system (b)

Parallel lines — no solution Inconsistent system (c)

FIGURE 1.58

Solution by Substitution

Solving systems of equations by graphing with a graphing utility is not always the easiest method to use because the equations must be solved for y to be entered in the utility and an appropriate window must be used. A second solution method for a system of linear equations is called the **substitution method**, where one equation is solved for a variable and that variable is replaced by the equivalent expression in the other equation.

EXAMPLE 3 Concerta

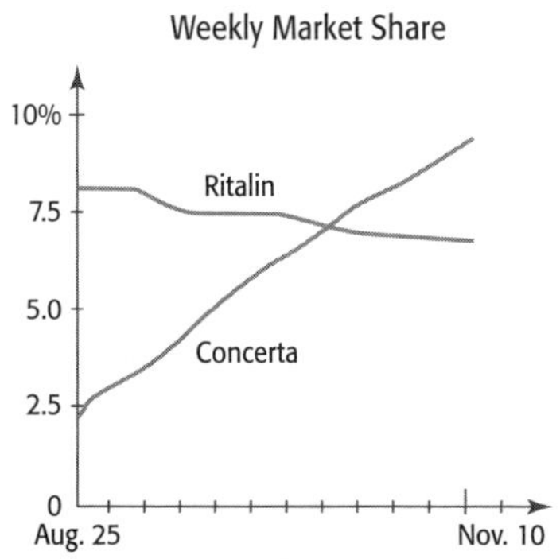

FIGURE 1.59

(Source: *Newsweek*, December 4, 2000)

The graphs that compare the weekly market share of the two drugs Concerta and Ritalin are shown in Figure 1.59. If the linear equation $y = 0.690x + 2.4$ models the percent of market share of Concerta and $y = -0.072x + 7.7$ models the percent of market share of Ritalin, where x is the number of weeks since Concerta was made available, find the number of weeks past the release date that the weekly market share of Concerta reached that of Ritalin.

Solution

To find the number of weeks (x) until the Concerta market share equals the Ritalin share, we solve the system

$$\begin{cases} y = 0.690x + 2.4 \\ y = -0.072x + 7.7 \end{cases}$$

We can solve this system by substitution. From the first equation we see that $0.690x + 2.4$ is equal to y, so we substitute this expression for y in the second equation. This substitution gives the equation

$$0.690x + 2.4 = -0.072x + 7.7$$

Solving this equation for x gives

$$0.762x = 5.3$$
$$x = 6.96 \approx 7 \text{ (weeks)}$$

Thus, the weekly market share of Concerta reached that of Ritalin 7 weeks after it was released. ■

The substitution method is also illustrated by the following example, which finds market equilibrium. The quantity of a product that is demanded by consumers is called the **demand** for the product and the quantity that is supplied is called the **supply**. In a free economy, both demand and supply are related to the price, and the price where the number of units demanded equals the number of units supplied is called the **equilibrium price**.

EXAMPLE 4 Market Equilibrium

Suppose the daily demand for a product is given by $p = 200 - 2q$, where q is the number of units demanded and p is the price per unit in dollars, and that the daily supply is given by $p = 60 + 5q$, where q is the number of units supplied and p is the price in dollars. If a price results in more units being supplied than demanded, we say there is a *surplus*, and if the price results in fewer units being supplied than demanded, we say there is a *shortfall*. **Market equilibrium** occurs when the supply quantity equals the demand quantity (and when the prices are equal); that is, when q and p both satisfy the system

$$\begin{cases} p = 200 - 2q \\ p = 60 + 5q \end{cases}$$

a. If the price is \$140, how many units are supplied and how many are demanded?

b. Does this price give a surplus or a shortfall of the product?

c. What price gives market equilibrium?

Solution

a. If the price is \$140, the number of units demanded satisfies $140 = 200 - 2q$, or $q = 30$, and the number of units supplied satisfies $140 = 60 + 5q$, or $q = 16$.

b. At this price, the quantity supplied is less than the quantity demanded, so a shortfall occurs.

c. Because market equilibrium occurs where q and p both satisfy the system

$$\begin{cases} p = 200 - 2q \\ p = 60 + 5q \end{cases}$$

we seek the solution to this system.

We can solve this system by substitution. Substituting $60 + 5q$ for p in the first equation gives the equation

$$60 + 5q = 200 - 2q$$

Solving this equation gives

$$\begin{aligned} 60 + 5q &= 200 - 2q \\ 7q &= 140 \\ q &= 20 \end{aligned}$$

Thus, market equilibrium occurs when the number of units is 20, and the equilibrium price is

$$p = 200 - 2(20) = 60 + 5(20) = 160 \text{ dollars per unit}$$

■

The substitution in Example 4 was not difficult because both equations were solved for p. In general, we use the following steps to solve systems of two equations in two variables by substitution.

SOLUTION OF SYSTEMS OF EQUATIONS BY SUBSTITUTION

1. Solve one of the equations for one of the variables in terms of the other variable.
2. Substitute the expression from step 1 into the other equation to give an equation in one variable.
3. Solve the linear equation for the variable.
4. Substitute this solution into the equation from step 1 or into one of the original equations and solve this equation for the second variable.
5. Check the solution in both original equations or check graphically.

EXAMPLE 5 Solution by Substitution

Solve the system $\begin{cases} 3x + 4y = 10 \\ 4x - 2y = 6 \end{cases}$ by substitution.

Solution

To solve this system by substitution, we can solve either equation for either variable and substitute the resulting expression into the other equation. Solving the second equation for y gives

$$\begin{aligned} 4x - 2y &= 6 \\ -2y &= -4x + 6 \\ y &= 2x - 3 \end{aligned}$$

Substituting this expression for y in the first equation gives

$$3x + 4(2x - 3) = 10$$

Solving this equation gives

$$\begin{aligned} 3x + 8x - 12 &= 10 \\ 11x &= 22 \\ x &= 2 \end{aligned}$$

Substituting $x = 2$ into $y = 2x - 3$ gives $y = 2(2) - 3 = 1$, so the solution to the system is $x = 2, y = 1$, or $(2, 1)$.

Checking shows that this solution satisfies both original equations. ■

Solution by Elimination

A second analytical method, called the **elimination method**, is frequently an easier method to use to solve a system of linear equations. The elimination method is based on rewriting one or both of the equations in an equivalent form that allows us to eliminate one of the variables by adding or subtracting the equations.

SOLVING A SYSTEM OF TWO EQUATIONS IN TWO VARIABLES BY ELIMINATION

1. If necessary, multiply one or both equations by a nonzero number that will make the coefficients of one of the variables in the equations equal, except perhaps for sign.
2. Add or subtract the equations to eliminate one of the variables.
3. Solve for the variable in the resulting equation.
4. Substitute the solution from step 3 into one of the original equations and solve for the second variable.
5. Check the solutions in the remaining original equation, or check graphically.

EXAMPLE 6 Solution by Elimination

Use the elimination method to solve the system

$$\begin{cases} 3x + 4y = 10 \\ 4x - 2y = 6 \end{cases}$$

and check the solution graphically.

Solution

The goal is to convert one of the equations into an equivalent equation of a form so that addition of the two equations will eliminate one of the variables. Notice that the coefficient of y in the second equation, -2, is a factor of the coefficient of y in the first equation, 4. If we multiply both sides of the second equation by 2 and add the two equations, this will eliminate the y-variable.

$$\begin{cases} 3x + 4y = 10 & (1) \\ 4x - 2y = 6 & (2) \end{cases}$$

Multiply 2 times Equation (2), getting equivalent Equation (3).

$$\begin{cases} 3x + 4y = 10 & (1) \\ 8x - 4y = 12 & (3) \end{cases}$$

Add Equations (1) and (3) to eliminate y. $\qquad 11x = 22$

Solve the new equation for x. $\qquad x = 2$

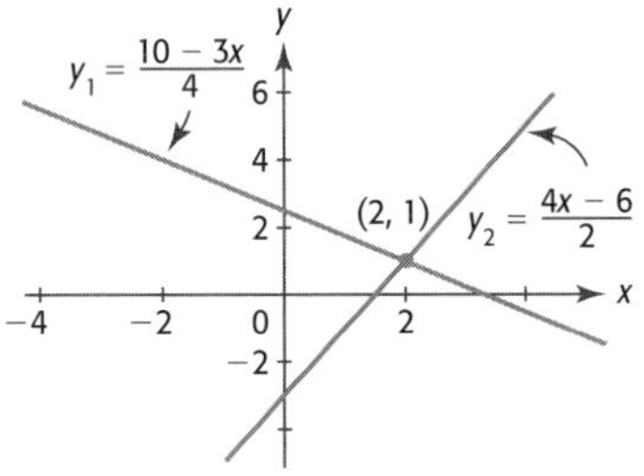

FIGURE 1.60

Substituting $x = 2$ in the first equation gives $3(2) + 4y = 10$, or $y = 1$. Thus, the solution to the system is $x = 2$, $y = 1$, or (2, 1).

To check graphically, we solve both equations for y, getting $y_1 = \frac{10 - 3x}{4}$ and $y_2 = \frac{4x - 6}{2}$, then graph the equations and find the point of intersection to be (2, 1) (Figure 1.60). ■

Modeling Systems of Linear Equations

Solving some real problems requires us to create two or more equations whose simultaneous solution is the solution to the problem. Consider the following examples.

EXAMPLE 7 Investments

An investor has \$300,000 to invest, part at 12% and the remainder in a less risky investment at 7%. If her investment goal is to have an annual income of \$27,000, how much should she put in each investment?

Solution

If we denote the amount invested at 12% as x and the amount invested at 7% as y, the sum of the investments is $x + y$, so we have the equation

$$x + y = 300{,}000$$

The annual income from the 12% investment is $0.12x$ and the annual income from the 7% investment is $0.07y$. Thus, the desired annual income from the two investments is

$$0.12x + 0.07y = 27{,}000$$

Thus, we can write the given information as a system of equations.

$$\begin{cases} x + \quad y = 300{,}000 \\ 0.12x + 0.07y = \quad 27{,}000 \end{cases}$$

To solve this system, we multiply the first equation by -0.12 and add the two equations. This results in an equation with one variable:

$$\begin{cases} -0.12x - 0.12y = -36{,}000 \\ \quad 0.12x + 0.07y = 27{,}000 \end{cases}$$

$$-0.05y = -9000$$

$$y = 180{,}000$$

Substituting 180,000 for y into the first original equation and solving for x gives $x = 120{,}000$. Thus, \$120,000 should be invested at 12%, and \$180,000 should be invested at 7%.

To check this solution, we see that the total investment is \$120,000 + \$180,000, which equals \$300,000. The interest earned at 12% is \$120,000(.12) = \$14,400, and the interest earned at 7% is \$180,000(.07) = \$12,600. The total interest is \$14,400 + \$12,600, which equals \$27,000. This agrees with the given information. ■

EXAMPLE 8 Medication

A nurse has two solutions that contain different concentrations of a certain medication. One is a 12% concentration, and the other is an 8% concentration. How many cubic centimeters (cc) of each should she mix together to obtain 20 cc of a 9% solution?

Solution

We begin by denoting the total amount of the first solution by x and the amount of the second solution by y. The total amount of solution is the sum of x and y, so

$$x + y = 20$$

The total medication in the combined solution is 9% of 20 cc, or $0.09(20) = 1.8$ cc, and the mixture is obtained by adding $0.12x$ and $0.08y$, so

$$0.12x + 0.08y = 1.8$$

We can use substitution to solve the system

$$\begin{cases} x + \quad y = 20 \\ 0.12x + 0.08y = 1.8 \end{cases}$$

Substituting $20 - x$ for y in $0.12x + 0.08y = 1.8$ gives $0.12x + 0.08(20 - x) = 1.8$, and solving this equation gives

$$0.12x + 0.08(20 - x) = 1.8$$

$$0.12x + 1.6 - 0.08x = 1.8$$

$$0.04x = 0.2$$

$$x = 5$$

Thus, combining 5 cc of the first solution with $20 - 5 = 15$ cc of the second solution gives 20 cc of the 9% solution. ■

Inconsistent and Dependent Systems

EXAMPLE 9 Systems with Nonunique Solutions

Use the elimination method to solve each of the following systems, if possible. Verify the solution graphically.

a. $\begin{cases} 2x - 3y = 4 \\ 6x - 9y = 12 \end{cases}$ **b.** $\begin{cases} 2x - 3y = 4 \\ 6x - 9y = 36 \end{cases}$

Solution

a. To solve $\begin{cases} 2x - 3y = 4 \\ 6x - 9y = 12 \end{cases}$ we multiply the first equation by -3 and add the equations, getting the following:

$$\begin{cases} -6x + 9y = -12 \\ 6x - 9y = 12 \end{cases}$$
$$0 = 0$$

This indicates that the graphs of the equations intersect when $0 = 0$, *which is always true*. Thus any value that satisfies one of these equations also satisfies the other, and there are *infinitely many* solutions. Figure 1.61(a) shows that the graphs of the equations lie on the same line. Notice that the second equation is a multiple of the first, so the equations are equivalent. This system is *dependent*.

b. To solve $\begin{cases} 2x - 3y = 4 \\ 6x - 9y = 36 \end{cases}$ we multiply the first equation by -3 and add the equations, getting the following:

$$\begin{cases} -6x + 9y = -12 \\ 6x - 9y = 36 \end{cases}$$
$$0 = 24$$

This indicates that the equations intersect when $0 = 24$, *which is never true*. Thus, no values of x and y satisfy both of the equations. Figure 1.61(b) shows that the graphs of the equations are parallel. This system is *inconsistent*. ■

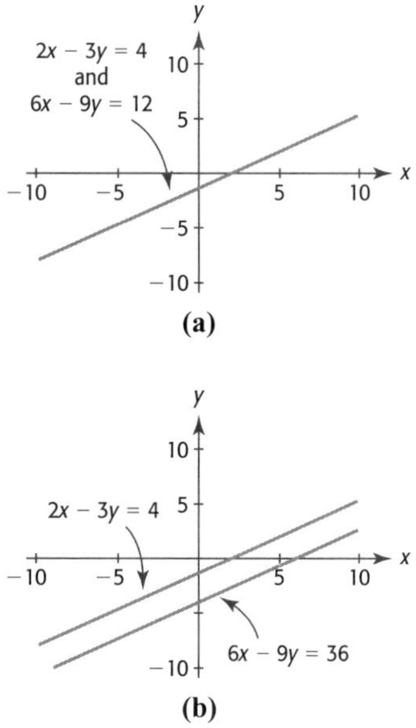

FIGURE 1.61

1.7 SKILLS CHECK

1. What are the coordinates of the point of intersection of $y = 3x - 2$ and $y = 3 - 2x$?

2. Give the coordinates of the point of intersection of $3x + 2y = 5$ and $5x - 3y = 21$.

In Exercises 3–6, solve the systems of equations graphically.

3. $\begin{cases} y = 3x - 12 \\ y = 4x + 2 \end{cases}$

4. $\begin{cases} 2x - 4y = 6 \\ 3x + 5y = 20 \end{cases}$

5. $\begin{cases} 4x - 3y = -4 \\ 2x - 5y = -4 \end{cases}$

6. $\begin{cases} 5x - 6y = 22 \\ 4x - 4y = 16 \end{cases}$

7. Does the system $\begin{cases} 2x + 5y = 6 \\ x + 2.5y = 3 \end{cases}$ have a unique solution, no solution, or many solutions? What does this mean graphically?

8. Does the system $\begin{cases} 6x + 4y = 3 \\ 3x + 2y = 3 \end{cases}$ have a unique solution, no solution, or many solutions? What does this mean graphically?

In Exercises 9–12, solve the systems of equations by substitution.

9. $\begin{cases} x - 5y = 12 \\ 3x + 4y = -2 \end{cases}$

10. $\begin{cases} 2x - 3y = 2 \\ 5x - y = 18 \end{cases}$

11. $\begin{cases} 2x - 3y = 5 \\ 5x + 4y = 1 \end{cases}$

12. $\begin{cases} 4x - 5y = -17 \\ 3x + 2y = -7 \end{cases}$

In Exercises 13–22, solve the systems of equations by elimination, if a solution exists.

13. $\begin{cases} x + 3y = 5 \\ 2x + 4y = 8 \end{cases}$

14. $\begin{cases} 4x - 3y = -13 \\ 5x + 6y = 13 \end{cases}$

15. $\begin{cases} 5x + 3y = 8 \\ 2x + 4y = 8 \end{cases}$

16. $\begin{cases} 3x + 3y = 5 \\ 2x + 4y = 8 \end{cases}$

17. $\begin{cases} 0.3x + 0.4y = 2.4 \\ 5x - 3y = 11 \end{cases}$

18. $\begin{cases} 8x - 4y = 0 \\ 0.5x + 0.3y = 2.2 \end{cases}$

19. $\begin{cases} 3x + 6y = 12 \\ 2x + 4y = 8 \end{cases}$

20. $\begin{cases} -4x + 6y = 12 \\ 10x - 15y = -30 \end{cases}$

21. $\begin{cases} 6x - 9y = 12 \\ 3x - 4.5y = -6 \end{cases}$

22. $\begin{cases} 4x - 8y = 5 \\ 6x - 12y = 10 \end{cases}$

In Exercises 23–32, solve the systems of equations by any convenient method, if a solution exists.

23. $\begin{cases} y = 3x - 2 \\ y = 5x - 6 \end{cases}$

24. $\begin{cases} y = 8x - 6 \\ y = 14x - 12 \end{cases}$

25. $\begin{cases} 4x + 6y = 4 \\ x = 4y + 8 \end{cases}$

26. $\begin{cases} y = 4x - 5 \\ 3x - 4y = 7 \end{cases}$

27. $\begin{cases} 2x - 5y = 16 \\ 6x - 8y = 34 \end{cases}$

28. $\begin{cases} 4x - y = 4 \\ 6x + 3y = 15 \end{cases}$

29. $\begin{cases} 3x - 7y = -1 \\ 4x + 3y = 11 \end{cases}$

30. $\begin{cases} 5x - 3y = 12 \\ 3x - 5y = 8 \end{cases}$

31. $\begin{cases} 4x - 3y = 9 \\ 8x - 6y = 16 \end{cases}$

32. $\begin{cases} 5x - 4y = 8 \\ -15x + 12y = -12 \end{cases}$

1.7 EXERCISES

33. *Break Even* A manufacturer of kitchen sinks has total revenue given by the function $R = 76.50x$ and has total cost given by $C = 2970 + 27x$, where x is the number of sinks produced and sold. Use graphical methods to find the number of units that gives break even for the product.

34. *Break Even* A jewelry maker has total revenue for her bracelets given by $R = 89.75x$ and incurs a total cost of $C = 23.50x + 1192.50$, where x is the number of bracelets produced and sold. Use graphical methods to find the number of units that gives break even for the product.

35. *Supply and Demand* A certain product has supply and demand functions given by $p = 5q + 20$ and $p = 128 - 4q$, respectively.
   a. If the price p is \$60, how many units q are supplied and how many are demanded?
   b. What price gives market equilibrium, and how many units are demanded and supplied at this price?

36. *Market Equilibrium* The demand for a brand of clock radio is given by $p + 2q = 320$, and the supply for these radios is given by $p - 8q = 20$, where p is the price and q is the quantity demanded at price p. Solve the system containing these two equations to find the price at which the quantity demanded equals the quantity supplied and the equilibrium quantity.

37. *Military* The number of active-duty U.S. Navy personnel is given by $y = -5.686x + 676.173$ thousand, and the number of active-duty U.S. Air Force personnel is given by $y = -11.997x + 847.529$ thousand, where x is the number of years from 1960.
   a. Use graphical methods to find the year in which the number of Navy personnel reaches the number of Air Force personnel.
   b. How many will be in each service when the numbers of personnel are equal?
   c. For what years does the Navy have more active-duty personnel than the Air Force?
   (Source: *World Almanac*)

38. *College Enrollment* Suppose the percent of males who enrolled in college within 12 months of high school graduation is given by $y = -0.126x + 55.72$ and the percent of females enrolled in college within 12 months of high school graduation is given by $y = 0.73x + 39.7$, where x is the number of years after 1960. Use graphical methods to find the year these models indicate that the percent of females equaled the percent of males.
(Source: *Statistical Abstract of the United States, 2000*)

39. *Snack Sales* The graphs below make it appear that mint sales surpassed gum sales in 1999. Using this data, the best-fitting line for mint sales has equation $y = 24.5x + 93.5$ (million dollars) and the best-fitting line for gum sales is $y = -0.2x + 1007$ (million dollars), where x is the number of years after 1990.

a. For what value of x does mint sales equal gum sales according to these models?

b. In what year do these models predict that mint sales will reach gum sales?

c. Are these graphs misleading even if they are correct? How?

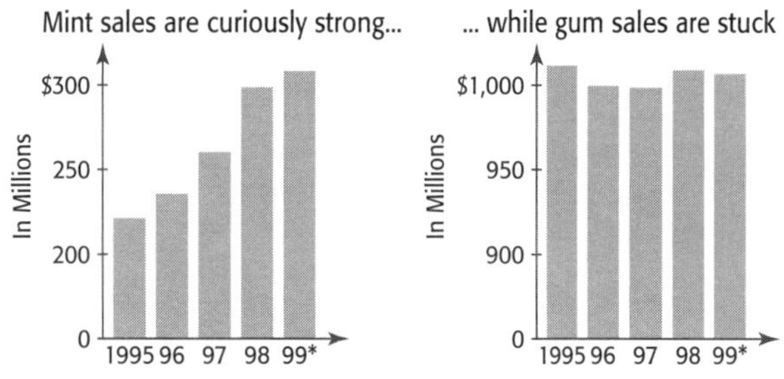

*52 weeks ended July 18, 1999 Source: Information Resources Inc.

(Source: *Newsweek*, November 1, 1999)

40. *Nursing Homes* The number of nursing homes in Massachusetts between 1993 and 1997 can be described by $m(x) = 3.37x - 6212.34$, and the number of nursing homes during that same time period in Illinois is given by $n(x) = 12.34x - 23{,}802.09$. In each equation, x represents the year. Assuming that the pattern indicated by these equations is valid prior to 1993, during what year did Massachusetts and Illinois have the same number of nursing homes? How many homes were there in each state when the number of nursing homes was equal?
(Source: Health Care Financing Administration)

41. *Stock Prices* The sum of the high and low prices of a share of stock for The Gap, Inc. in 2000 is \$83.50, and the difference of these two prices in 2000 is \$21.88. Find the high and low prices.
(Source: The Gap 2000 Annual Report)

42. *Revenue* The sum of the 1999 revenue and twice the 1998 revenue for Papa John's International, Inc. is \$2144.9 million. The difference in the 1999 and 1998 revenues is \$135.5 million. If Papa John's revenue between 1990 and 1999 is an increasing function, find the 1998 and 1999 revenues.
(Source: Papa John's 1999 Annual Report)

43. *Pricing* A concert promoter needs to make \$84,000 from the sale of 2400 tickets. The promoter charges \$30 for some tickets and \$45 for the others.

a. If there are x of the \$30 tickets sold and y of the \$45 tickets sold, write an equation that states that the sum of the tickets sold is 2400.

b. How much money is received from the sale of x tickets for \$30 each?

c. How much money is received from the sale of y tickets for \$45 each?

d. Write an equation that states that the total amount received from the sale is \$84,000.

e. Solve the equations simultaneously to find how many tickets of each type must be sold to yield the \$84,000.

44. *Rental Income* A woman has \$250,000 invested in two rental properties. One yields an annual return of 10% of her investment, and the other returns 12% per year on her investment. Her total annual return from the two investments is \$26,500. Let x represent the amount of the 10% investment and y represent the amount of the 12% investment.

a. Write an equation that states that the sum of the investments is \$250,000.

b. What is the annual return on the 10% investment?

c. What is the annual return on the 12% investment?

d. Write an equation that states that the sum of the annual returns is \$26,500.

e. Solve these two equations simultaneously to find how much is invested in each property.

45. *Investment* One safe investment pays 8% per year, and a more risky investment pays 12% per year.

a. How much must be invested in each account if the investor of \$100,000 would like a return of \$9000 per year?

b. Why might the investor use two accounts rather than put all money in the 12% investment?

46. *Investment* A woman invests \$52,000 in two different mutual funds, one that averages 10% per year and another that averages 14% per year. If her average annual return on the two mutual funds is \$5720, how much did she invest in each fund?

47. *Nutrition* A glass of skim milk supplies 0.1 mg of iron and 8.5 g of protein. A quarter pound of lean meat provides 3.4 mg of iron and 22 g of protein. If a person on a special diet is to have 7.1 mg of iron and 69.5 g of protein, how many glasses of skim milk and how many quarter-pound servings of meat will provide this?

48. *Nutrition* Each ounce of substance A supplies 6% of the nutrient a patient needs, and each ounce of substance B supplies 10% of the required nutrient. If the total number of ounces given to the patient is 14, and 100% of the nutrient is supplied, how many ounces of each substance were given?

49. *Medication* A nurse has two solutions that contain different concentrations of a certain medication. One is a 10% concentration and the other is a 5% concentration. How many cubic centimeters (cc) of each should he mix to obtain 20 cc of an 8% solution?

50. *Medication* A nurse has two solutions that contain different concentrations of a certain medication. One is a 30% concentration and the other is a 15% concentration. How many cubic centimeters (cc) of each should she mix to obtain 45 cc of a 20% solution?

51. *Supply and Demand* The table below gives the quantity of graphing calculators demanded and the quantity supplied for selected prices.

a. Find the linear equation that gives the price as a function of the quantity demanded.

b. Find the linear equation that gives the price as a function of the quantity supplied.

c. Use these equations to find the market equilibrium price.

Price ($)	Quantity Demanded (thousands)	Quantity Supplied (thousands)
50	210	0
60	190	40
70	170	80
80	150	120
100	110	200

52. *Market Analysis* The supply function and the demand function for a product are linear and are determined by the table that follows. Create the supply and demand functions and find the price that gives market equilibrium.

Supply Price	Function Quantity	Demand Price	Function Quantity
200	400	400	400
400	800	200	800
600	1200	0	1200

53. *Alcohol Use* The percent of the U.S. population who admitted to using alcohol at least once during the month prior to being asked about alcohol use is as shown in the table below.

Year	1994	1995	1996	1997
18- to 25-Year-Olds (%)	63	61	60	58
26- to 34-Year-Olds (%)	65	63	62	60

(Source: National Household Survey on Drug Abuse)

Letting x represent the number of years after 1990, the models are:

18- to 25-year-olds: $y = -1.6x + 69.3$

26- to 34-year-olds: $y = -1.6x + 71.3$

Solve this system of equations, if possible, to find when the percent of 18- to 25-year-olds who admitted to using alcohol equals the percent of 26- to 34-year-olds.

54. *Tax Rate* The federal tax rate is equivalent to a flat rate of 34% for taxable federal income between \$35,000 and \$1 million, and the Alabama tax rate is 5% of the taxable state income. Assume an income of \$1 million before either tax is paid, and use the fact that Alabama's income tax can be deducted from federal taxable income and that the federal tax can be deducted from Alabama's taxable income.

a. Create an equation that gives the federal tax x after the Alabama tax y is deducted.

b. Create an equation that gives the Alabama tax y after the federal tax x is deducted.

c. Write the equations created in parts (a) and (b) as a system of equations.

d. Solve the system of equations to find the amount paid in taxes.

55. *Social Agency* A social agency provides emergency food and shelter to two groups of clients. The first group has x clients who need an average of \$300 for emergencies, and the second group has y clients who need an average of \$200 for emergencies. The agency has \$100,000 to spend for these two groups.

a. Write an equation that describes the number of clients who can be served with the \$100,000.

b. If the first group has twice as many clients as the second group, how many clients are in each group if all the money is spent?

56. *Medication* Suppose combining x cubic centimeters (cc) of a 20% concentration of a medication and y cc of a 5% concentration of the medication gives $(x + y)$ cc of a 15.5% concentration. If 7 cc of the 20% concentration are added, how much must the amount of 5% concentration be increased to keep the same concentration?

57. *Market Equilibrium* A retail chain will buy 900 cordless phones if the price is \$10 each and 400 if the price is \$60. A wholesaler will supply 700 phones at \$30 each and 1400 at \$50 each. Assuming that the supply and demand functions are linear, find the market equilibrium point and explain what it means.

58. *Market Equilibrium* A retail chain will buy 800 televisions if the price is \$350 each and 1200 if the price is \$300. A wholesaler will supply 700 of these televisions at \$280 each and 1400 at \$385 each. Assuming that the supply and demand functions are linear, find the market equilibrium point and explain what it means.

CHAPTER 1 *Summary*

In this chapter, we studied the basic concept of a function and how to recognize functions with tables, graphs, and equations. We studied linear functions in particular—graphs, slope, model formation, equation writing, and solution of linear equations and systems of linear equations. We used graphing utilities to graph functions, to evaluate functions, and to solve linear equations. We solved business and economics problems involving linear functions, solved linear inequalities, and used graphing utilities to model linear functions.

Key Concepts and Formulas

1.1 Functions and Models

Scatter plot of data

A scatter plot is a graph of pairs of values that represent real-world information collected in numerical form.

Function definition

A function is a rule or correspondence that determines exactly one output for each input. The function may be defined by a set of ordered pairs, a table, a graph, or an equation.

Domain and range

The set of possible inputs for a function is called its domain, and the set of possible outputs is called its range. In general, if the domain of a function is not specified, we assume that it includes all real numbers except

- Values that result in a denominator of 0
- Values that result in an even root of a negative number

Independent variable and dependent variable

If x represents any element in the domain, then x is called the independent variable; if y represents an output of the function from an input x, then y is called the dependent variable.

Tests for functions

We can test for a function graphically, numerically, and analytically.

Vertical line test

A set of points in a coordinate plane is the graph of a function if and only if no vertical line intersects the graph in more than one point.

Functional notation We denote that y is a function of x using functional notation when we write $y = f(x)$.

Modeling The process of translating real-world information into a mathematical form so that it can be applied and then interpreted in the real-world setting is called modeling.

Mathematical model A mathematical model is a functional relationship (usually in the form of an equation) that includes not only the function rule but also descriptions of all involved variables and their units of measure.

1.2 Graphs of Functions

Point-plotting method The point-plotting method of sketching a graph means sketching the graph of a function by plotting enough points to determine the shape of the graph and then drawing a smooth curve through the points.

Complete graph A graph is a complete graph if it shows the basic shape of the graph and important points on the graph (including points where the graph crosses the axes and points where the graph turns), and suggests what the unseen portions of the graph will be.

Using a calculator to draw a graph After writing the function with x representing the independent variable and y representing the dependent variable, enter the function in the equation editor of the graphing utility. Activate the graph with ZOOM or GRAPH (Set the x- and y-boundaries of the viewing window before pressing GRAPH.)

Viewing windows The values that define the viewing window can be set manually or by using the ZOOM keys. The boundaries of a viewing window are

x_{min}: the smallest value on the x-axis (the left boundary of the window)

y_{min}: the smallest value on the y-axis (the bottom boundary of the window)

x_{max}: the largest value on the x-axis (the right boundary of the window)

y_{max}: the largest value on the y-axis (the top boundary of the window)

Spreadsheets Spreadsheets like Excel can be used to create accurate graphs, sometimes better-looking graphs than those created with graphing calculators.

Aligning data Using aligned inputs (input values that have been shifted horizontally to smaller numbers) rather than the actual data values results in smaller coefficients in models and less involved computations.

Evaluating functions with a calculator We can find (or estimate) the output of a function $y = f(x)$ at specific inputs with a graphing calculator by using TRACE and moving the cursor to (or close to) the value of the independent variable.

Graphing data points Graphing utilities can be used to create lists of numbers and to create graphs of the data stored in the lists.

1.3 Linear Functions

Linear functions A linear function is a function of the form $f(x) = ax + b$ where a and b are constants. The graph of a linear function is a line.

Constant function A special linear function that has the form $y = c$, where c is a real number, is called a constant function.

Identity function
The identity function $y = x$ is a linear function of the form $y = mx + b$ with slope $m = 1$ and y-intercept $b = 0$.

Intercepts
A point (or the x-coordinate of the point) where a graph crosses or touches the horizontal axis is called an x-intercept, and a point (or the y-coordinate of the point) where a graph crosses or touches the vertical axis is called a y-intercept. To find the x-intercept(s) of the graph of an equation, set $y = 0$ in the equation and solve for x. To find the y-intercept(s), set $x = 0$ and solve for y.

Slope of a line
The slope of a line is a measure of its steepness and direction. The slope is defined as

$$\text{slope} = \frac{\text{vertical change}}{\text{horizontal change}} = \frac{\text{rise}}{\text{run}}$$

If (x_1, y_1) and (x_2, y_2) are two points on a line, then the slope of the line is $m = \dfrac{y_2 - y_1}{x_2 - x_1}$.

Slope and y-intercept of a line
The slope of the graph of the equation $y = mx + b$ is m, and the y-intercept of the graph is b. (This is called slope-intercept form.)

Parallel and perpendicular lines
Two distinct nonvertical lines are parallel if they have the same slope. Two nonvertical and nonhorizontal lines are perpendicular if their slopes are negative reciprocals.

Constant rates of change
If a model is linear, the rate of change of the outputs with respect to the inputs will be constant, and the rate of change equals the slope of the line that is the graph of a linear function. Thus, we can determine if a linear model fits a set of real data by determining if the rate of change remains constant for the data.

1.4 Equations of Lines

Writing equations of lines

- **General form**
The general form of the equation of a line is $ax + by = c$, where a, b, and c are constants.

- **Slope-intercept form**
The slope-intercept form of the equation of a line with slope m and y-intercept b is $y = mx + b$.

- **Point-slope form**
The equation of the line with slope m and passing through a known point (x_1, y_1) is $y - y_1 = m(x - x_1)$.

- **Horizontal line**
The equation of a horizontal line is $y = b$, where b is a constant.

- **Vertical line**
The equation of a vertical line is $x = a$, where a is a constant.

Average rate of change
The average rate of change of a quantity over an interval describes how a change in the output of a function describing that quantity $f(x)$ is related to a change in the input x over that interval. The average rate of change of $f(x)$ with respect to x over the interval from $x = a$ to $x = b$ (where $a < b$) is calculated as

$$\text{average rate of change} = \frac{\text{change in } f(x) \text{ values}}{\text{corresponding change in } x \text{ values}} = \frac{f(b) - f(a)}{b - a}$$

Difference quotient
The average rate of change of a function from a point $(x, f(x))$ to the point $(x + h, f(x + h))$ is the difference quotient $\dfrac{f(x + h) - f(x)}{h}$.

Secant line
When a function is not linear, the average rate of change between two points is the slope of the line joining two points on the curve, which is called a secant line.

Approximately linear data Some sets of real data lie sufficiently close to a line that two points can be used to create a linear function that models the data.

1.5 Algebraic and Graphical Solution of Linear Equations

Algebraic solution of linear equations If a linear equation contains fractions, multiply both sides of the equation by a number that will remove all denominators in the equation. Next, remove any parentheses or other symbols of grouping and then perform any additions or subtractions to get all terms containing the variable on one side and all other terms on the other side of the equation. Combine like terms. Divide both sides of the equation by the coefficient of the variable. Check the solution by substitution in the original equation.

Solving real-world application problems To solve an application problem that is set in a real-world context, use the same solution methods. However, remember to include units of measure with your answer and check that your answer makes sense in the problem situation.

Zero of a function Any number a for which $f(a) = 0$ is called a zero of the function $f(x)$.

Solutions, zeros, and x-intercepts If a is an x-intercept of the graph of a function f, then a is a zero of the function f, and a is a solution to the equation $f(x) = 0$.

Graphical solution of linear equations

- **x-Intercept method** Rewrite the equation with 0 on one side, enter the nonzero side into the equation editor of a graphing calculator, and find the x-intercept of the graph. This is the solution to the equation.
- **Intersection method** Enter the left side of the equation into y_1, enter the right side of the equation into y_2, and find the point of intersection. The x-coordinate of the point of intersection is the solution to the equation.

Formulas; solving an equation for a specified linear variable To solve an equation for one variable if two or more variables are in the equation and if that variable is linear in the equation, we can solve for that variable by treating the other variables as constants and using the same steps that we used to solve a linear equation in one variable.

1.6 Fitting Lines to Data Points; Modeling Linear Functions

Fitting lines to data points When real-world information is collected as numerical information called *data*, technology can be used to determine the pattern exhibited by the data (provided that a recognizable pattern exists). These patterns can often be described by mathematical functions.

Linear regression We can determine the equation of the line that is the best fit for a set of points by using a procedure called linear regression (or the least-squares method), which defines the best-fit line as the line for which the sum of the squares of the vertical distances from the data points to the line is a minimum.

Modeling data We can model a set of data by entering the data into a graphing utility, obtaining a scatter plot, and using the graphing utility to obtain the linear equation that is the best fit for the data. The equation and/or numerical results should be reported in a way that makes sense in the context of the problem, with the appropriate units and with the variables identified.

Discrete versus continuous We use the term *discrete* to describe data or a function that is presented in the form of a table or in a scatter plot. We use the term *continuous* to describe a function or graph when the inputs can be any real number or any real number between two specified values.

Applying models Using a model to find an output for an input between two given data points is called *interpolation*. When a model is used for a prediction using an input outside the given data points, the output is called *extrapolation*.

Goodness of fit The goodness of fit of a linear model can be observed from a graph of the model and the data points and measured with the correlation coefficient.

Constant first differences If the first differences of data outputs are constant (for equally spaced inputs), a linear model can be found that fits the data exactly. If the first differences are "nearly constant," a linear model can be found that is an approximate fit for the data.

1.7 Systems of Linear Equations in Two Variables

System of equations A system of linear equations is a set of equations in two or more variables. A solution of the system must satisfy every equation in the system.

Solving a system of linear equations in two variables

- **Graphing** Graph the equations and find their point of intersection.
- **Substitution** Solve one of the equations for one variable and substitute that expression into the other equation, thus giving an equation in one variable.
- **Elimination** Rewrite one or both equations in a form that allows us to eliminate one of the variables by adding or subtracting the equations.

Break-even analysis A company is said to break even from the production and sale of a product if the total revenue equals the total cost—that is, if the profit for that product is zero.

Supply, demand, and market equilibrium *Market equilibrium* is said to occur when the quantity of a commodity demanded is equal to the quantity supplied. The price at this point is called the *equilibrium price*, and the quantity at this point is called the *equilibrium quantity*.

Possible solutions to a system of linear equations in two variables

- **Unique solution** Graphs are intersecting lines.
- **No solution** Graphs are parallel lines; system is *inconsistent*.
- **Many solutions** Graphs are the same line; system is *dependent*.

Modeling systems of equations Solution of real problems sometimes requires us to create two or more equations whose simultaneous solution is the solution to the problem.

Chapter 1 Skills Check

Use the values in the table below in Exercises 1–4.

x	-3	-1	1	3	5	7	9	11	13
y	9	6	3	0	-3	-6	-9	-12	-15

1. Explain why the relationship shown by the table describes y as a function of x.

2. State the domain and range of the function.

3. If the function defined by the table is denoted by f, so that $y = f(x)$, what is $f(3)$?

4. Do the outputs in this table indicate that a linear function fits the data? If so, write the equation of the line.

5. If $C(s) = 16 - 2s^2$, find
a. $C(3)$ **b.** $C(-2)$ **c.** $C(-1)$

6. For each of the functions $y = f(x)$ described below, find $f(-3)$.

a.

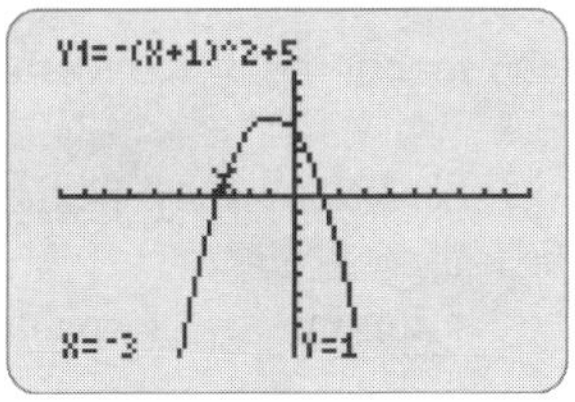

b.

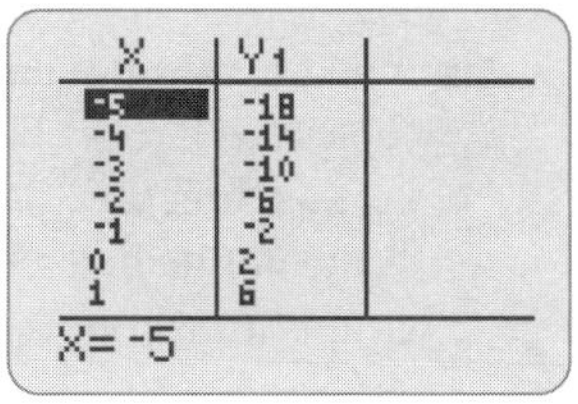

7. Graph the function $f(x) = -2x^3 + 5x$ with a graphing utility using a standard viewing window.

8. Graph the function $y = 3x^2$ with a graphing utility using a standard viewing window.

9. Graph $y = -10x^2 + 400x + 10$ on a standard window and on a window with $x_{min} = 0$, $x_{max} = 40$, $y_{min} = 0$, $y_{max} = 5000$. Which window gives a better view of the graph of the function?

Use the table of data below in Exercises 10–13.

x	1	3	6	8	10
y	-9	-1	5	12	18

10. Use a graphing utility to graph the points (x, y) from the table, using a viewing window with $x_{min} = 0$, $x_{max} = 12$, $y_{min} = -12$, and $y_{max} = 20$.

11. Find the linear function that is the best fit for the data in the table.

12. Use a graphing utility to graph the function found in Exercise 11 on the same set of axes as the scatter plot in Exercise 10, with $x_{min} = 0$, $x_{max} = 12$, $y_{min} = -12$, and $y_{max} = 20$.

13. Do the data points in the table fit exactly on the graph of the function from Exercise 12? Should they?

14. Find the slope of the line through $(-4, 6)$ and $(8, -16)$.

15. Given the equation $2x - 3y = 12$, (a) find the x- and y-intercepts of the graph and (b) graph the equation.

16. What is the slope of the graph of the function given in Exercise 15?

17. Find the slope and y-intercept of the graph of $y = -6x + 3$.

18. Find the rate of change of the function whose equation is given in Exercise 17.

19. Write the equation of the line that has slope $\frac{1}{3}$ and y-intercept 3.

20. Write the equation of a line that has slope $\frac{-3}{4}$ and passes through $(4, -6)$.

21. Write the equation of the line that passes through $(-1, 3)$ and $(2, 6)$.

Solve the systems of linear equations in Exercises 22–27, if possible.

22. $\begin{cases} 3x + 2y = 0 \\ 2x - y = 7 \end{cases}$

23. $\begin{cases} 3x + 2y = -3 \\ 2x - 3y = 3 \end{cases}$

24. $\begin{cases} -4x + 2y = -14 \\ 2x - y = 7 \end{cases}$

25. $\begin{cases} -6x + 4y = 10 \\ 3x - 2y = 5 \end{cases}$

26. $\begin{cases} 2x + 3y = 9 \\ -x - y = -2 \end{cases}$

27. $\begin{cases} 2x + y = -3 \\ 4x - 2y = 10 \end{cases}$

For the functions given in Exercises 28 and 29, find

a. $f(x + h)$ **b.** $f(x + h) - f(x)$ **c.** $\dfrac{f(x + h) - f(x)}{h}$

28. $f(x) = 5 - 4x$ **29.** $f(x) = 10x - 50$

In Exercises 30 and 31, solve the equations for x algebraically and with a graphing utility.

30. $3x + 22 = 8x - 12$

31. $\dfrac{3(x - 2)}{5} - x = 8 - \dfrac{x}{3}$

32. For the function $y = x^2$, compute the average rate of change between $x = 0$ and $x = 3$.

33. Solve $4x - 3y = 6$ for y and graph it on a graphing utility with a standard window.

In Exercises 34–36, solve the inequalities both algebraically and graphically.

34. $3x + 8 < 4 - 2x$ **35.** $3x - \dfrac{1}{2} \leq \dfrac{x}{5} + 2$

36. $18 \leq 2x + 6 < 42$

Chapter 1 Review

37. *Voters* The table gives the percent p of African American voters who have supported Democratic candidates for president for the years 1960–1996.
a. Is the percent p a function of the year y?
b. Let $p = f(y)$ denote that p is a function of y. Find $f(1992)$ and explain what it means.
c. What is y if $f(y) = 94$? What does this mean?

Year	Democrat (%)	GOP (%)
1960	68	32
1964	94	6
1968	85	15
1972	87	13
1976	85	15
1980	86	12
1984	89	9
1992	82	11
1996	84	12

(Source: Joint Center for Political and Economic Studies)

38. *Voters*
a. What is the domain of the function defined by the table in Exercise 37?
b. Is this function defined for 1982? Why is this value not included in the table?
c. Is this function discrete or continuous? Explain.

39. *Voters* Graph the function defined by the table in Exercise 37 on the window [1955, 2000] by [60, 100].

40. *Voters*
a. Find the slope of the line joining points (1968, 85) and (1996, 84).
b. Use the answer from part (a) to find the average annual rate of change of the percent of African American voters who voted for the Democratic candidate for president between the years 1968 and 1996, inclusive.
c. Is the average annual rate of change from 1968 to 1980 equal to the average annual rate of change from 1968 to 1996?
d. Does a linear function model this data exactly?

When money is borrowed to purchase an automobile, the amount borrowed A determines the monthly payment P. In particular, if a dealership offers a 5-year loan at 2.9% interest, then the amount borrowed for the car determines the payment according to the following table. Use the table to define the function $P = f(A)$ in Exercises 41–44.

Amount Borrowed ($)	Monthly Payment ($)
10,000	179.25
15,000	268.87
20,000	358.49
25,000	448.11
30,000	537.73

(Source: Sky Financial)

41. *Car Loans*

a. Explain why the monthly payment P is a linear function of the amount borrowed A.

b. Find $f(25{,}000)$ and interpret its meaning.

c. If $f(A) = 358.49$, what is A?

42. *Car Loans*

a. What are the domain and range of the function $P = f(A)$ defined by the table?

b. Is this function $P = f(A)$ defined for a \$12,000 loan?

c. Is the function defined by the table discrete or continuous?

43. *Car Loans*

a. Are the first differences of the outputs in the table constant?

b. Is there a line on which these data points fit exactly?

44. *Car Loans*

a. Write the equation $P = f(A)$ of the line that fits on the data points in the table.

b. Use the unrounded linear model found in part (a) to find $P = f(28{,}000)$ and explain what it means.

c. Can the function f be used to find the monthly payment for any dollar amount A of a loan if the interest rate and length of loan are unchanged?

d. Determine the amount of a loan that will keep the payment less than or equal to \$500.

45. *Life Expectancy* Figure (a) gives the number of years the average woman is estimated to live beyond age 65 during selected years between 1950 and 2030. Let x represent the year and let y represent the expected number of years a woman will live past age 65. Write $y = f(x)$ and answer the following questions:

a. What is $f(1960)$ and what does it mean?

b. What is the life expectancy for the average woman in 2010?

c. In what year is the average woman expected to live 19 years past age 65?

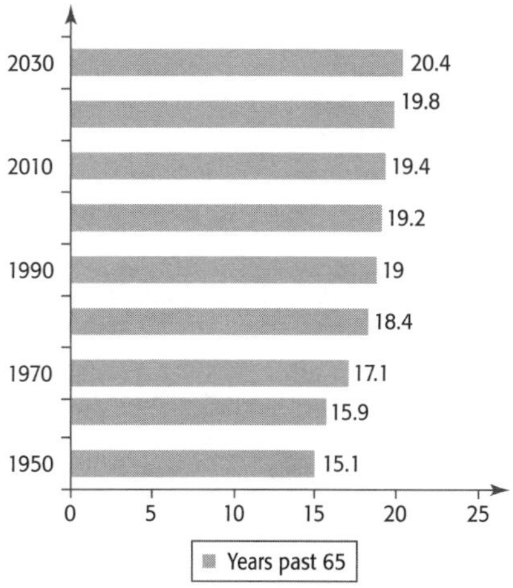

(a) Women

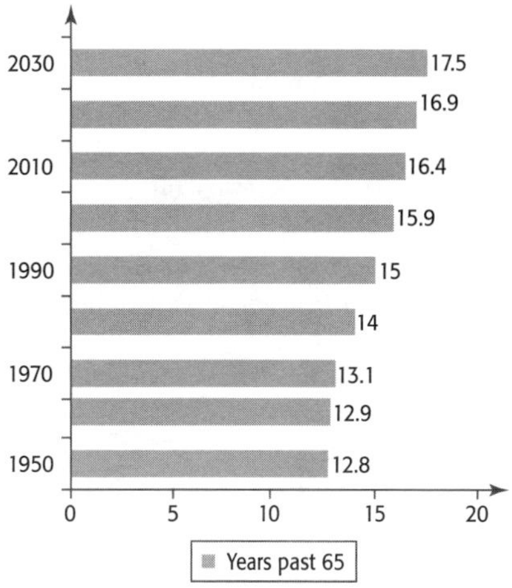

(b) Men

46. *Life Expectancy* Figure (b) in Exercise 45 gives the number of years the average man is estimated to live beyond age 65 during selected years between 1950 and 2030. Let x represent the year and let y represent the expected number of years a man will live past age 65. Write $y = g(x)$ and answer the following questions:

a. What is $g(2020)$ and what does it mean?

b. What was the life expectancy for the average man in 1950?

c. Write a functional expression that indicates that the average man in 1990 has a life expectancy of 80 years.

47. *Teacher Salaries* The average classroom teacher salary in the United States is given by $f(t) = 982.06t + 32{,}903.77$, where t is the number of years from 1990.

a. What was the average salary in 2000? Write this in functional notation.
b. Find $f(15)$ and interpret it.
c. Are the salaries increasing or decreasing?
(Source: www.ors2.state.sc.us/abstract)

48. *Teacher Salaries* The average classroom teacher salary in the United States is given by the function $f(t) = 982.06t + 32{,}903.77$, where t is the number of years from 1990.
a. Graph this function for values of t between 0 and 15.
b. What years do these values represent?
(Source: www.ors2.state.sc.us/abstract)

49. *Drug Users* The number of current users of illicit drugs increased from 12 million in 1992 to 14.5 million in 1999.
a. Find the slope of the line connecting the points (1992, 12.0) and (1999, 14.5), rounded to three decimal places.
b. Find the average annual rate of change of current users between 1992 and 1999.

50. *Fuel* The table below shows data for the number of gallons of gas purchased each day of a certain week by the 250 taxis owned by the Inner City Transportation Taxi Company. Write the equation that models this data.

Days from first day	0	1	2	3	4	5	6
Gas used (gal)	4500	4500	4500	4500	4500	4500	4500

51. *Work Hours* The average weekly hours worked by production and nonsupervisory workers on private nonfarm payrolls, seasonally adjusted, are given by the data in the table below.

Month and Year	Average Weekly Hours	Month and Year	Average Weekly Hours
June 1998	34.6	Oct. 1998	34.6
July 1998	34.6	Nov. 1998	34.6
Aug. 1998	34.6	Dec. 1998	34.6
Sept. 1998	34.6	Jan. 1999	34.6
		Feb. 1999	34.6

(Source: Bureau of Labor Statistics)

a. Write the equation of a function that describes the average weekly hours using an input equal to the number of months past May 1998.
b. Is this function a constant function?

52. *Job Selection* A job candidate is given the choice of two positions, one paying \$2100 per month and one paying \$1000 per month plus a 5% commission on all sales made during the month.
a. How much (in dollars) must the employee sell in a month for the second position to pay as much as the first?
b. To be sure that the second position will pay more than the first, how much (in dollars) must the employee sell each month?

53. *Marketing* A car dealer purchased 12 automobiles for \$24,000 each. If she sells 8 of them with an average profit of 12%, for how much must she sell the remaining 4 to obtain an average profit of 10% on all 12?

54. *Investment* A retired couple has \$420,000 to invest. They chose one relatively safe investment fund that has an annual yield of 6% and another riskier investment that has a 10% annual yield. How much should they invest in each fund to earn \$30,000 per year?

55. *Writing Scores* The average writing scores of 11th-graders on the National Assessment of Educational Progress tests have changed over the years since 1984, with the average score given by $y = -0.629x + 293.871$, where x is the number of years from 1980. For what year does this model give an average score of 285?
(Source: U.S. Dept of Education)

56. *Revenue* A company has revenue given by $R(x) = 564x$ dollars and total costs given by $C(x) = 40{,}000 + 64x$ dollars, where x is the number of units produced and sold.
a. What is the revenue when 120 units are produced?
b. What is the cost when 120 units are produced?
c. What is the marginal cost and what is the marginal revenue for this product?
d. What is the slope of the graph of $C(x) = 40{,}000 + 64x$?
e. Graph $R(x)$ and $C(x)$ on the same set of axes.

57. *Profit* A company has revenue given by $R(x) = 564x$ dollars and total cost given by $C(x) = 40{,}000 + 64x$ dollars, where x is the number of units produced and sold. The profit can be found by forming the function $P(x) = R(x) - C(x)$.
a. Write the profit function.

b. Find the profit when 120 units are produced and sold.
c. How many units give break even?
d. What is the marginal profit for this product?
e. How is the marginal profit related to the marginal revenue and the marginal cost?

58. *Depreciation* A business property can be depreciated for tax purposes by using the formula $y + 3000x = 300{,}000$, where y is the value of the property x years after it was purchased.
a. Find the y-intercept of the graph of this function. Interpret this value.
b. Find the x-intercept. Interpret this value.

59. *Marginal Profit* A company has determined that its profit for a product can be described by a linear function. The profit from the production and sale of 150 units is \$455 and the profit from 250 units is \$895.
a. What is the average rate of change of the profit for this product when between 150 and 250 units are sold?
b. What is the slope of the graph of this profit function?
c. Write the equation of the profit function for this product.
d. What is the marginal profit for this product?
e. How many units give break even for this product?

60. *Life Expectancy*
a. Find a linear function $y = f(x)$ that models the data shown in Figure (a) of Exercise 45 with x equal to the number of years after 1950 and y equal to the number of years the average woman is estimated to live beyond age 65.
b. Graph the data and the model on the same set of axes.
c. Use the model to estimate $f(104)$ and explain what it means.
d. Determine an inequality to represent the time period (in years) for which the average woman can expect to live at least 84 years.

61. *Life Expectancy*
a. Find a linear function $y = g(x)$ that models the data shown in Figure (b) of Exercise 45 with x equal to the number of years after 1950 and y equal to the number of years the average man is estimated to live beyond age 65.
b. Graph the data and the model on the same set of axes.
c. Use the model to estimate $g(130)$ and explain what it means.
d. In what year would the average man expect to live to age 90?
e. Determine an inequality to represent the time period for which the average man could expect to live no more than 81 years.

62. *Education Spending* The following table gives the amount (in billions of dollars) spent on education during the years 1996–2001.
a. Graph the data points *or* find the first differences to determine if a linear equation is a reasonable model for this data.
b. If it is reasonable, find the linear model that is the best fit for this data, with x equal to the number of years after 1990.
c. Use the unrounded model to predict the spending in 2002.

Years	Amount Spent (\$ billions)
1996	23
1997	26.6
1998	29.9
1999	33.5
2000	35.6
2001	40.1

63. *Population Growth* The resident population of Florida (in thousands) is given in the table below. Let x equal the number of years after 1980 and y equal the number of thousands of residents.
a. Graph the data points to determine if a linear equation is a reasonable model for this data.
b. If it is reasonable, find the linear model that is the best fit for this data, with x equal to the number of years after 1980.
c. What does this unrounded model predict that the population will be in 2002?

Years	1980	1985	1990	1995
Population (thousands)	9746	11,351	12,938	14,180

Years	1996	1997	1998
Population (thousands)	14,425	14,677	14,916

(Source: U.S. Census Bureau)

64. *Earnings per Share* The table below gives the earnings ($ thousands) per share (EPS) for ACS stock for the years 1996–2000. Let x equal the number of years after 1995 and y equal the EPS, in thousands of dollars.

a. Graph the data points to determine if a linear equation is a reasonable model for this data.

b. If it is reasonable, find the linear model that is the best fit for this data.

c. Graph the data points and the function on the same axes and discuss the goodness of fit.

Years	1996	1997	1998	1999	2000
EPS ($ thousands)	.85	1.05	1.29	1.66	2.05

(Source: 2000 Report of ACS)

65. *Marriage Rate* According to data from the Index of Leading Cultural Indicators, the marriage rate (the number of marriages per 1000 women) can be described by $y = -0.763x + 85.284$, where x is the number of years after 1950. For what years does this model indicate that the rate

a. Was above 48.66 per 1000 women?

b. Will be below 41.03 per 1000 women?

66. *Marijuana Use* The percent p of high school seniors who use marijuana daily is given by the equation $30p - 19x = 1$, where x is the number of years after 1990. If this model is accurate, what is the range of years for which the percent of use is 3.2%–7%? (Source: Index of Leading Cultural Indicators)

67. *Prison Sentences* The mean time in prison y for a crime can be found as a function of the mean sentence length x, using $y = 0.554x - 2.886$, where x and y are in months. If a judge sentences a convicted criminal to serve between 3 and 5 years, how many months would we expect the criminal to serve? (Source: Index of Leading Cultural Indicators)

68. *Investment* A retired couple has $240,000 to invest. They chose one relatively safe investment fund that has an annual yield of 8% and another riskier investment that has a 12% annual yield. How much should they invest in each fund to earn $23,200 per year?

69. *Break Even* A computer manufacturer has a new product with daily total revenue given by $R = 565x$ and daily total cost given by $C = 6000 + 325x$. How many units per day must be produced and sold to give break even for the product?

70. *Medication* Medication A is given six times per day and medication B is given twice per day. For a certain patient, the total intake of the two medications is limited to 25.2 mg per day. If the ratio of dosage of medication A to the dosage of medication B is 2 to 3, how many milligrams are in each dosage?

71. *Market Equilibrium* The demand for a certain brand of women's shoes is given by $3q + p = 340$, and the supply for these shoes is given by $p - 4q = -220$, where p is the price in dollars and q is the number of pairs demanded at price p. Solve the system containing these two equations to find the equilibrium price and the equilibrium quantity.

72. *Market Analysis* Suppose for a certain product, the supply and demand functions are $p = \frac{q}{10} + 8$ and $10p + q = 1500$, respectively, where p is in dollars and q is in units. Find the equilibrium price and quantity.

73. *Pricing* A concert promoter needs to make $120,000 from the sale of 2600 tickets. The promoter charges $40 for some tickets and $60 for the others.

a. If there are x of the $40 tickets and y of the $60 tickets, write an equation that states that the total number of the tickets sold is 2600.

b. How much money is made from the sale of x tickets for $40 each?

c. How much money is made from the sale of y tickets for $60 each?

d. Write an equation that states that the total amount made from the sale is $120,000.

e. Solve the equations simultaneously to find how many tickets of each type must be sold to yield the $120,000.

74. *Rental Income* A woman has $500,000 invested in two rental properties. One yields an annual return of 12% of her investment, and the other returns 15% per year on her investment. Her total annual return from the two investments is $64,500. If x represents the 12% investment and y represents the 15% investment,

a. Write an equation that states that the sum of the investments is $500,000.

b. What is the annual return on the 12% investment?

c. What is the annual return on the 15% investment?

d. Write an equation that states that the sum of the annual returns is $64,500.

e. Solve these two equations simultaneously to find how much is invested in each property.

GROUP ACTIVITIES/EXTENDED APPLICATIONS

1. Body Mass Index

Obesity is a risk factor for the development of medical problems, including high blood pressure, high cholesterol, heart disease, and diabetes. Of course, how much a person can safely weigh depends on his or her height. One way of comparing weights that account for height is the *body mass index (BMI)*. The table below gives the BMI for a variety of heights and weights of people. Roche Pharmaceuticals, manufacturers of Xenical, state that a BMI of 30 or greater can create an increased risk of developing medical problems associated with obesity.

Describe how to assist a group of people in using the information. Some things you would want to include in your description follow.

a. How a person uses the table to determine his or her BMI.

b. How a person determines the weight that will put him or her at medical risk.

c. How a person whose weight or height is not in the table can determine if his or her BMI is 30. To answer this question, develop a formula to find the weight that would give a BMI of 30 for a person of a given height, including how you would:

1. Pick the points from the table that correspond to a BMI of 30 and create a table of these heights and weights. Change the units of measurement to simplify the data.

2. Determine whether a linear model is an exact or approximate fit for the data.

3. Create the linear model if it is reasonable.

4. Test the model with existing data.

5. Explain how to use the model to test for obesity.

Body Mass Index for Specified Height (FT/IN.) and Weight (LB)

Height/Weight	120	130	140	150	160	170	180	190	200	210	220	230	240	250
5′0″	23	25	27	29	31	33	35	37	39	41	43	45	47	49
5′1″	23	25	27	28	30	32	34	36	38	40	42	44	45	47
5′2″	22	24	26	27	29	31	33	35	37	38	40	42	44	46
5′3″	21	23	25	27	28	30	32	34	36	37	39	41	43	44
5′4″	21	22	24	26	28	29	31	33	34	36	38	40	41	43
5′5″	20	22	23	25	27	28	30	32	33	35	37	38	40	42
5′6″	19	21	23	24	26	27	29	31	32	34	36	37	39	40
5′7″	19	20	23	24	25	27	28	30	31	33	35	36	38	39
5′8″	18	20	21	23	24	26	27	29	30	32	34	35	37	38
5′9″	18	19	21	22	24	25	27	28	30	31	33	34	36	37
5′10″	17	19	20	22	23	24	25	27	28	29	31	33	35	36
5′11″	17	18	20	21	22	24	25	27	28	29	31	32	34	35
6′0″	16	18	19	20	22	23	24	26	27	29	30	31	33	34
6′1″	16	17	19	20	21	22	24	25	26	28	29	30	32	33

(Source: Roche Pharmaceuticals)

2. Research

Linear functions can be used to model many types of real data. Graphs displaying linear growth are frequently displayed in periodicals such as *Newsweek* and *Time*, in newspapers such as *USA Today* and the *Wall Street Journal*, and on numerous Web sites on the Internet. Tables of data can also be found in these sources, especially in federal and state government Web sites, such as www.census.gov. (This Web site is the source of the Florida resident population data used in the Chapter Review, for example.)

Your mission is to find a company sales, stock price, biological growth, or sociological trend over a period of years (using at least four points) that is linear or "nearly" linear, and to find a linear function that is a model for this data.

A linear model will be a good fit for the data

1. If the data is presented as a graph that is linear or "nearly linear."
2. If the data is presented in a table and the plot of the data points lies near some line.
3. If the data is presented in a table and the first differences of the outputs are nearly constant for equally spaced inputs.

After you have created the model, you should test the goodness of fit of the model to the data and discuss uses that you could make of the model.

Your completed project should include

a. A complete citation of the source of the data you are using.
b. An original copy or photocopy of the data being used.
c. A scatter plot of the data.
d. The equation that you have created.
e. A graph containing the scatter plot and the modeled equation.
f. A statement about how the model could be used to make estimations or predictions about the trend you are observing.

Some helpful hints:

1. If you decide to use a relation determined by a graph that you have found, read the graph very carefully to determine the data points or contact the source of the data to get the data from which the graph was drawn.
2. Align the independent variable by letting x represent the number of years from some convenient year and then enter the data into a graphing utility and create a scatter plot.
3. Use your graphing utility to create the equation of the function that is the best fit for the data. Graph this equation and the data points on the same axes to see if the equation is reasonable.

CHAPTER 2

Quadratic and Other Nonlinear Functions

Revenue and profit from the sale of products frequently cannot be modeled by linear functions because they increase at rates that are not constant. In this chapter, we use nonlinear functions, including quadratic and power functions, to model numerous applications in business, economics, and the life and social sciences.

SECTIONS	TOPICS	APPLICATIONS
2.1 Quadratic Functions; Parabolas	Graphing quadratic functions; finding vertices of parabolas; finding intercepts of parabolas	Maximum revenue from sales, teenage drug use, height of a ball
2.2 Solving Quadratic Equations	Solving by factoring; solving graphically; combining graphs and factoring; solving with the root method; completing the square; the quadratic formula; complex solutions	Hospital admissions, profit, marijuana use
2.3 A Library of Functions	Graphing and applying power, root, reciprocal, piecewise-defined, and absolute value functions	Allometric relationships, average cost, postage, residential power costs
2.4 Quadratic and Power Models	Modeling with quadratic functions; comparing linear and quadratic models; modeling with power functions; comparing power and quadratic models	Cocaine use in the United Kingdom, height of a rocket, aid to dependent children, auto noise, cohabiting households, voting

18 9-2.

$3x^2 + 9x - 2x - 6 = 0$

$3x(x+3) - 2(x+3) = 0$

$(3x-2)(x+3) = 0$

$x = \frac{2}{3}$ or -3

2.1 Quadratic Functions; Parabolas

Key Concepts
- Quadratic function
- Parabola
- Vertex
- Axis of symmetry
- Maxima and minima
- Intercepts

Section Preview Revenue

Suppose the monthly revenue from the sale of Carlson 32-inch televisions is given by the function

$$R(x) = -0.1x^2 + 600x \text{ dollars}$$

where x is the number of TVs sold. In this case, the revenue of this product is represented by a **second-degree**, or **quadratic**, **function**. A quadratic function is a function that can be written in the form

$$f(x) = ax^2 + bx + c$$

where a, b, and c are real numbers with $a \neq 0$.

The graph of the quadratic function $f(x) = ax^2 + bx + c$ is a **parabola** with a "turning point" called the **vertex**. Figure 2.1 shows the graph of the function $R(x) = -0.1x^2 + 600x$, which is a parabola that opens downward, and the vertex occurs where the function has its maximum value. Notice also that the graph of this revenue function is symmetric about a vertical line through the vertex. This vertical line is called the **axis of symmetry**.

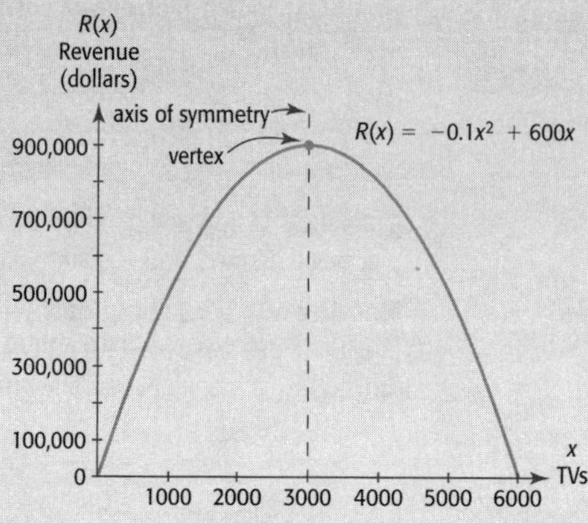

FIGURE 2.1

Knowing where the maximum value of $R(x)$ occurs can show the company how many units must be sold to obtain the largest revenue. Knowing the maximum output of $R(x)$ helps the company plan its sales campaign. In this section, we graph and apply quadratic functions.

Parabolas

The graph of every quadratic function has the distinctive shape known as a parabola. The graph of a quadratic function is determined by the location of the vertex and whether the parabola opens upward or downward.

Consider the basic quadratic function $y = x^2$. Each output y is obtained by squaring an input x (Table 2.1). The graph of $y = x^2$, shown in Figure 2.2, is a parabola that opens upward with the **vertex** (turning point) at the origin, (0, 0).

TABLE 2.1

x	y_1
−4	16
−3	9
−2	4
−1	1
0	0
1	1
2	4
3	9
4	16

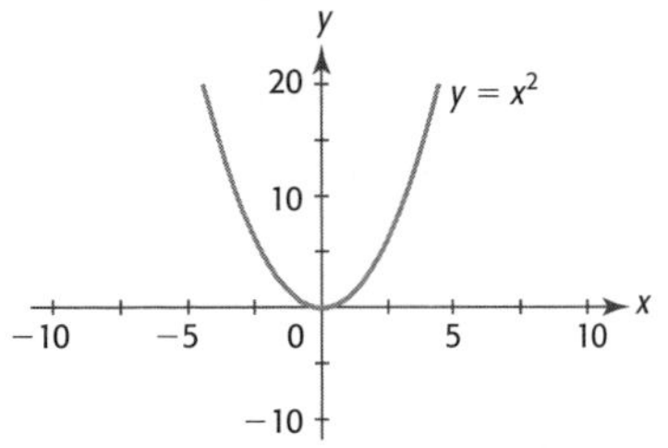

FIGURE 2.2

Observe that for $x > 0$, the graph of $y = x^2$ rises as it moves left from to right (that is, as the x-values increase), so the function $y = x^2$ is **increasing** for $x > 0$. For values of $x < 0$, the graph falls as it moves from left to right (as the x-values increase), so the function $y = x^2$ is **decreasing** for $x < 0$.

The quadratic function $y = -x^2$ has the form $y = ax^2$ with $a < 0$, and its graph is a parabola that opens downward with vertex at (0, 0) (Figure 2.3(a)).

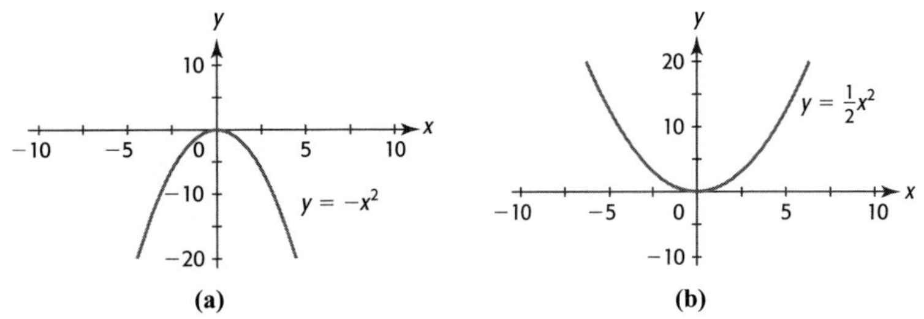

FIGURE 2.3

The function $y = \frac{1}{2}x^2$ has the form $y = ax^2$ with $a > 0$, and its graph is a parabola opening upward (Figure 2.3(b)). In general, the graph of a quadratic function of the form $y = ax^2$ is a parabola that opens upward (is **concave up**) if a is positive and opens downward (is **concave down**) if a is negative.* The **vertex**, which is the point where the parabola turns, is a **minimum point** if a is positive and is a **maximum point** if a is negative. The vertical line through the vertex is called the **axis of symmetry** because this line divides the graph into two halves that are reflections of each other (Figure 2.4).

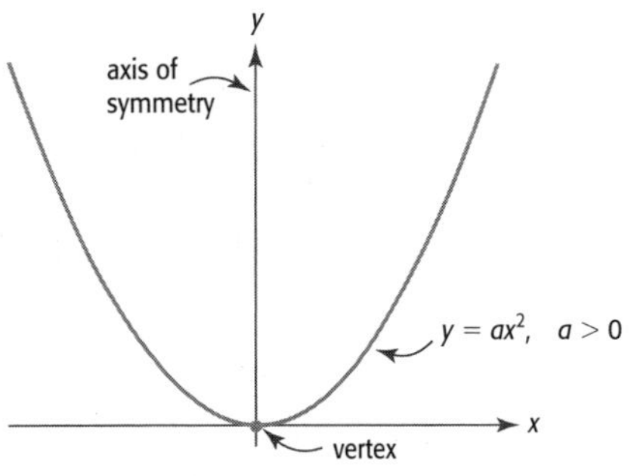

FIGURE 2.4

*A parabola that is concave up appears that it will "hold water," and a parabola that is concave down will appear to "shed water."

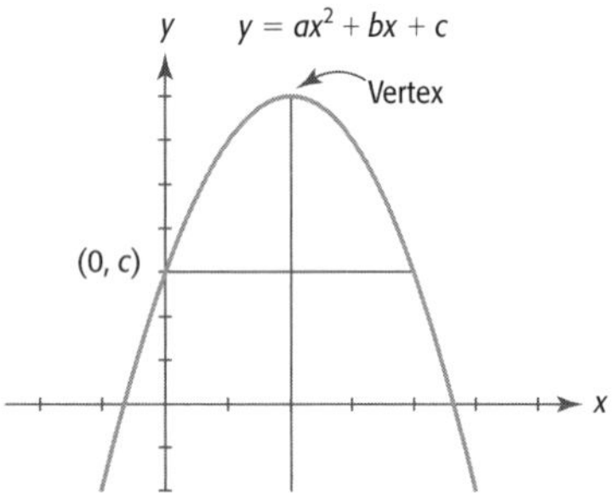

FIGURE 2.5

We can find the x-coordinate of the vertex of the graph of $y = ax^2 + bx + c$ by using the fact that the axis of symmetry of a parabola passes through the vertex. As Figure 2.5 shows, the y-intercept of the graph of $y = ax^2 + bx + c$ is $(0, c)$, and there is another point on the graph with y-coordinate c.

The x-coordinates of the points on this graph with y-coordinate c satisfy

$$c = ax^2 + bx + c$$

Solving this equation gives

$$0 = ax^2 + bx$$

$$0 = x(ax + b)$$

$$x = 0 \qquad \text{or} \qquad x = \frac{-b}{a}$$

The x-coordinate of the vertex is on the axis of symmetry, which is halfway from $x = 0$ to $x = \frac{-b}{a}$, so it is at $x = \frac{-b}{2a}$. The y-coordinate of the vertex can be found by evaluating the function at the x-coordinate of the vertex.

GRAPH OF A QUADRATIC FUNCTION

The graph of the function

$$f(x) = ax^2 + bx + c$$

is a parabola that opens upward, and the vertex is a minimum, if $a > 0$. The parabola opens downward, and the vertex is a maximum, if $a < 0$.

Its vertex is at the point $\left(\frac{-b}{2a}, f\left(\frac{-b}{2a}\right)\right)$ (Figure 2.6).

The axis of symmetry of the parabola has equation $x = \frac{-b}{2a}$.

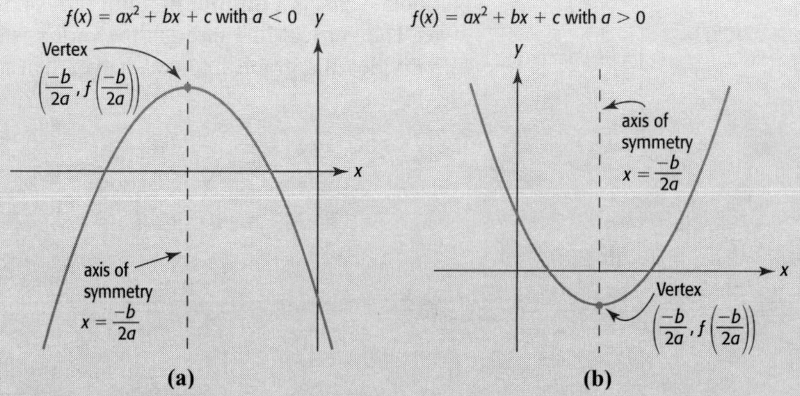

FIGURE 2.6

If we know the location of the vertex and the direction in which the parabola opens, we can make a good sketch of the graph by plotting just a few more points.

EXAMPLE 1 Vertices of Parabolas

Suppose the profit from the sale of sweatshirts sold at a school fund-raiser is

$$P(x) = -2x^2 + 60x$$

where $P(x)$ is the profit when each sweatshirt sells for x dollars.

a. Find the vertex and the axis of symmetry of the graph of this function.

b. Determine if the vertex represents a maximum or minimum point.

c. Interpret the vertex in the context of the application.

d. Graph this function.

Solution

a. The function is a quadratic function with $a = -2$, $b = 60$, and $c = 0$. The x-coordinate of the vertex is $\frac{-b}{2a} = \frac{-60}{2(-2)} = 15$, and the axis of symmetry is the line $x = 15$. The y-coordinate of the vertex is

$$P(15) = -2(15)^2 + 60(15) = 450$$

b. Because $a < 0$, the parabola opens downward, so the vertex is a maximum point.

c. The x-coordinate of the vertex gives the price that should be charged to maximize the profit, so charging \$15 will result in the maximum profit. The maximum profit, \$450, is the y-coordinate of the vertex.

d. A table of some values that satisfy $P(x) = -2x^2 + 60x$ and the graph are shown in Figure 2.7(a).

x	P
0	0
10	400
15	450
20	400
30	0

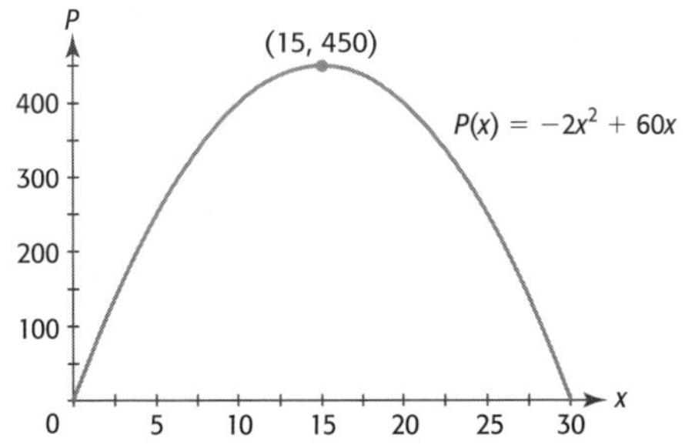

FIGURE 2.7(a)

Even when using a graphing utility to graph a quadratic function, it is important to recognize that the graph is a parabola and to locate the vertex. Determining which way the parabola opens and the x-coordinate of the vertex is very useful in setting the viewing window so that a complete graph (that includes the vertex and the intercepts) is shown. For example, suppose we want to graph the sweatshirt function with technology. Graphing this function with the standard viewing window yields a graph that appears to be a line very close to the y-axis (Figure 2.7(b)); this graph gives no indication of the location of the vertex or which way the parabola opens. We know that the vertex of the graph of this function is (15, 450) and that it is a maximum, so we can choose a window with $x = 15$ near the center and $y = 450$ near the top. We use a window with $x_{min} = 0$, $x_{max} = 30$, $y_{min} = 0$, and $y_{max} = 460$ to get a better view of the graph (Figure 2.7(c)).

$y = -2x^2 + 60x$

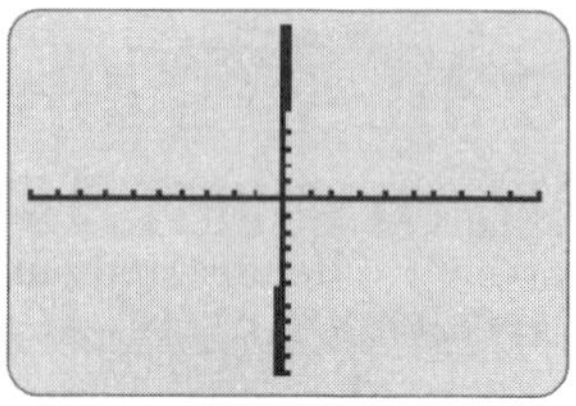

[−10, 10] by [−10, 10]

FIGURE 2.7(b)

$y = -2x^2 + 60x$

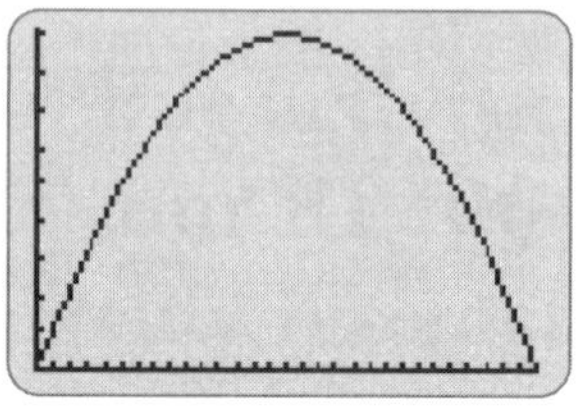

[0, 30] by [0, 460]

FIGURE 2.7(c)

EXAMPLE 2 Teenage Drug Use

The percent of 12th-graders using any drug during the years 1980–1998 can be modeled by the equation $y = 0.181x^2 - 4.093x + 68.771$, where x is the number of years after 1980.

a. During what year does the model indicate that this usage was a minimum?

b. What is the minimum percent of usage?
(Source: Index of Leading Cultural Indicators)

Solution

a. This equation is in the form $f(x) = ax^2 + bx + c$, so $a = 0.181$. Because $a > 0$, the parabola opens upward, the vertex of this parabola is a minimum, and the x-coordinate of the vertex is where the minimum usage occurs.

$$x = \frac{-b}{2a} = \frac{-(-4.093)}{2(0.181)} \approx 11.3$$

So the minimum usage occurs in the 12th year, in 1992.

b. The minimum for the model is found by evaluating the function at $x = 11.3$. This value, $y = 45.6$, can be found with TRACE, CALC VALUE, or TABLE, or with direct evaluation (Figure 2.8). So the minimum percent is 45.6%.

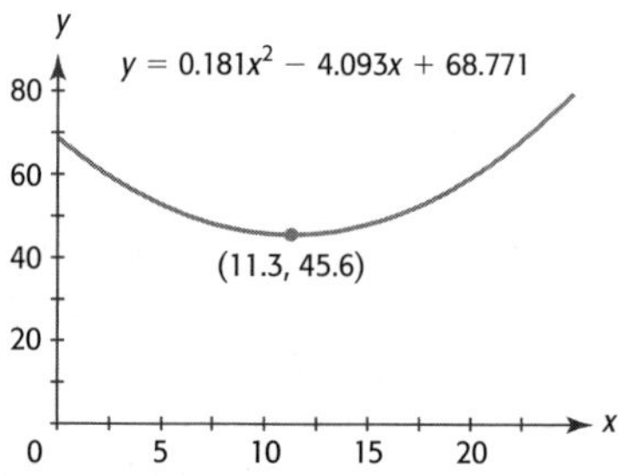

FIGURE 2.8

■

Quadratic Functions in the Form $f(x) = a(x - h)^2 + k$

When a quadratic function is written in the form $f(x) = ax^2 + bx + c$, we can calculate the coordinates of the vertex. But if a quadratic function is written in the form

$$y = a(x - h)^2 + k$$

the vertex of the parabola is at (h, k) (Figure 2.9(a)). For example, the graph of $y = (x - 2)^2 + 3$ is a parabola opening upward with vertex (2, 3) (Figure 2.9(b)).

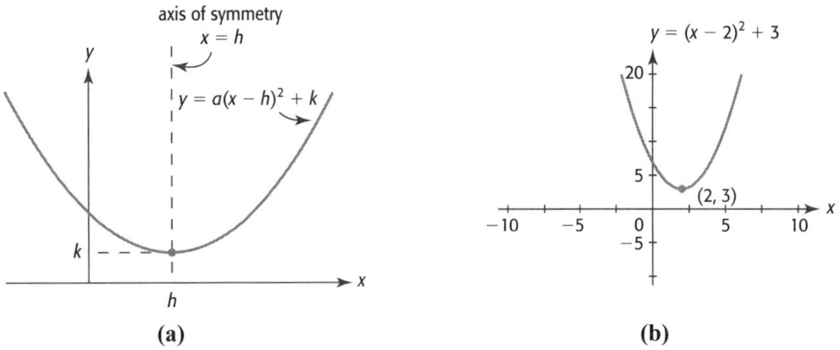

FIGURE 2.9

> **GRAPH OF A QUADRATIC FUNCTION**
>
> In general, the graph of the function
>
> $$y = a(x - h)^2 + k$$
>
> is a parabola with its vertex at the point (h, k).
>
> The parabola opens upward if $a > 0$, and the vertex is a minimum.
> The parabola opens downward if $a < 0$, and the vertex is a maximum.
> The axis of symmetry of the parabola has equation $x = h$.

EXAMPLE 3 Vertex of a Parabola

a. Find the vertex of the function $R(x) = -0.1(x - 3000)^2 + 900{,}000$.

b. Is the vertex a maximum or a minimum?

c. Graph the function for positive values of $R(x)$.

Solution

a. This function is in the form $y = a(x - h)^2 + k$ with $h = 3000$ and $k = 900{,}000$. Thus, the vertex of $R(x) = -0.1(x - 3000)^2 + 900{,}000$ is (3000, 900,000).

b. Because $a = -0.1$, which is negative, the vertex is a maximum.

c. Using a window with $x = 3000$ at the center and $y = 900{,}000$ near the top gives the graph of this function, shown in Figure 2.10 on the next page.

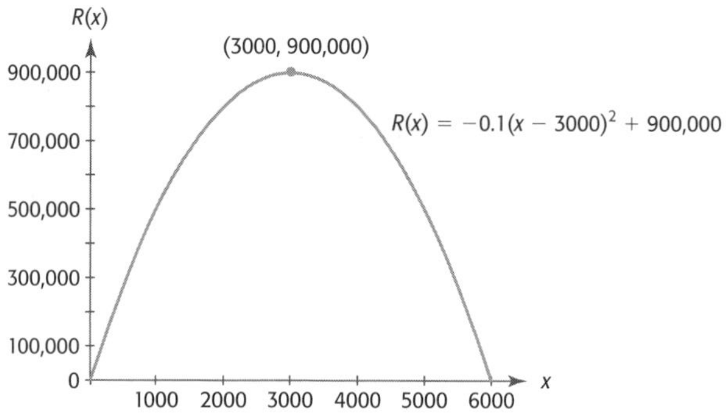

FIGURE 2.10

Intercepts

As with a linear function, we can use technology to find the y-intercept and x-intercept(s) of the graph of a quadratic function. The graph of a quadratic function may intersect the horizontal axis in two, one, or zero points (Figure 2.11).

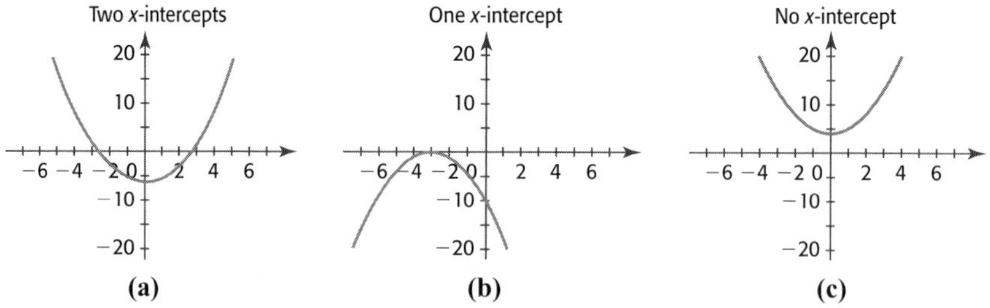

FIGURE 2.11

EXAMPLE 4 Height of a Ball

The height above ground of a ball thrown at 64 feet per second from the top of an 80-foot-high building is modeled by $S = 80 + 64t - 16t^2$ feet, where t is the number of seconds after the ball is thrown.

a. Describe the graph of the model.

b. Find the t-coordinate and S-coordinate of the vertex of the graph of this quadratic function.

c. Explain the meaning of the coordinates of the vertex for this model.

d. Over what time interval is this model increasing? What does this mean in the context of the model?

e. Over what time interval is this model decreasing, and what does this mean?

Solution

a. The function is quadratic, and the coefficient of the second-degree term $(-16t^2)$ is negative, so the graph is a parabola that opens down. The graph is shown in Figure 2.12(a).

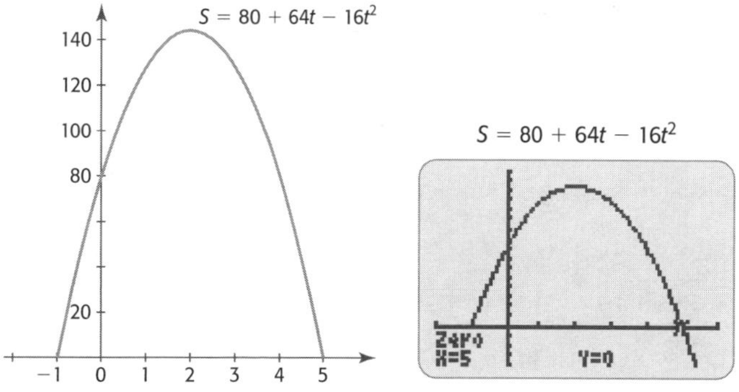

(a) (b)

FIGURE 2.12

b. The height S is a function of the time t, and the t-coordinate of the vertex is

$$t = \frac{-b}{2a} = \frac{-64}{2(-16)} = 2$$

The S-coordinate of the vertex is the value of S at $t = 2$, and $S(2) = 80 + 64(2) - 16(2^2) = 144$. Thus, the vertex is (2, 144).

c. The graph is a parabola that opens down, so the vertex is the highest point on the graph and the function has its maximum there. The t-coordinate of the vertex, 2, is the time (in seconds) at which the ball reaches its maximum height, and the S-coordinate, 144, is the maximum height (in feet) that the ball reaches.

d. The function is increasing until $t = 2$, where the graph reaches its vertex. Because the model does not apply for negative values of t (the time), we consider only non-negative values of t. Thus the ball is climbing for $0 < t < 2$; that is, for the first 2 seconds after it is thrown.

e. The function $S = 80 + 64t - 16t^2$ decreases for all $t > 2$, but the ball will only fall until it reaches the ground, so the *model* is decreasing from $t = 2$ to the time when $S = 0$. To find this value of t, we find the x-intercepts of the graph of the function. The graph of this function, with x representing the time t and y representing the height S, is shown in Figure 2.12(b) with a positive x-intercept of 5. The negative zero is not relevant to the problem, so the ball strikes the ground at $t = 5$ seconds. Thus, the model is decreasing for $2 < t < 5$. This means that the ball is falling after it has reached its maximum height; this occurs between 2 seconds and 5 seconds after it is thrown. ■

2.1 SKILLS CHECK

In Exercises 1–6, (a) determine if the function is quadratic. If it is, (b) determine if the graph is concave up or concave down. (c) Determine if the vertex of the graph is a maximum point or a minimum point.

1. $y = 2x^2 - 8x + 6$

2. $y = 4x - 3$

3. $y = 2x^3 - 3x^2$

4. $f(x) = x^2 + 4x + 4$

5. $g(x) = -5x^2 - 6x + 8$

6. $h(x) = -2x^2 - 4x + 6$

In Exercises 7–14, (a) graph each quadratic function on $[-10, 10]$ *by* $[-10, 10]$*. (b) Does this window give a complete graph?*

7. $y = 2x^2 - 8x + 6$
8. $f(x) = x^2 + 4x + 4$
9. $g(x) = -5x^2 - 6x + 8$
10. $h(x) = -2x^2 - 4x + 6$
11. $y = x^2 + 8x + 19$
12. $y = x^2 - 4x + 5$
13. $y = 0.01x^2 - 8x$
14. $y = 0.1x^2 + 8x + 2$

In Exercises 15–22, (a) give the coordinates of the vertex of the graph of each function. (b) Graph each function on a window that includes the vertex.

15. $y = (x - 1)^2 - 3$
16. $y = (x + 10)^2 - 6$
17. $y = (x + 8)^2 + 8$
18. $y = (x - 12)^2 + 1$
19. $y = 12x - 3x^2$
20. $y = 3x + 18x^2$
21. $y = 3x^2 + 18x - 3$
22. $y = 5x^2 + 75x + 8$

For Exercises 23–26, (a) Find the x-coordinate of the vertex of the graph. (b) Set the viewing window so that the x-coordinate of the vertex is near the center of the window and the vertex is visible, and then graph the given equation.

23. $y = 2x^2 - 40x + 10$
24. $y = -3x^2 - 66x + 12$
25. $y = -0.2x^2 - 32x + 2$
26. $y = 0.3x^2 + 12x - 8$

In Exercises 27–32, sketch complete graphs of the functions.

27. $y = x^2 + 24x + 144$
28. $y = x^2 - 36x + 324$
29. $y = -x^2 - 100x + 1600$
30. $y = -x^2 - 80x - 2000$
31. $y = 2x^2 + 10x - 600$
32. $y = 2x^2 - 75x - 450$

Use the graph of each function in Exercises 33–38 to estimate the x-intercepts.

33. $y = 2x^2 - 8x + 6$
34. $f(x) = x^2 + 4x + 4$
35. $y = x^2 - x - 110$
36. $y = x^2 + 9x - 36$
37. $g(x) = -5x^2 - 6x + 8$
38. $h(x) = -2x^2 - 4x + 6$

2.1 EXERCISES

39. *Profit* The daily profit for a product is given by $P = 32x - 0.01x^2 - 1000$, where x is the number of units produced and sold.
    a. Graph this function for x between 0 and 3200.
    b. Describe what happens to the profit for this product when the number of units produced is between 1 and 1600.
    c. What happens after 1600 units are produced?

40. *Profit* The daily profit for a product is given by $P = 420x - 0.1x^2 - 4100$ dollars, where x is the number of units produced and sold.
    a. Graph this function for x between 0 and 4200.
    b. Is the graph of the function concave up or down?

41. *World Population* A low-projection scenario of world population for the years 1995–2150 by the United Nations is given by the function $y = -0.36x^2 + 38.52x + 5822.86$, where x is the number of years after 1990 and the world population is measured in millions of people.
    a. Graph this function for $x = 0$ to $x = 120$.
    b. What will be the world population in 2010 if the projections made using this model are accurate?
    (Source: *World Population Prospects: 1998 Revision*, United Nations, 1999)

42. *Crimes* When x represents the number of years after 1950, the number of arrests of juveniles per 100,000 juveniles for crimes is given by $f(x) = -0.027x^2 + 5.630x + 51.148$.
    a. Graph this function for the years 1950–2010.
    b. What does this model estimate the number of arrests to be in 2003?
    (Source: Federal Bureau of Investigation)

43. *Tourism Spending* The equation $y = 0.442x^2 + 23.126x + 181.178$, with x equal to the number of years from 1987 to 1999, models the global spending (in billions of dollars) on travel and tourism.
    a. Graph this function for $x = 0$ to $x = 12$.
    b. If $x = 0$ in 1987, find the spending projected by this model in 2002.

c. Is the value in part (b) an interpolation or an extrapolation?
(Source: *Statistical Abstract of the United States, 1999*)

44. *Prison Population* The average number of inmates in the Beaufort County, South Carolina, jail is modeled by $y = 0.161x^2 + 7.582x + 82.157$, where x is the number of years after 1991.
a. Graph this function from $x = 0$ to $x = 20$.
b. Does this model indicate that the population will increase or decrease in years 1991–2011?
c. How would this information be useful to government officials planning for the future?
(Source: *The Island Packet,* November 1999)

45. *Flight of a Ball* If a ball is thrown upward at 96 feet per second from the top of a building that is 100 feet high, the height of the ball can be modeled by $S = 100 + 96t - 16t^2$ feet, where t is the number of seconds after the ball is thrown.
a. Describe the graph of the model.
b. Find the t-coordinate and S-coordinate of the vertex of the graph of this quadratic function.
c. Explain the meaning of the coordinates of the vertex for this model.

46. *Flight of a Ball* If a ball is thrown upward at 39.2 meters per second from the top of a building that is 30 meters high, the height of the ball can be modeled by $S = 30 + 39.2t - 9.8t^2$ meters, where t is the number of seconds after the ball is thrown.
a. Find the t-coordinate and S-coordinate of the vertex of the graph of this quadratic function.
b. Explain the meaning of the coordinates of the vertex for this function.
c. Over what time interval is the function increasing? What does this mean in relation to the ball?

47. *Photosynthesis* The rate of photosynthesis R for a certain plant depends on the intensity of light x, in lumens, according to $R(x) = 270x - 90x^2$.
a. Sketch the graph of this function on a meaningful window.
b. Determine the intensity x that gives the maximum rate of photosynthesis.

48. *Workers and Output* The weekly output of a certain product is $Q(x) = 200x + 6x^2$. Graph this function for values of x and Q that make sense in this application, if x is the number of weeks, $x \le 10$.

49. *Profit* The profit for a product can be described by the function $P(x) = 40x - 3000 - 0.01x^2$ dollars, where x is the number of units produced and sold.
a. To maximize profit, how many units must be produced and sold?
b. What is the maximum possible profit?

50. *Profit* The profit for a product can be described by the function $P(x) = 840x - 75.6 - 0.4x^2$ million dollars, where x is the number of units produced and sold.
a. To maximize profit, how many units must be produced and sold?
b. What is the maximum possible profit?

51. *Revenue* The annual total revenue for a product is given by $R(x) = 1500x - 0.02x^2$ dollars, where x is the number of units sold.
a. To maximize the annual revenue, how many units must be sold?
b. What is the maximum possible annual revenue?

52. *Revenue* The monthly total revenue for a product is given by $R(x) = 300x - 0.01x^2$ dollars, where x is the number of units sold.
a. To maximize the monthly revenue, how many units must be sold?
b. What is the maximum possible monthly revenue?

53. *Area* If 200 feet of fence are used to enclose a rectangular pen, the resulting area of the pen is $A = x(100 - x)$, where x is the width of the pen.
a. Is A a quadratic function of x?
b. What is the maximum possible area of the pen?

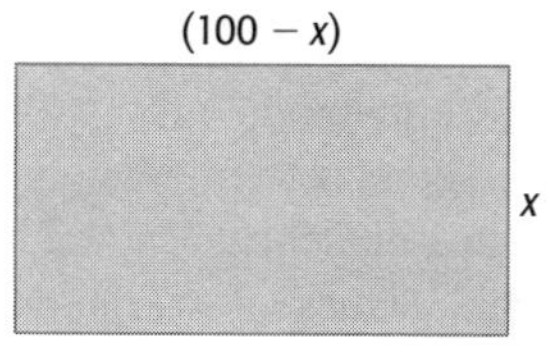

54. *Area* If 25,000 feet of fence are used to enclose a rectangular field, the resulting area of the field is $A = (12{,}500 - x)x$, where x is the width of the pen. What is the maximum possible area of the pen?

55. *Satellite Dishes* The number of U.S. schools with satellite dishes between 1992 and 1997 can be modeled by the function $S(x) = -0.877x^2 + 10.867x - 17.041$ thousand schools, where x is the number of years after 1990. Did the number of schools with satellite dishes reach a maximum or a minimum value between 1992 and 1997? If so, give the maximum or minimum number of dishes.
(Source: *Statistical Abstract of the United States, 1998*)

56. *U.S. Visitors* The number of international visitors to the United States, in millions, can be modeled by

$y = -0.233x^2 + 7.238x - 9.167$, where x is the number of years after 1980.

a. Does the model estimate that a maximum or minimum number of visitors will occur during this period? How can you tell without graphing the equation?

b. Find the input and output at the vertex.

c. Interpret the results of part (b).

(Source: *Statistical Abstract of the United States, 1999*)

57. *Foreign-Born Population* The percent of the U.S. population that is foreign born can be modeled by the equation $y = 0.003x^2 - 0.419x + 19.858$, where x is the number of years after 1900.

a. Is the vertex of the graph of this function a maximum or a minimum?

b. Assuming that the pattern indicated by this model remains the same, find the input and output at the vertex.

c. Interpret the results of part (b).

d. Use the vertex to set the window and graph the model.

(Source: Bureau of the Census, U.S. Department of Commerce)

58. *Abortions* Using data from 1973 to 1996, the number of abortions in the United States can be modeled by the function $y = -4.408x^2 + 149.930x + 383.373$, where x is the number of years after 1970 and y is measured in thousands.

a. In what year does this model indicate the number will be a maximum?

b. What is the maximum number of abortions in this year if this model is accurate?

c. For what values of x can we be sure that this model no longer applies?

(Source: Alan Guttmacher Institute)

59. *Wind and Pollution* The amount of particulate pollution p in the air depends on the wind speed s, among other things, with the relationship between p and s approximated by $p = 25 - 0.01s^2$, where p is in ounces per cubic yard and s is in miles per hour.

a. Sketch the graph of this model with s on the horizontal axis and with nonnegative values of s and p.

b. Is the function increasing or decreasing on this domain?

c. What is the p-intercept of the graph?

d. What does the p-intercept mean in the context of this application?

60. *Drug Sensitivity* The sensitivity S to a drug is related to the dosage size x by $S = 1000x - x^2$.

a. Sketch the graph of this model using a domain and range with nonnegative x and S.

b. Is the function increasing or decreasing for x between 0 and 500?

c. What is the positive x-intercept of the graph?

d. Why is this x-intercept important in the context of this application?

61. *Falling Object* A tennis ball is thrown into a swimming pool from the top of a tall hotel. The height of the ball from the pool is given by $D(t) = -16t^2 - 4t + 210$ feet where t is the time, in seconds, after the ball was thrown. Graphically find the t-intercepts for this function. Interpret the value(s) that make sense in this problem context.

62. *Break Even* The profit for a product is given by $P = 1600 - 100x + x^2$, where x is the number of units produced and sold. Graphically find the x-intercepts of the graph of this function to find how many units will give break even (that is, return a profit of zero).

63. *Apartment Rental* The owner of an apartment building can rent all 100 apartments if he charges \$1200 per apartment per month, but the number of apartments rented is reduced by 2 for every \$40 increase in the monthly rent.

a. Construct a table that gives the revenue if the rent charged is \$1240, \$1280, and \$1320.

b. Does $R(x) = (1200 + 40x)(100 - 2x)$ model the revenue from these apartments if x represents the number of \$40 increases?

c. What monthly rent gives the maximum revenue for the apartments?

64. *Rink Rental* The owner of a skating rink rents the rink for parties at \$720 if 60 or fewer skaters attend, so that the cost is \$12 per person if 60 attend. For each 6 skaters above 60, she reduces the price per skater by \$.50.

a. Construct a table that gives the revenue if the number attending is 66, 72, and 78.

b. Does the function $R(x) = (60 + 6x)(12 - 0.5x)$ model the revenue from the party if x represents the number of increases of 6 people each?

c. How many people should attend for the rink's revenue to be a maximum?

65. *World Population* A low-projection scenario of world population for 1995–2150 by the United Nations is given by the function $y = -0.36x^2 + 38.52x + 5822.86$, where x is the number of years after 1990 and the world population is measured in millions of people.

a. Find the input and output at the vertex of the graph of this model.

b. Interpret the values from part (a).

c. For what years after 1995 does this model predict that the population will increase?
(Source: *World Population Prospects: 1998 Revision*, United Nations, 1999)

2.2 Solving Quadratic Equations

Key Concepts

- Solving quadratic equations
- Factoring; zero product property
- The square root method
- Completing the square
- The quadratic formula
- Graphical solution methods
  x-Intercept method
  Intersection method
- Combining graphical and factoring methods
- Combining graphical and numerical methods
- Equations with complex solutions

Section Preview Admissions

The number of admissions at all hospitals in the United States between 1985 and 1997 can be described by the function

$$A(x) = 38.228x^2 - 1069.600x + 40{,}698.547 \text{ thousand people}$$

where x is the number of years after 1985. Viewing and understanding the pattern indicated by this model can help the medical industry, the government, medical training facilities, and consumers to better understand current and future needs for hospital space and facilities and for hospital personnel. (Source: Based on data from the American Hospital Association: *Hospital Statistics*, 1986, 1991–1999 editions)

To find the year after 1985 that is predicted by this model to give 33,563,000 (that is, 33,563 thousand) admissions, we solve the **quadratic equation**

$$38.228x^2 - 1069.600x + 40{,}698.547 = 33{,}563 \quad \text{or}$$
$$38.228x^2 - 1069.600x + 7135.547 = 0$$

The values of x that satisfy this equation are called the **solutions** of the equation; they are also zeros of the function

$$y = 38.228x^2 - 1069.600x + 7135.547$$

and they are the x-intercepts of the graph of this function. In this section, we learn how to solve quadratic equations by using factoring methods, graphical methods, the square root property, completing the square, and the quadratic formula.

Factoring Methods

An equation that can be written in the form $ax^2 + bx + c = 0$, with $a \neq 0$, is called a **quadratic equation**. Solutions to some quadratic equations can be found exactly by factoring; other quadratic equations require different types of solution methods to find or to approximate solutions.

Solution by factoring is based on the following property of real numbers.

ZERO PRODUCT PROPERTY

For real numbers a and b, the product $ab = 0$ if and only if either $a = 0$ or $b = 0$ or both a and b are zero.

To use this property to solve a quadratic equation by factoring, we must first make sure that the equation is written in a form with zero on one side. If the resulting nonzero expression is factorable, we factor it and use the zero product property to convert the equation into two linear equations that are easily solved.* Before applying this technique to real-world applications, we consider the following example.

EXAMPLE 1 Solving a Quadratic Equation by Factoring

Solve the equation $3x^2 + 7x = 6$.

Solution

We first subtract 6 from both sides of the equation to rewrite the equation with 0 on one side:

$$3x^2 + 7x - 6 = 0$$

To begin factoring the trinomial $3x^2 + 7x - 6$, we seek factors of $3x^2$ (that is, $3x$ and x) as the first terms of two binomials and factors of -6 as the last terms of the binomials. The factorization whose inner and outer products combine to $7x$ is $(3x - 2)(x + 3)$, so we have

$$(3x - 2)(x + 3) = 0$$

Using the zero product property gives

$$3x - 2 = 0 \quad \text{or} \quad x + 3 = 0$$

Solving these linear equations gives the two solutions to the original equation.

$$x = \frac{2}{3} \quad \text{or} \quad x = -3$$

■

EXAMPLE 2 Profit

Suppose the daily profit from the production and sale of x units of a product is given by

$$P(x) = -0.01x^2 + 20x - 500 \text{ dollars}$$

What levels of production and sales will give a daily profit of \$1400?

Solution

To answer this question, note that \$1400 is the amount of profit. Thus, we solve the equation

$$1400 = -0.01x^2 + 20x - 500$$

to find the number of units x that gives a \$1400 profit.

To find the solution by factoring, we use the following steps:

$1400 = -0.01x^2 + 20x - 500$	
$0 = -0.01x^2 + 20x - 1900$	Subtract 1400 from both sides.
$0 = -0.01(x^2 - 2000x + 190{,}000)$	Factor -0.01 from the expression.
$0 = -0.01(x - 1900)(x - 100)$	Factor the trinomial into binomial factors.
$x - 1900 = 0 \quad \text{or} \quad x - 100 = 0$	Set the binomial factors equal to 0.
$x = 1900 \quad \text{or} \quad x = 100$	Solve for x.

*For a review of factoring methods, see the Algebra Toolbox in Appendix B.

Thus, producing and selling either 100 units or 1900 units give the daily profit of \$1400. ■

Graphical Methods

Factoring the right-hand side of the quadratic equation in Example 2 was not especially easy. In cases where factoring $f(x)$ is difficult or impossible, graphing $y = f(x)$ after the equation is in the form $f(x) = 0$ can be helpful in finding the solution. Recall that if a is a real number, the following three statements are equivalent:

- a is a solution to the equation $f(x) = 0$.
- a is a zero of the function f.
- a is an x-intercept of the graph of $y = f(x)$.

It is important to remember that the above three statements are equivalent because connecting these concepts allows us to use different methods for solving equations. Sometimes graphical methods are the easiest way to find or approximate solutions to real data problems. If the x-intercepts of the graph of $y = f(x)$ are easily found, then graphical methods may be helpful in finding the solutions. Note that if the graph of $y = f(x)$ does not cross or touch the x-axis, there are no real solutions to the equation $f(x) = 0$.

We can also find solutions or decimal approximations of solutions to quadratic equations by using the intersection method with a graphing utility.

EXAMPLE 3 Profit (Revisited)

Consider again the daily profit from the production and sale of x units of a product, given by

$$P(x) = -0.01x^2 + 20x - 500 \text{ dollars}$$

a. Use a graph to find the levels of production and sales that give a daily profit of \$1400.

b. Is it possible for the profit to be greater than \$1400?

Solution

a. To find the level of production and sales, x, that gives a daily profit of 1400 dollars, we solve

$$1400 = -0.01x^2 + 20x - 500$$

To solve this equation by the intersection method, we graph

$$y_1 = -0.01x^2 + 20x - 500 \quad \text{and} \quad y_2 = 1400$$

To find the appropriate window for this graph, we note that the graph of the function $y_1 = -0.01x^2 + 20x - 500$ is a parabola with the vertex at

$$x = \frac{-b}{2a} = \frac{-20}{2(-0.01)} = 1000 \quad \text{and} \quad y = P(1000) = 9500$$

We use a viewing window containing this point, with $x = 1000$ near the center, to graph the function (Figure 2.13). Using the intersection method, we see that (100, 1400) and (1900, 1400) are the points of intersection of the graphs of the two functions.

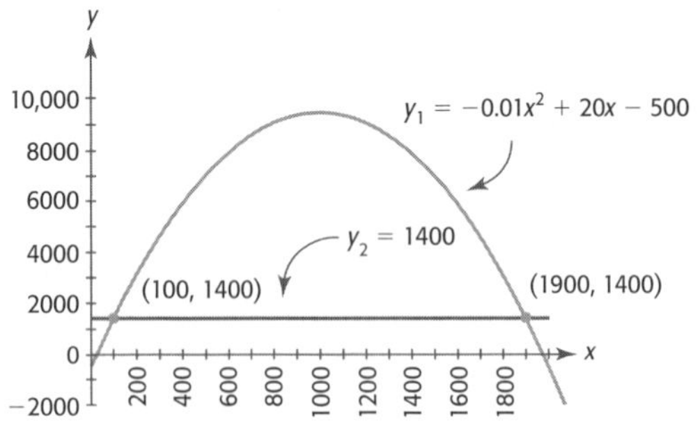

FIGURE 2.13

Thus, $x = 100$ and $x = 1900$ are solutions to the equation $1400 = -0.01x^2 + 20x - 500$, and the profit is \$1400 when $x = 100$ units or $x = 1900$ units of the product are produced and sold. Note that this is the same solution as the one found by factoring in Example 2.

b. We can see from the graph in Figure 2.13 that the profit is more than \$1400 for many values of x. Because the graph of this profit function is a parabola that opens down, the maximum profit occurs at the vertex of the graph. As we found in part (a), the vertex occurs at $x = 1000$ and the maximum possible profit is $P(1000) =$ \$9500, which is more than \$1400. ■

Combining Graphs and Factoring

Examples 2 and 3 indicate that the solutions found graphically agree with the solutions found by factoring. So if the factors of a quadratic function are not easily found, its graph may be helpful in finding the factors. Note once more the important correspondences that exist among solutions, zeros, and x-intercepts: the x-intercepts of the graph of $y = P(x)$ are the real solutions of the equation $0 = P(x)$ and the zeros of $P(x)$. In addition, the factors of $P(x)$ are also related to the zeros of $P(x)$. We saw in Example 2 that $(x - 100)$ is a factor of $-0.01x^2 + 20x - 1900$ and that $x = 100$ is a solution of $0 = -0.01x^2 + 20x - 1900$, so that $P(100) = 0$. This relationship among the factors, solutions, and zeros is true for any polynomial function f and can be generalized by the following theorem.

FACTOR THEOREM

The polynomial function f has a factor $(x - a)$ if and only if $f(a) = 0$. Thus, $(x - a)$ is factor of $f(x)$ if and only if $x = a$ is a solution to $f(x) = 0$.

This means that we can verify our factorization of f and the solutions to $0 = f(x)$ by graphing $y = f(x)$ and observing where the graph crosses the x-axis. We can also sometimes use our observation of the graph to assist us in the factorization of a quadratic function:

> If one solution can be found *exactly* from the graph, it can be used to find one of the factors of the function. The second factor can then be found easily, leading to the second solution.

EXAMPLE 4 Graphing and Factoring Methods Combined

Solve $0 = 3x^2 - x - 10$ by using the following steps.

a. Graphically find one of the x-intercepts of $y = 3x^2 - x - 10$.

b. Algebraically verify that the zero found in part (a) is an exact solution to $0 = 3x^2 - x - 10$

c. Use the method of factoring to find the other solution to $0 = 3x^2 - x - 10$.

Solution

a. The vertex of the graph of $y = 3x^2 - x - 10$ is at $x = \frac{1}{6}$ and $y \approx -10.08$, and because the parabola opens up, we set a viewing window that includes this point near the bottom center of the window. Graphing the function and using the x-intercept method, we find that the graph crosses the x-axis at $x = 2$ (Figure 2.14). This means that 2 is a zero of $f(x) = 3x^2 - x - 10$.

b. Because $x = 2$ was obtained graphically, it may be an approximation of the exact solution. To verify that it is exact, we show that $x = 2$ makes the equation $0 = 3x^2 - x - 10$ a true statement:

$$3(2)^2 - 2 - 10 = 12 - 2 - 10 = 0$$

c. Because $x = 2$ is a solution of $0 = 3x^2 - x - 10$, one factor of $3x^2 - x - 10$ is $(x - 2)$. Thus, we use $(x - 2)$ as one factor and seek a second binomial factor so that the product of the two binomials is $3x^2 - x - 10$:

$$3x^2 - x - 10 = 0$$

$$(x - 2)(\quad) = 0$$

$$(x - 2)(3x + 5) = 0$$

The remaining factor is $3x + 5$, which we set equal to 0 to obtain the other solution:

$$3x = -5$$

$$x = -\frac{5}{3}$$

Thus, the two solutions to $0 = 3x^2 - x - 10$ are $x = 2$ and $x = -\frac{5}{3}$.

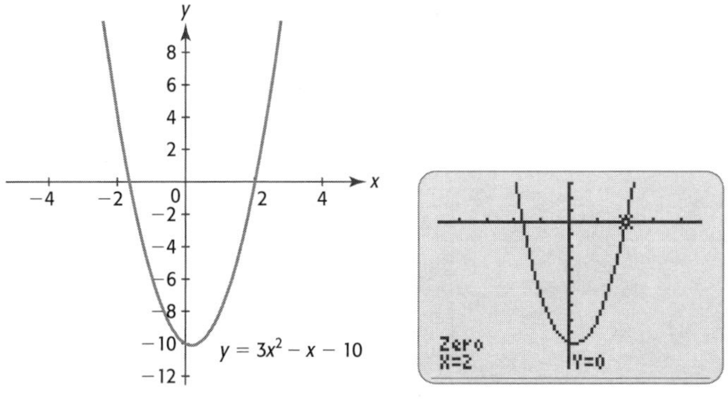

FIGURE 2.14 ■

Graphical and Numerical Methods

When approximate solutions to quadratic equations are sufficient, graphical and/or numerical solution methods can be used. These methods of solving are illustrated in Example 5.

EXAMPLE 5 Marijuana Use

For the years from 1975 to 1991, the percent p of high school seniors who have tried marijuana can be considered as a function of time t according to the equation

$$p = -0.2286t^2 + 2.7783t + 49.8138$$

where t is the number of years after 1975. Use the indicated method to find the year after 1975 during which the percent predicted by the function reached 33%.

a. Graphical method

b. Numerical method

(Source: National Institute on Drug Abuse)

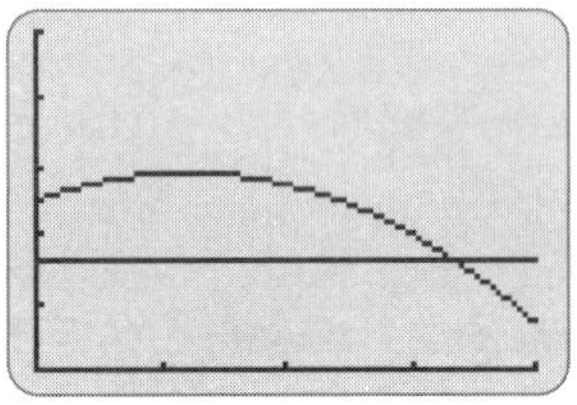

[0, 20] by [0, 100]

(a)

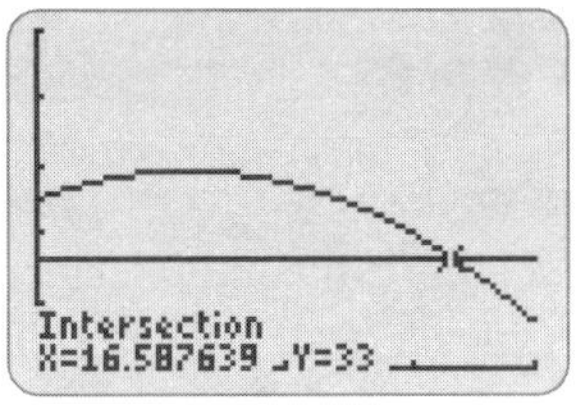

(b)

FIGURE 2.15

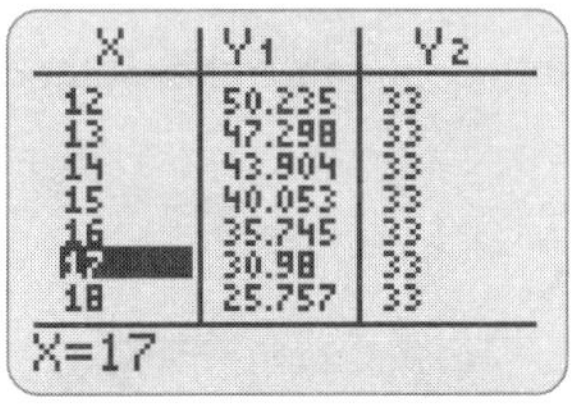

X	Y1	Y2
12	50.235	33
13	47.298	33
14	43.904	33
15	40.053	33
16	35.745	33
17	30.98	33
18	25.757	33

X=17

FIGURE 2.16

Solution

Note that the output p is measured in percent, so for the output 33%, $p = 33$. Thus, to find where the percent is 33, we solve the equation

$$33 = -0.2286t^2 + 2.7783t + 49.8138$$

a. We could use the x-intercept method and find the zeros of

$$y = -0.2286t^2 + 2.7783t + 16.8138$$

Instead, we choose to use the intersection method. Figure 2.15(a) shows the graphs of $y_1 = -0.2286t^2 + 2.7783t + 49.8138$ and $y_2 = 33$, and Figure 2.15(b) shows the point of intersection. We find that t is approximately 16.59.

Because 16 years after 1975 is the end of 1991, the percent reaches 33 in 1992. Note that the graphs also intersect at $t \approx -4.43$, but this value would not give a year after 1975.

b. Figure 2.16 shows a table of inputs (years after 1975) and outputs (percent of high school seniors who have tried marijuana). The table shows that $p = 35.745\%$ at the end of the 16th year after 1975, or 1991, and the percent reached 33 (and fell below it) during 1992. ■

The Square Root Method

We have solved quadratic equations by factoring and by graphical methods. Another method can be used to solve quadratic equations that are written in a particular form. In general, we can find the solutions of quadratic equations of the form $x^2 = C$ by taking the square root of both sides. For example, to solve $x^2 = 25$, we take the square root of both sides of the equation, getting $x = \pm 5$. Note that there are two solutions because $5^2 = 25$ and $(-5)^2 = 25$.

> **SQUARE ROOT METHOD**
>
> The solutions of the quadratic equation $x^2 = C$ are $x = \pm\sqrt{C}$. Note that, when we take the square root of both sides, we use a $\pm$ symbol because there are both a positive and a negative value that, when squared, give C.

We can also use this method to solve quadratic equations of the form $(ax + b)^2 = C$.

EXAMPLE 6 Square Root Solution Method

Solve the following equations by using the square root method.

a. $2x^2 - 16 = 0$ **b.** $(x - 6)^2 = 18$

Solution

a. This equation can be written in the form $x^2 = C$, so the square root method can be used. We want to rewrite the equation so that the x^2-term is isolated and its coefficient is 1 and then take the square root of both sides.

$$\begin{aligned} 2x^2 - 16 &= 0 \\ 2x^2 &= 16 \\ x^2 &= 8 \\ x &= \pm\sqrt{8} = \pm 2\sqrt{2} \end{aligned}$$

The exact solutions to $2x^2 - 16 = 0$ are $x = 2\sqrt{2}$ and $x = -2\sqrt{2}$.

b. The left side of this equation is a square, so we can take the square root of both sides to find x.

$$\begin{aligned} (x - 6)^2 &= 18 \\ x - 6 &= \pm\sqrt{18} \\ x &= 6 \pm \sqrt{18} \\ &= 6 \pm \sqrt{9}\sqrt{2} \\ &= 6 \pm 3\sqrt{2} \end{aligned}$$

■

Completing the Square

When we convert one side of a quadratic equation to a perfect binomial square and use the square root method to solve the equation, the method used is called **completing the square**.

EXAMPLE 7 Completing the Square

Solve the equation $x^2 - 12x + 7 = 0$ by completing the square.

Solution

Because the left side of this equation is not a perfect square, we rewrite it in a form where we can more easily get a perfect square. We subtract 7 from both sides, getting

$$x^2 - 12x = -7$$

We complete the square on the left side of this equation by adding the appropriate number to both sides of the equation to make the left side a perfect square trinomial. Because $(x + a)^2 = x^2 + 2ax + a^2$, the constant term of a perfect square trinomial will be the *square of half the coefficient of* x. Using this rule with the equation $x^2 - 12x = -7$, we would need to take half the coefficient of x and add the square of this number to get a perfect square trinomial. Half of -12 is -6, so adding

$(-6)^2 = 36$ to $x^2 - 12x$ gives the perfect square $x^2 - 12x + 36$. We also have to add 36 to the other side of the equation (to preserve the equality), giving

$$x^2 - 12x + 36 = -7 + 36$$

Factoring the perfect square trinomial gives

$$(x - 6)^2 = 29$$

We can now solve the equation with the square root method.

$$\begin{aligned} (x - 6)^2 &= 29 \\ x - 6 &= \pm\sqrt{29} \\ x &= 6 \pm \sqrt{29} \end{aligned}$$

■

The Quadratic Formula

We can generalize the method of completing the square to derive a general formula that gives the solution to any quadratic equation. We use this method to find the general solution to $ax^2 + bx + c = 0, a \neq 0$.

$$ax^2 + bx + c = 0 \qquad \text{Standard form}$$

$$ax^2 + bx = -c \qquad \text{Subtract } c \text{ from both sides.}$$

$$x^2 + \frac{b}{a}x = -\frac{c}{a} \qquad \text{Divide both sides by } a.$$

We would like to make the left side of the last equation a perfect square trinomial. Half the coefficient of x is $\frac{b}{2a}$, and squaring this gives $\frac{b^2}{4a^2}$. Hence, adding $\frac{b^2}{4a^2}$ to both sides of the equation gives a perfect square trinomial on the left side, and we can continue with the solution.

$$x^2 + \frac{b}{a}x + \frac{b^2}{4a^2} = \frac{b^2}{4a^2} - \frac{c}{a} \qquad \text{Add } \frac{b^2}{4a^2} \text{ to both sides of the equation.}$$

$$\left(x + \frac{b}{2a}\right)^2 = \frac{b^2 - 4ac}{4a^2} \qquad \text{Factor the left side and combine the fractions on the right side.}$$

$$x + \frac{b}{2a} = \pm\sqrt{\frac{b^2 - 4ac}{4a^2}} \qquad \text{Take the square root of both sides.}$$

$$x = -\frac{b}{2a} \pm \frac{\sqrt{b^2 - 4ac}}{2|a|} \qquad \text{Subtract } \frac{b}{2a} \text{ from both sides and simplify.}$$

$$x = -\frac{b}{2a} \pm \frac{\sqrt{b^2 - 4ac}}{2a} \qquad \pm 2|a| = \pm 2a$$

$$x = \frac{-b \pm \sqrt{b^2 - 4ac}}{2a} \qquad \text{Combine the fractions.}$$

The formula we have developed is called the **quadratic formula**.

QUADRATIC FORMULA

The solutions of the quadratic equation $ax^2 + bx + c = 0$ are given by the formula

$$x = \frac{-b \pm \sqrt{b^2 - 4ac}}{2a}$$

Note that a is the coefficient of x^2, b is the coefficient of x, and c is the constant term.

Because of the $\pm$ sign, there are two solutions to the equation:

$$x = \frac{-b + \sqrt{b^2 - 4ac}}{2a} \quad \text{and} \quad x = \frac{-b - \sqrt{b^2 - 4ac}}{2a}$$

We can use the quadratic formula to solve all quadratic equations exactly, but it is especially useful for finding exact solutions to those equations for which factorization is difficult or impossible. For example, the solutions to $33 = -0.2286t^2 + 2.7783t + 49.8138$ were found approximately by graphical methods in Example 5. If we need to find the exact solutions, we could use the quadratic formula.

EXAMPLE 8 Solving Using the Quadratic Formula

Solve $6 - 3x^2 + 4x = 0$ using the quadratic formula.

Solution

The equation $6 - 3x^2 + 4x = 0$ can be rewritten as $-3x^2 + 4x + 6 = 0$, so $a = -3$, $b = 4$, and $c = 6$. The two solutions to this equation are

$$x = \frac{-4 \pm \sqrt{4^2 - 4(-3)(6)}}{2(-3)} = \frac{-4 \pm \sqrt{88}}{-6} = \frac{-4 \pm 2\sqrt{22}}{-6} = \frac{2 \pm \sqrt{22}}{3}$$

Thus, the exact solutions are the irrational numbers $x = \frac{2 + \sqrt{22}}{3}$ and $x = \frac{2 - \sqrt{22}}{3}$. Three-place decimal approximations for these solutions are $x \approx -0.897$ and $x \approx 2.230$. ■

Decimal approximations of irrational solutions found with the quadratic formula will often suffice as answers to an applied problem. The quadratic formula is especially useful when the coefficients of a quadratic equation are decimal values that make factorization impractical. This occurs in many applied problems, such as the one discussed at the beginning of this section. If the graph of the quadratic function $y = f(x)$ does not intersect the x-axis at "nice" values of x, the solutions may be irrational numbers, and using the quadratic formula allows us to find these solutions exactly.

EXAMPLE 9 Hospital Admissions

The annual number of admissions at all hospitals in the United States between 1985 and 1997 can be described by the function $A(x) = 38.228x^2 - 1069.600x + 40{,}698.547$ thousand people, where x is the number of years after 1985. If the model is valid

until 2003, find the years after 1985 that this model estimates the admissions to be 33,563,000.
(Source: American Hospital Association: *Hospital Statistics*, 1986, 1991–1999 editions)

Solution

Note that the output of A is measured in thousands of admissions, so we must first write 33,563,000 admissions as 33,563 thousand admissions. Then, to answer the given question, we solve (using an algebraic, numerical, or graphical method) the quadratic equation

$$38.228x^2 - 1069.600x + 40{,}698.547 = 33{,}563$$

We choose to use the quadratic formula (even though an approximate answer is all that is needed). We first write $38.228x^2 - 1069.600x + 40{,}698.547 = 33{,}563$ with 0 on one side:

$$38.228x^2 - 1069.600x + 40{,}698.547 - 33{,}563 = 0$$
$$38.228x^2 - 1069.600x + 7135.547 = 0$$

This gives $a = 38.228$, $b = -1069.6$, and $c = 7135.547$. Substituting in the quadratic formula gives

$$x = \frac{-(-1069.6) \pm \sqrt{(-1069.6)^2 - 4(38.228)(7135.547)}}{2(38.228)}$$
$$= \frac{1069.6 \pm \sqrt{52{,}933.39714}}{76.456}$$
$$x \approx 16.999 \quad \text{or} \quad x \approx 10.981$$

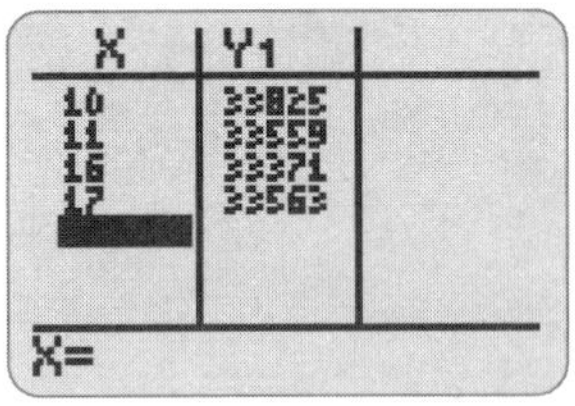

FIGURE 2.17

Thus, 33,563,000 admissions occur in the 17th year from the end of 1985, in 2002, and in 1985 + 11 = 1996. Figure 2.17 shows the outputs of

$$y_1 = 38.228x^2 - 1069.600x + 40{,}698.547$$

at integer values near the x-values that are the results from the quadratic formula, which confirms our conclusions above. ■

The Discriminant

We can also determine the type of solutions a quadratic equation has by looking at the expression $b^2 - 4ac$, which is called the **discriminant** of the quadratic equation $ax^2 + bx + c = 0$. The discriminant is the expression inside the radical in the quadratic formula $x = \frac{-b \pm \sqrt{b^2 - 4ac}}{2a}$, so it determines if the quantity inside the radical is positive, zero, or negative. Thus, we have

- If $b^2 - 4ac > 0$, there are two different real solutions.
- If $b^2 - 4ac = 0$, there is one real solution.
- If $b^2 - 4ac < 0$, there is no real solution.

For example, the equation $3x^2 + 4x + 2 = 0$ has no real solution because $4^2 - 4(3)(2) = -8 < 0$, and the equation $x^2 + 4x + 2 = 0$ has two different real solutions because $4^2 - 4(1)(2) = 8 > 0$.

Aids for Solving Quadratic Equations

The x-intercepts of the graph of the quadratic function $y = f(x)$ can be used to determine how to solve the quadratic equation $f(x) = 0$. Table 2.2 summarizes these ideas with suggested methods for solving, and the graphical representations of the solutions.

TABLE 2.2 Connections Between Graphs of Quadratic Functions and Solution Methods

Graph	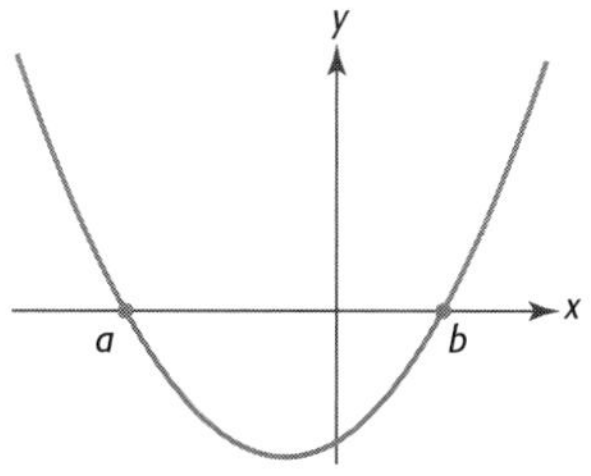	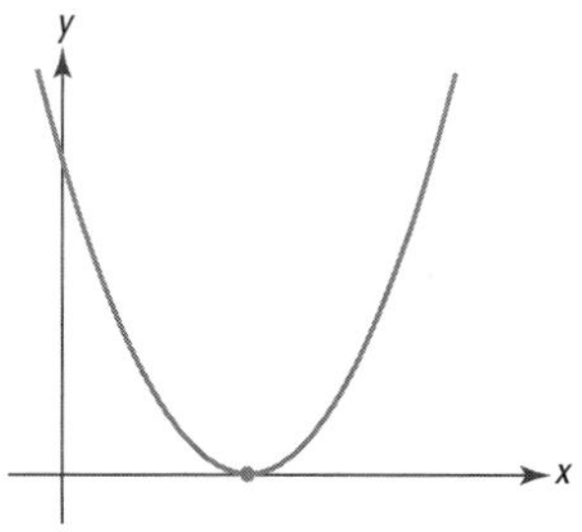	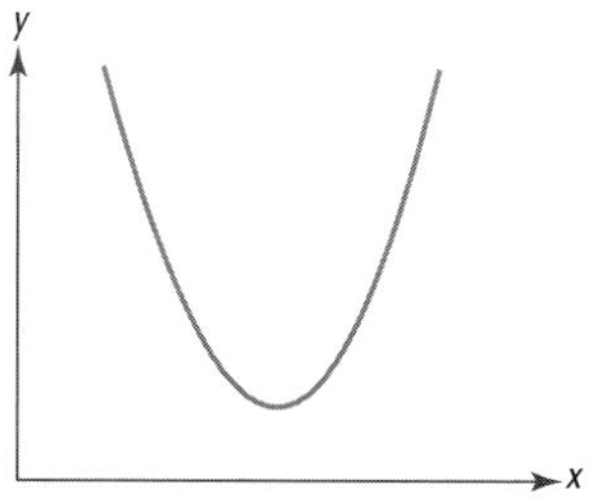
Type of *x*-Intercepts	Graph crosses *x*-axis twice.	Graph touches but does not cross *x*-axis.	Graph does not cross the *x*-axis.
Type of Solutions	Equation has two real solutions.	Equation has one real solution.	Equation has no real solutions.
Suggested Solution Methods*	Use factoring, graphing, or quadratic formula.	Use factoring, graphing, or the square root method.	Use quadratic formula to verify that solutions are not real.

*Solution methods other than the suggested methods may also be successful.

Equations with Complex Solutions

Recall that the solutions of the quadratic equation $x^2 = C$ are $x = \pm\sqrt{C}$, so the solutions of the equation $x^2 = -a$ for $a > 0$ are

$$x = \pm\sqrt{-a} = \pm\sqrt{-1}\sqrt{a} = \pm i\sqrt{a}$$

EXAMPLE 10 Solution Using the Root Method

Solve the equations:

a. $x^2 = -9$ **b.** $3x^2 + 24 = 0$

Solution

a. Taking the square root of both sides of the equation gives the solution of $x^2 = -9$:

$$x = \pm\sqrt{-9} = \pm\sqrt{-1}\sqrt{9} = \pm 3i$$

b. We solve $3x^2 + 24 = 0$ using the root method, as follows:

$$3x^2 = -24$$
$$x^2 = -8$$
$$x = \pm\sqrt{-8} = \pm\sqrt{-1}\sqrt{4 \cdot 2} = \pm 2i\sqrt{2}$$

■

We used the root method to solve the equations in Example 10 because neither equation contained a first-degree term (that is, a term containing x to the first power). We can also find complex solutions by using the quadratic formula.* Recall that the solutions of the quadratic equation $ax^2 + bx + c = 0$ are given by the formula

$$x = \frac{-b \pm \sqrt{b^2 - 4ac}}{2a}$$

EXAMPLE 11 Solution with the Quadratic Formula

Solve the equations:

a. $x^2 - 3x + 5 = 0$

b. $3x^2 + 4x = -3$

Solution

a. Using the quadratic formula, with $a = 1$, $b = -3$, and $c = 5$, gives

$$x = \frac{-(-3) \pm \sqrt{(-3)^2 - 4(1)(5)}}{2(1)} = \frac{3 \pm \sqrt{-11}}{2} = \frac{3 \pm i\sqrt{11}}{2}$$

Note that the solutions can also be written in the form $\frac{3}{2} \pm \frac{\sqrt{11}}{2}i$.

Thus, the solutions are the complex numbers $\frac{3}{2} + \frac{\sqrt{11}}{2}i$ and $\frac{3}{2} - \frac{\sqrt{11}}{2}i$.

b. Writing $3x^2 + 4x = -3$ in the form $3x^2 + 4x + 3 = 0$ gives $a = 3$, $b = 4$, and $c = 3$, so the solution is

$$x = \frac{-4 \pm \sqrt{(4)^2 - 4(3)(3)}}{2(3)} = \frac{-4 \pm \sqrt{-20}}{6} = \frac{-4 \pm \sqrt{-1}\sqrt{4}\sqrt{5}}{6}$$

$$= \frac{-4 \pm 2i\sqrt{5}}{6} = \frac{2(-2 \pm i\sqrt{5})}{2 \cdot 3} = \frac{-2 \pm i\sqrt{5}}{3}$$

Thus,

$$x = -\frac{2}{3} + \frac{\sqrt{5}}{3}i \quad \text{and} \quad x = -\frac{2}{3} - \frac{\sqrt{5}}{3}i$$

■

*The solutions could also be found by completing the square. Recall that the quadratic formula was developed by completing the square on the quadratic equation $ax^2 + bx + c = 0$.

2.2 SKILLS CHECK

In Exercises 1–10, use factoring to solve the equations.

1. $x^2 - 3x - 10 = 0$ **2.** $x^2 - 9x + 18 = 0$

3. $x^2 - 11x + 24 = 0$ **4.** $x^2 + 3x - 10 = 0$

5. $2x^2 + 2x - 12 = 0$ **6.** $2s^2 + s - 6 = 0$

7. $0 = 2t^2 - 11t + 12$ **8.** $6x^2 - 13x + 6 = 0$

9. $6x^2 + 10x = 4$ **10.** $10x^2 + 11x = 6$

Use a graphing utility to find or to approximate the x-intercepts of the graph of each function in Exercises 11–16.

11. $y = x^2 - 3x - 10$ **12.** $y = x^2 + 4x - 32$

13. $y = 3x^2 - 8x + 4$ **14.** $y = 2x^2 + 8x - 10$

15. $y = 2x^2 + 7x - 4$ **16.** $y = 5x^2 - 17x + 6$

Use a graphing utility as an aid in factoring to solve the equations in Exercises 17–22.

17. $2w^2 - 5w - 3 = 0$ **18.** $3x^2 - 4x - 4 = 0$

19. $x^2 - 40x + 256 = 0$ **20.** $x^2 - 32x + 112 = 0$

21. $2s^2 - 70s = 1500$ **22.** $3s^2 - 130s = -1000$

In Exercises 23–26, use the square root method to solve the quadratic equations.

23. $4x^2 - 9 = 0$ **24.** $x^2 - 20 = 0$

25. $x^2 - 32 = 0$ **26.** $5x^2 - 25 = 0$

In Exercises 27–30, complete the square to solve the quadratic equations.

27. $x^2 - 4x - 9 = 0$ **28.** $x^2 - 6x + 1 = 0$

29. $x^2 - 3x + 2 = 0$ **30.** $2x^2 - 9x + 8 = 0$

In Exercises 31–34, use the quadratic formula to solve the equations.

31. $x^2 - 5x + 2 = 0$ **32.** $3x^2 - 6x - 12 = 0$

33. $5x + 3x^2 = 8$ **34.** $3x^2 - 30x - 180 = 0$

In Exercises 35–40, use a graphing utility to find or approximate solutions of the equations.

35. $2x^2 + 2x - 12 = 0$ **36.** $2x^2 + x - 6 = 0$

37. $0 = 6x^2 + 5x - 6$ **38.** $10x^2 = 22x - 4$

39. $4x + 2 = 6x^2 + 3x$ **40.** $(x - 3)(x + 2) = -4$

In Exercises 41–48, find the exact solutions to $f(x) = 0$ in the complex numbers and confirm that the solutions are not real by showing that the graph of $y = f(x)$ does not cross the x-axis.

41. $x^2 + 25 = 0$ **42.** $2x^2 + 40 = 0$

43. $(x - 1)^2 = -4$ **44.** $(2x + 1)^2 + 7 = 0$

45. $x^2 + 4x + 8 = 0$ **46.** $x^2 - 5x + 7 = 0$

47. $2x^2 - 8x + 9 = 0$ **48.** $x^2 - x + 1 = 0$

2.2 EXERCISES

In Exercises 49–60, solve analytically and then check graphically.

49. *Flight of a Ball* If a ball is thrown upward at 96 feet per second from the top of a building that is 100 feet high, the height of the ball can be modeled by $S = 100 + 96t - 16t^2$ feet, where t is the number of seconds after the ball is thrown.

a. How long after the ball is thrown is the height 228 feet?

b. For what time interval is the ball higher than 228 feet?

50. *Falling Object* A tennis ball is thrown into a swimming pool from the top of a tall hotel. The height of the ball above the pool is modeled by $D(t) = -16t^2 - 4t + 200$ feet, where t is the time, in seconds, after the ball was thrown. How long after the ball is thrown is it 44 feet above the pool?

51. *Break-Even* The profit for a product is given by $P(x) = -12x^2 + 1320x - 21,600$, where x is the number of units produced and sold. How many units give break-even (that is, give zero profit) for this product?

52. *Break-Even* The profit for a product is given by $P(x) = -15x^2 + 180x - 405$ thousand dollars, where x is the number of tons of product produced and sold. How many tons give break-even (that is, give zero profit) for this product?

53. *Break Even* The total revenue function for a product is given by $R = 550x$ dollars, and the total cost function for this same product is given by $C = 10{,}000 + 30x + x^2$, where C is measured in dollars. For both functions, the input x is the number of units produced and sold.

a. Form the profit function for this product from the two given functions.

b. What is the profit when 18 units are produced and sold?

c. What is the profit when 32 units are produced and sold?

d. How many units must be sold to break even on this product?

54. *Break Even* The total revenue function for a product is given by $R = 266x$, and the total cost function for this same product is $C = 2000 + 46x + 2x^2$, where R and C are each measured in thousands of dollars and x is the number of units produced and sold.

a. Form the profit function for this product from the two given functions.

b. What is the profit when 55 units are produced and sold?

c. How many units must be sold to break even on this product?

55. *Wind and Pollution* The amount of particulate pollution p from a power plant in the air above the plant depends on the wind speed s, among other things, with the relationship between p and s approximated by $p = 25 - 0.01s^2$, with s in miles per hour.

a. Find the value(s) of s that will make $p = 0$.

b. What does $p = 0$ mean in this application?

c. What solution to $0 = 25 - 0.01s^2$ makes sense in the context of this application?

56. *Drug Sensitivity* The sensitivity S to a drug is related to the dosage size x by $S = 100x - x^2$, where x is the dosage size in milliliters.

a. What dosage(s) give zero sensitivity?

b. Explain what your answers in part (a) might mean.

57. *Velocity of Blood* Because of friction from the walls of an artery, the velocity of a blood corpuscle in an artery is greatest at the center of the artery and decreases as the distance r from the center increases. The velocity of the blood in the artery can be modeled by the function

$$v = k(R^2 - r^2)$$

where R is the radius of the artery and k is a constant that is determined by the pressure, viscosity of the blood, and the length of the artery. In the case where $k = 2$ and $R = 0.1$ centimeter, the velocity is $v = 2(0.01 - r^2)$ centimeters per second (cm/sec).

a. What distance r would give a velocity of 0.02 cm/sec?

b. What distance r would give a velocity of 0.015 cm/sec?

c. What distance r would give a velocity of 0 cm/sec? Where is the blood corpuscle?

58. *Body-Heat Loss* The model for body-heat loss depends on the coefficient of convection K, which depends on wind speed s according to the equation $K^2 = 16s + 4$, where s is in miles per hour. Find the positive coefficient of convection when the wind speed is

a. 20 mph. **b.** 60 mph.

c. What is the change in K for a change in speed from 20 mph to 60 mph?

59. *Market Equilibrium* Suppose the demand for artificial Christmas trees is given by the function

$$p = 109.70 - 0.10q$$

and that the supply of these trees is given by

$$p = 0.01q^2 + 5.91$$

where p is the price of a tree in dollars and q is the quantity of trees that are demanded/supplied in hundreds. Find the price that gives market equilibrium price and the number of trees that will be sold/bought at this price.

60. *Market Equilibrium* The demand for a product is given by $p = 7000 - 2x$ dollars, and the supply for this product is given by $p = 0.01x^2 + 2x + 1000$ dollars, where x is the number of units demanded and supplied when the price per unit is p dollars. Find the equilibrium quantity and equilibrium price.

61. *Foreign-Born Population* The percent of the U.S. population that is foreign born can be modeled by the equation $y = 0.003x^2 - 0.42x + 19.8$, where x is the number of years after 1900.

a. Verify graphically that one solution to $7.8 = 0.003x^2 - 0.42x + 19.8$ is $x = 40$.

b. Use this information and factoring to find the year or years when the percent of the U.S. population that is foreign born is 7.8.

c. During what years does this model indicate that the percent of the US population that is foreign born is below 7.8?
(Source: U.S. Census Bureau)

62. *Smoking Cessation* Using data from 1965 to 1990, the percent of people over 19 years of age who have ever smoked and quit is given by the equation $y = 0.010x^2 + 0.344x + 28.74$, where x is the number of years after 1960.

a. One solution to the equation $58.5 = 0.010x^2 + 0.344x + 28.74$ is $x = 40$. What does this mean?

b. To find when after 1960 the percent of people over 19 years of age who have ever smoked and quit is 58.5, do we need to find the second solution to this equation? Why or why not?

c. Graphically verify that $x = 40$ is a solution to $58.5 = 0.010x^2 + 0.344x + 28.74$.
(Source: *Substance Abuse*, Princeton, N.J., 1993)

63. *Baggage Complaints* The number of consumer complaints about baggage against U.S. airlines between 1989 and 1997 is given by

$$C(x) = 41.018x^2 - 428.194x + 1662.14$$

where x is the number of years after 1989.
(Source: *Statistical Abstract of the United States, 1999*)

a. Describe how to set the viewing window so that you can see the important points for this model for the years 1989–1997. Graph the function $y = C(x)$.

b. When did the minimum number of baggage complaints against U.S. airlines occur?

c. Graphically determine the year after 1990 when the number of baggage complaints reached 1482.

d. Is the result in part (c) an interpolation or an extrapolation?

64. *Internet Use* An equation that models the number of users of the Internet is

$$y = 11.786x^2 - 142.214x + 493 \text{ million users}$$

where x is the number of years after 1990.

a. If the pattern indicated by the model is based on data collected between 1990 and 1999, when does this model predict there will be 812 million users?

b. Of what significance in this real-world situation is the answer to part (a)?
(Source: *CyberAtlas*, 1999)

65. *Tourism Spending* The global spending on travel and tourism (in billions of dollars) can be described by the equation $y = 0.442x^2 + 23.126x + 181.178$, where x equals the number of years after 1987. Graphically find the year in which spending is projected to reach \$820.5 billion.
(Source: *Statistical Abstract of the United States, 1999*)

66. *Prison Population* The average daily number of inmates in the Beaufort County, South Carolina, prison can be modeled by

$$I(x) = 0.161x^2 + 7.582x + 82.157$$

where x is the number of years after 1991.

a. How would you set the viewing window if you wanted a complete graph of the function? Graph the function.

b. Graphically estimate when this model predicts that the average daily inmate population will be 185.
(Source: *The Island Packet*, November 1999)

67. *Hospital Admissions* The number of admissions to all types of hospitals in the U.S. between 1985 and 1997 can be described by the function

$$A(t) = 38.228t^2 - 1069.60t + 40{,}698.547$$

thousand people where t is the number of years after 1980.

a. By how much did the number of hospital admissions change between 1990 and 1997?

b. Use this model and graphical methods to find in what year after 1980 the number of admissions was equal to 33,217,000.

c. Use the model to estimate the number of hospital admissions in 2005. What assumptions are you making when you make this estimation?
(Source: American Hospital Association, *Hospital Statistics*)

68. *Smoking Cessation* Using data from 1965 to 1990, the percent of people over 19 years of age who have ever smoked and quit is given by the equation $y = 0.010x^2 + 0.344x + 28.74$, where x is the number of years after 1960. In what year will this model predict y to be over 100 percent, making it certain that the model is invalid?
(Source: *Substance Abuse*, Princeton, N.J., 1993)

69. *Student Aid* Federal funds providing for student financial assistance through programs funded by congressional appropriations between 1980 and 2000 can be described by the model

$$f(x) = 9.032x^2 + 99.970x + 3645.90$$

million dollars where x is the number of years after 1980.

a. Use the model to compare the federal aid available to students in 1990 and 2000.
b. When did the available federal aid reach and exceed $8,000,000,000?
(Source: U.S. Department of Education, National Center for Educational Statistics)

70. *Bomb Threats* The number of bomb threats received against U.S. aircraft between 1980 and 1996 can be approximated by the function $B(t) = 3.268t^2 - 45.733t + 277.910$ threats, where t is the number of years after 1980.
a. What does this model estimate the number of bomb threats to be in 1997?
b. Use graphical or numerical methods with this model to estimate when the number of bomb threats will reach 1177 if this model remained valid.
(Source: *Statistical Abstract of the United States, 1998*)

71. *Quitting Smoking* Using data from 1965 to 1990, the percent of people over 19 years of age who have ever smoked and quit is given by the equation $y = 0.010x^2 + 0.344x + 28.74$, where x is the number of years since 1960.
a. Graph this function for values of x representing 1960–2002.
b. Assuming that the pattern indicated by this model continued through 2002, what would be the percent in 2002?
c. When does this model indicate that the percent reached 50%?
(Source: *Substance Abuse*, Princeton, N.J., 1993)

72. *Unemployment* The unemployment as a percent of the labor force in South Carolina for the years 1982–1992 is given by the table below. The best-fitting quadratic model, with x representing the number of years after 1980, is $y = 0.144x^2 - 2.48x + 15.47$. In what year after 1990 does this model indicate that the unemployment percent is 5.6 percent?

Year	Total Unemployment as a Percent of Labor Force
1982	10.8
1983	10.1
1984	7.1
1985	6.9
1986	5.6
1987	5.6
1988	4.6
1989	4.7
1990	4.7
1991	6.2
1992	6.2

(Source: South Carolina Employment Security Commission, September 1993)

73. *World Population* One projection of the world population by the United Nations (a low-projection scenario) is given in the table below. This data can be modeled by $y = -0.36x^2 + 38.52x + 5822.86$ million people, where x is the number of years from 1990. In what year past 1990 does this model predict the world population will first reach 6,702,000,000?

Year	Projected Population (millions)
1995	5666
2000	6028
2025	7275
2050	7343
2075	6402
2100	5153
2125	4074
2150	3236

(Source: *World Population Prospects: 1998 Revision*, United Nations, 1999)

2.3 A Library of Functions

Key Concepts
- Power functions
- Cubing function
- Root functions
- Reciprocal function
- Vertical asymptote
- Horizontal asymptote
- Piecewise-defined functions
- Step functions
- Absolute value function

Section Preview Advertising

The data in Table 2.3 give the total U.S. cigarette advertising and promotional expenditures, in millions of dollars, for selected years from 1975 to 1993. These data can be modeled by the **power function** $y = 33.036x^{1.623}$, where x is the number of years after 1970 and y is in millions of dollars. The data points and the function are graphed in Figure 2.18.

Power functions are used in applications in the life, social, and physical sciences as well as in business. Power functions are one of a group of special functions that are the basic building blocks for many mathematical applications. In this section, we see how to graph, evaluate, and apply power functions and other nonlinear functions, including root, reciprocal, absolute value, and piecewise-defined functions.

TABLE 2.3

Year	Cigarette Advertising ($ millions)	Year	Cigarette Advertising ($ millions)
1975	491.3	1991	4650.1
1980	1242.3	1992	5231.9
1985	2476.4	1993	6035.4
1990	3992.0		

(Source: *Federal Trade Commission Report to Congress* for 1990–1996)

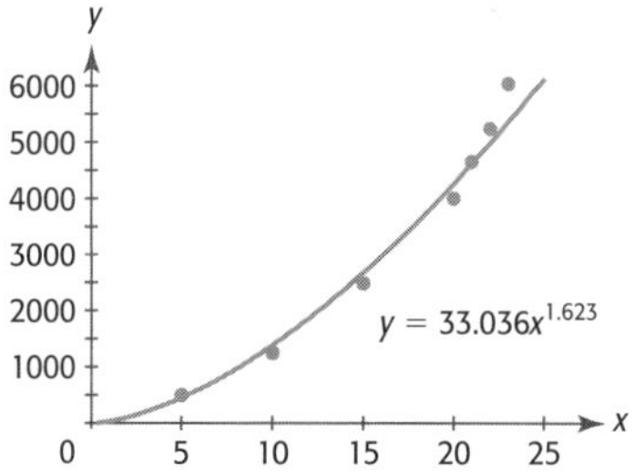

FIGURE 2.18

Power Functions

We define power functions as follows.

POWER FUNCTIONS

A **power function** is a function of the form $y = ax^b$, where a and b are real numbers, $b \neq 0$.

We know that the area of a rectangle is $A = lw$ square units, where l and w are the length and the width, respectively, of the rectangle. Because $l = w$ when the rectangle is a square, the area of a square can be found with the formula $A = x^2$ if each side is x units long. Table 2.4 gives the area of a square for the selected lengths of sides.

TABLE 2.4

Side Length x (units)	Area of Square $A = x^2$ (square units)
1	1
2	4
3	9
4	16
5	25
6	36
7	49

The graph of these data points is shown in Figure 2.19. Because the formula for area applies for any (positive) length of a side, we can graph the model for the area of a square for any nonnegative value of x (Figure 2.20). This is the function $A = x^2$ for $x \geq 0$. A related function is $y = x^2$, defined on the set of all real numbers; this is a power function with power 2. This function is also a basic quadratic function, and its complete graph is a parabola (Figure 2.21).

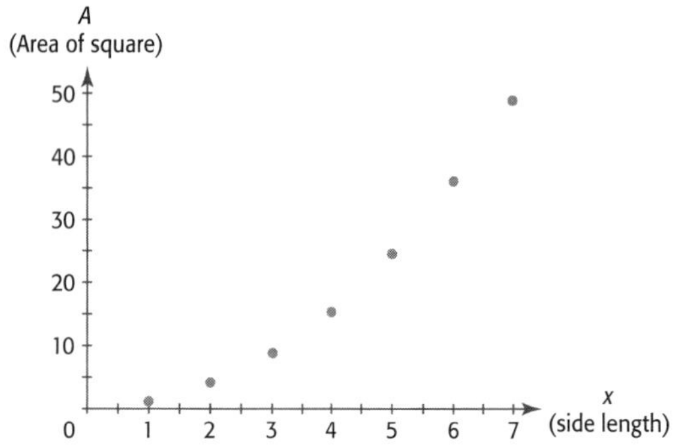

FIGURE 2.19

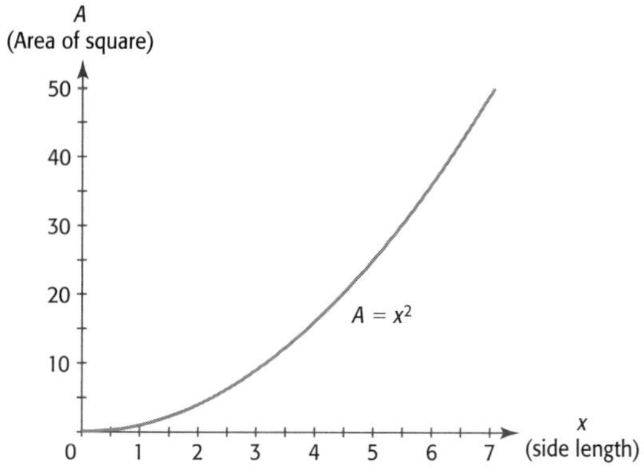

FIGURE 2.20

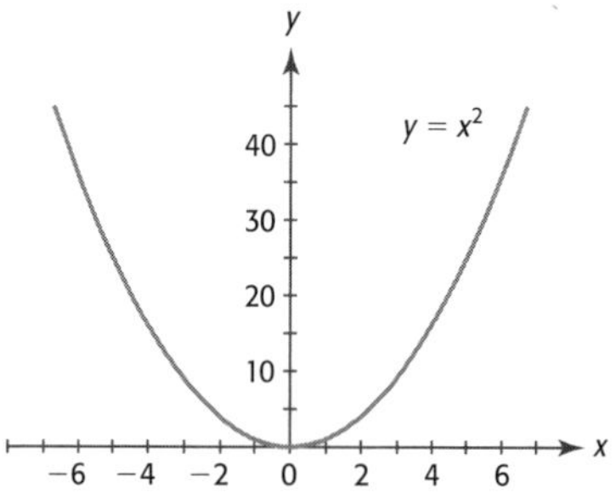

FIGURE 2.21

If we fill a larger cube with cubes that are 1 unit on each edge and have a volume of 1 cubic unit, the number of these small cubes that fit gives the volume of the larger cube. Table 2.5 contains the results of measuring the edges and the volumes of cubes of different sizes and shows the relationship between the edge length and the volume for cubes. It is easily observed that if the length of an edge of a cube is x units, its volume is x^3 cubic units.

TABLE 2.5

Edge Length x (units)	Volume of Cube $V = x^3$ (cubic units)
1	1
2	8
3	27
4	64
5	125
6	216

The graph of these data points is shown in Figure 2.22(a), and the graph of $y = x^3$ for any nonnegative value of x is shown in Figure 2.22(b).

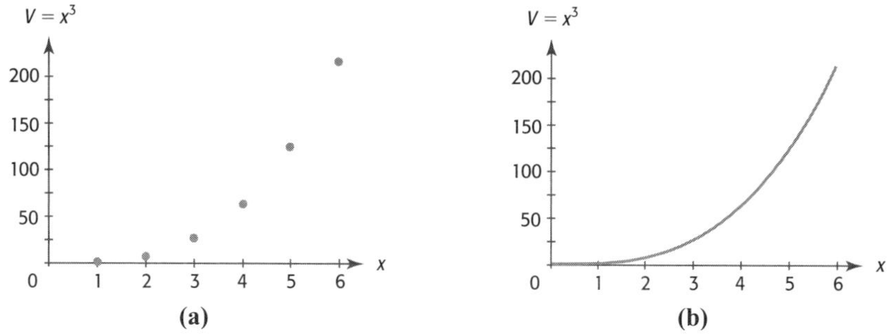

FIGURE 2.22

The related function $y = x^3$, defined for all real numbers x, is called the **cubing function**. A complete graph of this function is shown in Figure 2.23.

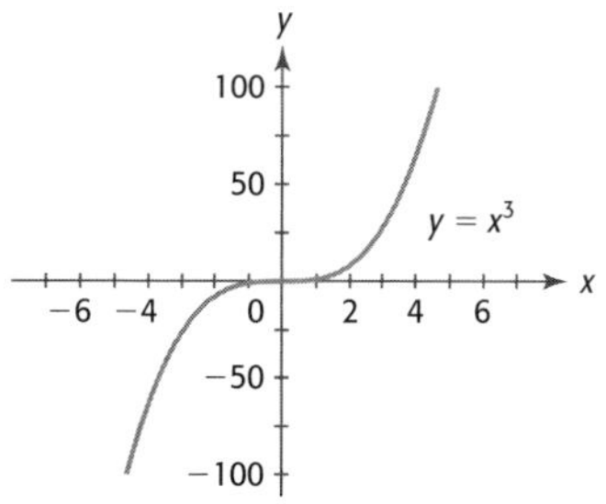

FIGURE 2.23

The functions $y = x, y = x^2$, and $y = x^3$ are examples of power functions with positive integer powers. Additional examples of power functions include functions with noninteger powers of x, like the function in the following example.

EXAMPLE 1 Allometric Relationships

For fiddler crabs, data shows that the relationship between the weight C of the claw and the weight W of the body is given by

$$C = 0.11W^{1.54}$$

with W and C in grams.

(Source: d'Arcy Thompson, *On Growth and Form*, Cambridge University Press, 1961)

a. Graph this function in the context of this application.

b. Use this function to compute the weight of a claw if the weight of the crab is 5 grams.

Solution

a. It is appropriate to graph this model only in the first quadrant because $C > 0$ and $W > 0$ in the problem context. The graph is shown in Figure 2.24.

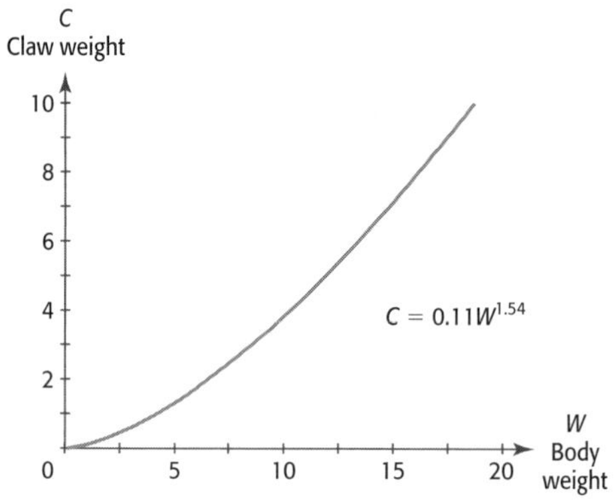

FIGURE 2.24

b. We find the weight of the claw by evaluating the function with an input of 5, giving an output of $C = 0.11(5)^{1.54} \approx 1.312$. Thus, the weight of the claw is approximately 1.3 grams. ■

Note that the power in $C = 0.11W^{1.54}$ is 1.54, which can be written in the form $\frac{154}{100}$ and is therefore a rational number. The power function $y = x^{2/3}$ has the rational exponent 2/3; its graph is shown in Figure 2.25.

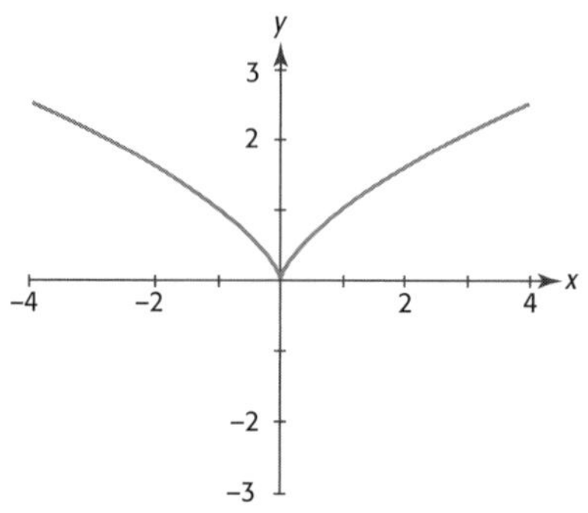

FIGURE 2.25

Functions with rational powers can also be written with radicals. For example, $y = x^{1/3}$ can be written in the form $y = \sqrt[3]{x}$, and $y = x^{1/2}$ can be written in the form $y = \sqrt{x}$. Functions like $y = \sqrt[3]{x}$ and $y = \sqrt{x}$ are special power functions called **root functions**.

ROOT FUNCTIONS

A root function is a function of the form $y = ax^{1/n}$, or $y = a\sqrt[n]{x}$, where n is an integer, $n \geq 2$.

The graphs of $y = \sqrt{x}$ and $y = \sqrt[3]{x}$ are shown in Figures 2.26 and 2.27. Note that the domain of $y = \sqrt{x}$ is $x \geq 0$ because $\sqrt{x}$ is undefined for negative values of x. Note also that these root functions can be written as power functions with powers less than 1. The graphs of these functions increase in the first quadrant but not as fast as $y = x$ does for values of x greater than 1, so we say that they are *concave down* in the first quadrant.

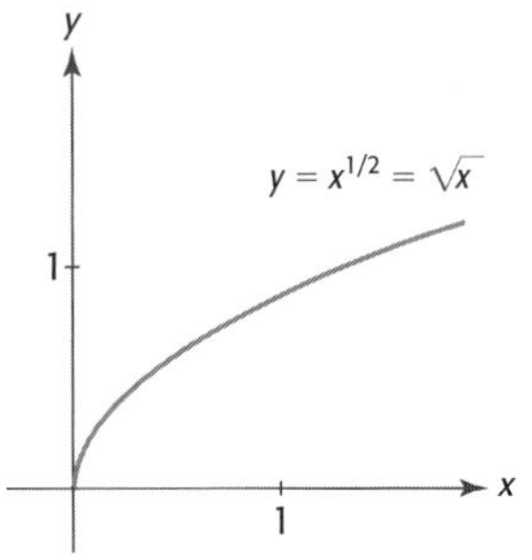

FIGURE 2.26

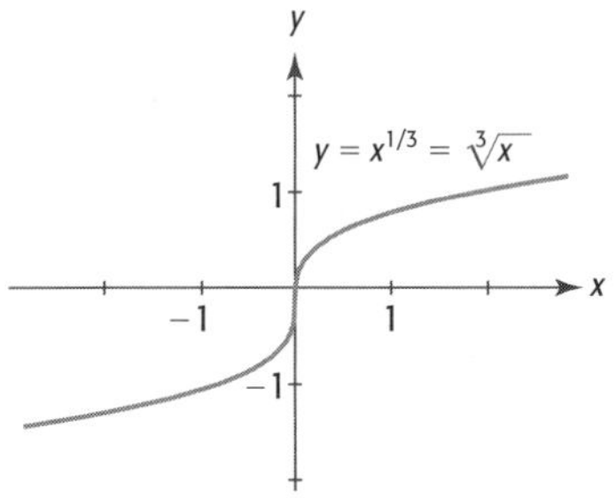

FIGURE 2.27

In general, if $a > 0$ and $x > 0$, the graph of $y = ax^b$ is concave up if $b > 1$ and concave down if $0 < b < 1$. Figures 2.28(a) and 2.28(b) show the first-quadrant portion of typical graphs of $y = ax^b$ for $a > 0$ and different values of b.

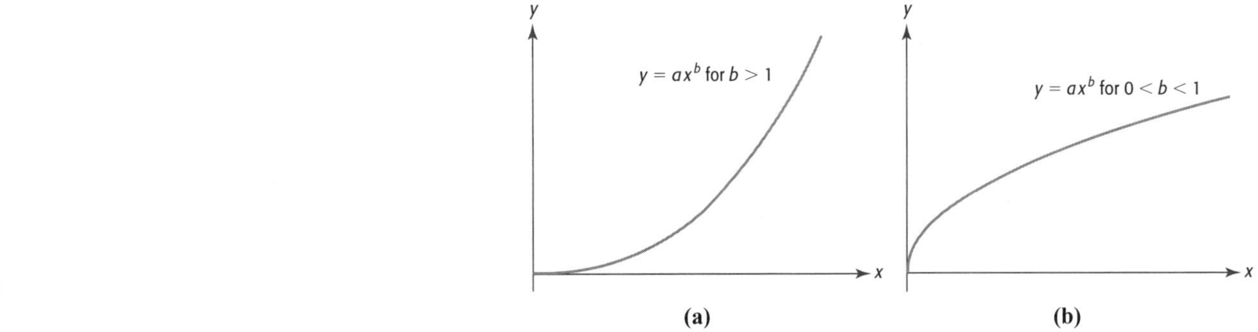

FIGURE 2.28

Reciprocal Function

The function formed by the quotient of $f(x) = 1$ and the identity function $g(x) = x$ is the **reciprocal function** with equation

$$y = \frac{1}{x}$$

The graph of the reciprocal function, shown in Figure 2.29, is called a **rectangular hyperbola**. Note that x cannot equal 0, so 0 is not in the domain of this function. As the

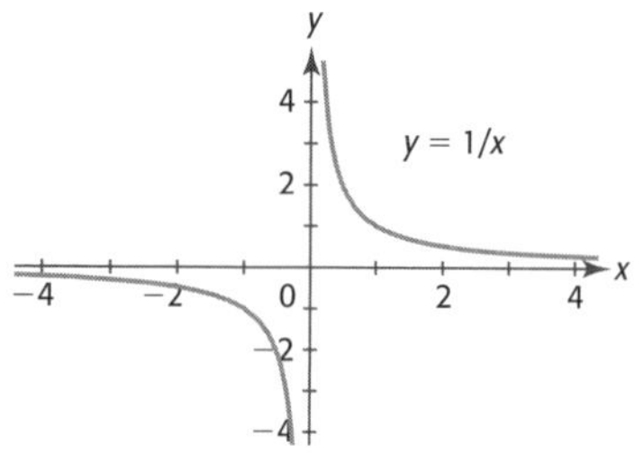

FIGURE 2.29

graph of $y = \frac{1}{x}$ shows, as x gets close to 0, $|y|$ gets large and the graph approaches but does not touch the y-axis. We say that the y-axis is a **vertical asymptote** of this graph.

> **VERTICAL ASYMPTOTE**
>
> If $|y|$ gets large without bound as x approaches a, then $x = a$ is a vertical asymptote of the graph of $y = f(x)$. We can denote this by
>
> $$f(x) \to \infty \text{ or } f(x) \to -\infty \text{ as } x \to a \text{ from the left or right.}$$

Note also that the values of $y = \frac{1}{x}$ get very small as $|x|$ gets large, and the graph approaches but does not touch the x-axis. We say that the x-axis is a **horizontal asymptote** for this graph.

> **HORIZONTAL ASYMPTOTE**
>
> If the values of y approach some finite number b as $|x|$ becomes very large, we say that the graph of $y = f(x)$ has a horizontal asymptote at $y = b$. We can denote this as
>
> $$f(x) \to b \text{ as } x \to \infty \text{ or } x \to -\infty.$$

In the following example, we consider an application of a reciprocal function.

EXAMPLE 2 Average Cost

The average cost per hat for producing baseball hats is $\overline{C}(x) = 4.50 + \frac{1}{x}$ dollars, where x is the number of units produced.

a. Determine the domain of this function without concern for the context of the application and graph the function on a standard window.

b. Use knowledge of the context of the problem to graph the function on a window that applies to the application.

c. What is the horizontal asymptote of the graph? What does this tell you about how low the average cost can be?

Solution

a. All values except $x = 0$ result in real values for the function. Thus, the domain of $\overline{C}(x)$ is all real numbers except 0. The graph, shown in Figure 2.30(a), has the same shape as the graph of $y = \frac{1}{x}$, shown in Figure 2.29.

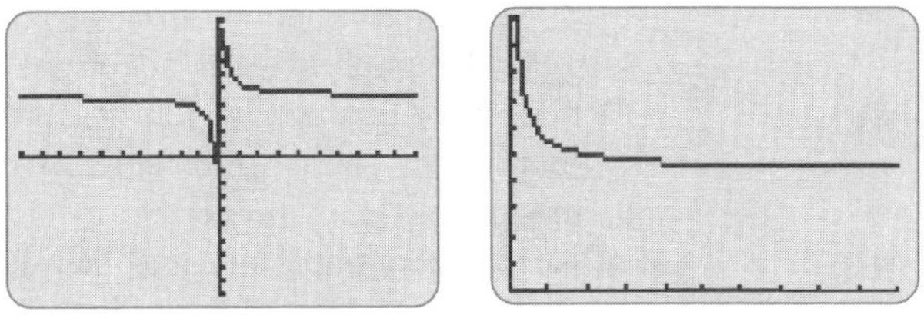

[−10, 10] by [−10, 10]
(a)

[0, 10] by [0, 10]
(b)

FIGURE 2.30

b. Because x represents the number of units produced, the viewing window is set for values of $x \geq 0$. The average cost cannot be negative, so the window is set with $\overline{C}(x) \geq 0$. Keeping the upper limits at 10 gives the graph shown in Figure 2.30(b).

c. The horizontal asymptote is $y = 4.50$. This means that the average cost approaches \$4.50 as the number of hats produced increases. ■

TABLE 2.6

Weight x (oz)	Postage y (¢)
$0 < x \leq 1$	39
$1 < x \leq 2$	63
$2 < x \leq 3$	87
$3 < x \leq 4$	111

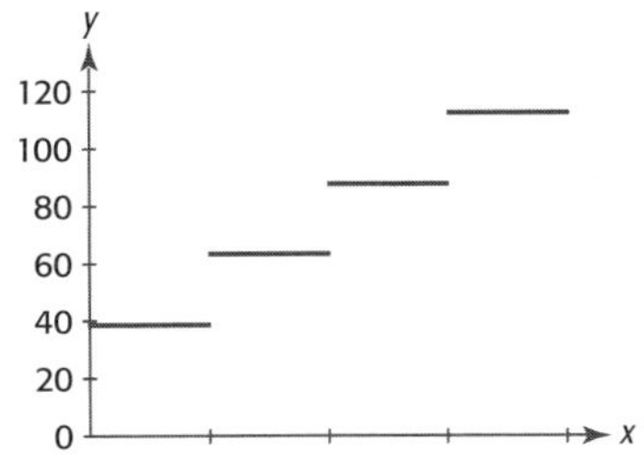

FIGURE 2.31

Piecewise-Defined Functions

It is possible that a set of data cannot be modeled with a single equation. For example, the postage paid for a first-class letter "jumps" by 24¢ for each ounce after the first ounce but does not increase until the weight increases by 1 ounce. (See Table 2.6, which gives the postage for letters up to 4 ounces.)

The function that models the price of postage for the first 4 ounces is

$$\begin{cases} 39 & \text{if } 0 < x \leq 1 \\ 63 & \text{if } 1 < x \leq 2 \\ 87 & \text{if } 2 < x \leq 3 \\ 111 & \text{if } 3 < x \leq 4 \end{cases}$$

This is an example of a **piecewise-defined function**, so named because it is defined with different pieces for different parts of its domain rather than one equation. As its graph shows in Figure 2.31, each output is a constant with discontinuous "steps" at 1, 2, and 3. This is a special piecewise-defined function, called a **step function**. Other piecewise-defined functions can be defined by polynomial functions over limited domains.

EXAMPLE 3 Piecewise-Defined Function

Graph the function

$$f(x) = \begin{cases} 5x + 2 & \text{if } 0 \leq x < 3 \\ x^3 & \text{if } 3 \leq x \leq 5 \end{cases}$$

Solution

To graph this function, we can construct tables of values for each of the pieces. Table 2.7(a) gives outputs of the function for some sample inputs x on the interval $[0, 3)$; on this interval, $f(x)$ is defined by $f(x) = 5x + 2$. Table 2.7(b) gives outputs for some sample inputs on the interval $[3, 5]$; on this interval $f(x)$ is defined by $f(x) = x^3$.

TABLE 2.7(a)

x	0	1	2	2.99
$f(x)$	2	7	12	16.95

TABLE 2.7(b)

x	3	4	5
$f(x)$	27	64	125

Plotting the points from Table 2.7(a) and connecting them with a smooth curve gives the graph of $y = f(x)$ on the x-interval $[0, 3)$ (Figure 2.32). The open circle on this piece of the graph indicates that this piece of the function is not defined for $x = 3$. Plotting the points from Table 2.7(b) and connecting them with a smooth curve gives the graph of $y = f(x)$ on the x-interval $[3, 5]$ (Figure 2.32). The closed circle on this piece of the graph indicates that this piece of the function is defined for $x = 3$. The graph of $y = f(x)$, shown in Figure 2.32, consists of these two pieces.

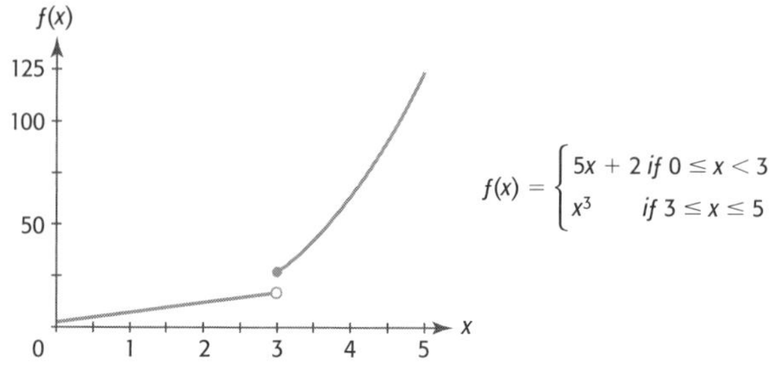

FIGURE 2.32

EXAMPLE 4 Residential Power Costs

Excluding fuel adjustment costs and taxes, Georgia Power Company charges its residential power customers for electricity during the months of June through September according to Table 2.8.

TABLE 2.8

Monthly Kilowatt-Hours (kWh)	Monthly Charge
0 to 650	\$7.50 plus \$0.04783 per kWh
More than 650, up to 1000	\$38.59 plus \$0.07948 per kWh above 650
More than 1000	\$66.41 plus \$0.08184 per kWh above 1000

a. Write the piecewise-defined function C that gives the monthly charge for residential electricity, with input x equal to the monthly number of kilowatt-hours.

b. Find $C(950)$ and explain what it means.

c. Find the charge for using 1560 kWh in a month.

Solution

a. The monthly charge is given by the function

$$C(x) = \begin{cases} 7.50 + 0.04783x & \text{if } 0 \le x \le 650 \\ 38.59 + 0.07948(x - 650) & \text{if } 650 < x \le 1000 \\ 66.41 + 0.08184(x - 1000) & \text{if } x > 1000 \end{cases}$$

where $C(x)$ is the charge in dollars for x kWh of electricity.

b. To evaluate $C(950)$, we must determine which "piece" defines the function when $x = 950$. Because 950 is between 650 and 1000, we use the "middle piece" of the function:

$$C(950) = 38.59 + 0.07948(950 - 650) = 62.434$$

Companies regularly round charges *up* to the next cent if any part of a cent is due. This means that if 950 kWh are used in a month, the bill is \$62.44.

c. To find the charge for 1560 kWh, we evaluate $C(1560)$, using the "bottom piece" of the function because $1560 > 1000$. Evaluating this gives

$$C(1560) = 66.41 + 0.08184(1560 - 1000) = 112.2404$$

so the charge for the month is \$112.25. ■

Absolute Value Function

We can construct a new function by combining two functions into a special piecewise-defined function. For example, we can write a function with the form

$$f(x) = \begin{cases} x & \text{if } x \geq 0 \\ -x & \text{if } x < 0 \end{cases}$$

This function is called the **absolute value function**, which is denoted by $f(x) = |x|$ and is derived from the definition of the absolute value of a number. Recall that the definition of the absolute value of a number is

$$|x| = \begin{cases} x & \text{if } x \geq 0 \\ -x & \text{if } x < 0 \end{cases}$$

To graph $f(x) = |x|$, we graph the portion of the line $y_1 = x$ for $x \geq 0$ (Figure 2.33(a)) and the portion of the line $y_2 = -x$ for $x < 0$ (Figure 2.33(b)). When these pieces are joined on the same graph (Figure 2.33(c)), we have the graph of $y = |x|$.

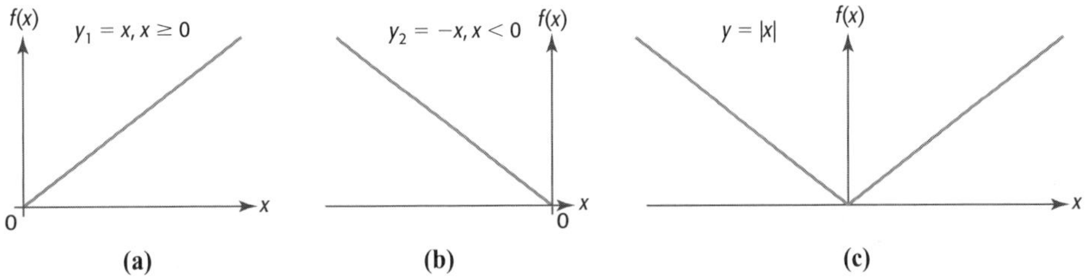

FIGURE 2.33

2.3 SKILLS CHECK

In Exercises 1–10, sketch the graph of each function using the window $[-5, 5]$ *by* $[-5, 5]$.

1. $y = x^3$

2. $y = x^4$

3. $y = x^{1/3}$

4. $y = x^{4/3}$

5. $y = \sqrt{x} + 2$

6. $y = \sqrt[3]{x} - 2$

7. $y = \dfrac{1}{x} - 3$

8. $y = 4 - \dfrac{1}{x}$

9. $y = \begin{cases} -1 & \text{if } x < 0 \\ 1 & \text{if } x \geq 0 \end{cases}$

10. $y = \begin{cases} 2 & \text{if } x \geq 2 \\ 6 & \text{if } x < 2 \end{cases}$

11. a. Graph $f(x) = \begin{cases} 4x - 3 & \text{if } x \leq 3 \\ x^2 & \text{if } x > 3 \end{cases}$.

b. Find $f(2)$ and $f(4)$.

c. State the domain of the function.

12. a. Graph $f(x) = \begin{cases} 3 - x & \text{if } x \le 2 \\ x^2 & \text{if } x > 2 \end{cases}$.
 b. Find $f(2)$ and $f(3)$.
 c. State the domain of the function.

13. a. Graph $f(x) = |x|$.
 b. Find $f(-2)$ and $f(5)$.
 c. State the domain of the function.

14. a. Graph $f(x) = |x - 4|$.
 b. Find $f(-2)$ and $f(5)$.
 c. State the domain of the function.

For each of the functions in Exercises 15–26, find the value of (a) $f(-1)$ and (b) $f(3)$, if possible.

15. $f(x) = (x - 2)^3$

16. $f(x) = x^2 + 3$

17. $f(x) = \sqrt{x - 2}$

18. $f(x) = \sqrt[3]{x^2 - 2}$

19. $y = \dfrac{1}{x - 2}$

20. $y = \dfrac{1}{x + 5}$

21. $y = \dfrac{1}{x - 4} - 9$

22. $y = \dfrac{1}{x + 3} + 6$

23. $y = \begin{cases} 5 & \text{if } x \le 1 \\ 6 & \text{if } x > 1 \end{cases}$

24. $y = \begin{cases} -2 & \text{if } x < -1 \\ 4 & \text{if } x \ge -1 \end{cases}$

25. $y = \begin{cases} x^2 - 1 & \text{if } x \le 0 \\ x^3 + 2 & \text{if } x > 0 \end{cases}$

26. $y = \begin{cases} 3x + 1 & \text{if } x < 3 \\ x^2 & \text{if } x \ge 3 \end{cases}$

27. Determine if the function $y = 4x^3$ is increasing or decreasing for
 a. $x < 0$. b. $x > 0$.

28. Determine if the function $y = -3x^4$ is increasing or decreasing for
 a. $x < 0$. b. $x > 0$.

For each of the functions in Exercises 29–32, determine if the function is concave up or concave down in the first quadrant.

29. $y = x^{1/2}$

30. $y = x^{3/2}$

31. $y = x^{1.4}$

32. $y = x^{0.6}$

33. Graph $f(x) = \begin{cases} x & \text{if } x \ge 0 \\ -x & \text{if } x < 0 \end{cases}$.

34. Compare the graph in Exercise 33 with the graph in Exercise 13(a).

2.3 EXERCISES

35. *Suicides* The number of suicides can be modeled by the function $f(x) = 20{,}000x^{0.11}$, where x is the number of years after 1960.
 a. What type of function is this?
 b. What is $f(5)$? What does this mean?
 c. How many suicides does this model estimate for 1982?
 (Source: National Center for Health Statistics, for selected years 1962–1996)

36. *Single-Parent Families* The percent of all families that were single-parent families between 1960 and 1998 was found to be modeled by $y = 1.053x^{0.888}$, with $x = 0$ in 1950.
 a. What is $f(45)$? What does this mean?
 b. Does this model indicate that the percent of single-parent families increased or decreased during the period from 1960 to 1998?

37. *Taxi Miles* The Inner City Taxi Company estimated, on the basis of collected data, that the number of taxi miles driven each day can be modeled by the function $Q = 489L^{0.6}$, when they employ L drivers per day.
 a. Graph this function for $0 \le L \le 35$.
 b. How many taxi miles are driven each day if there are 32 drivers employed?
 c. Does this model indicate that the number of taxi miles increases or decreases as the number of drivers increases? Is this reasonable?

38. *Single-Parent Families* The percent of all families that were single-parent families between 1960 and 1998 can be modeled by $y = 1.053x^{0.888}$, with $x = 0$ in 1950.
 a. According to the model, what percent of all families were estimated to have been single-parent families in 1975?
 b. What percent of all families did the model predict would be single-parent families in 2005? Comment on the reliability of your answer.

c. Is this function concave up or concave down during this period of time?

d. Use numerical or graphical methods to find when the model predicts that the percent will be 40.
(Source: U.S. Census Bureau)

39. *Mutual Funds* The household assets in mutual funds, measured in billions of dollars, for the years 1995–1999 can be modeled by $f(x) = 105.095x^{1.5307}$ billion dollars, where x is the number of years after 1990.

a. Is this function increasing or decreasing from 1995 to 1999?

b. Is this function concave up or concave down during this period of time?

c. Use numerical or graphical methods to find when the model predicts that the assets will reach \$4,000,000,000,000.
(Source: Federal Reserve, *Time*, April 3, 2000)

40. *U.S. Population* The U.S. population can be modeled by the function $y = 165.6x^{1.345}$, where y is in thousands and x is the number of years after 1800.

a. What was the population in 1960, according to this model?

b. Is the graph of this function concave up or concave down?

c. Use numerical or graphical methods to find when the model estimates the population to be 92,370,000.

41. *Production Output* The monthly output of a product (in units) is given by $P = 1200x^{5/2}$, where x is the capital investment in thousands of dollars.

a. Graph this function for x from 0 to 10 and P from 0 to 200,000.

b. Is the graph concave up or concave down?

42. *Harvesting* A farmer's main cash crop is tomatoes, and the tomato harvest begins in the month of May. The number of bushels of tomatoes harvested on the xth day of May is given by the equation $B(x) = 6(x + 1)^{3/2}$. How many bushels did the farmer harvest on May 8?

43. *Voter Turnout* The function $y = 92.239x^{-0.1585}$ gives the percent of voter turnout during presidential election years, with x representing the number of years after 1950.

a. Graph the function for $x > 0$.

b. Does this model indicate that the percent of voter turnout is increasing or decreasing?

c. Use the graph to determine the percent in 1996.

d. What percent voter turnout did this model estimate for the 2000 election?

e. Did the 2000 turnout agree with the prediction from the model?
(Source: Federal Election Commission)

44. *Trust in the Government* The percent of people who say they trust the government in Washington always or most of the time is given by $y = 154.131x^{-0.492}$, with x equal to the number of years after 1960.

a. Graph the function.

b. Does this model indicate that trust in the government is increasing or decreasing?

c. Use the graph to estimate the percent of people who trust the government in 1998.
(Source: Pew Research Center)

45. *Average Cost* The monthly average cost of producing 27-inch television sets is $C(x) = 105 + \dfrac{50{,}000}{x}$, where x is the number of sets produced per month. What is the average cost per set if 2000 sets are produced?

46. *Cost Benefit* Suppose for a certain city the cost C of obtaining drinking water that contains $p\%$ impurities (by volume) is given by

$$C = \frac{120{,}000}{p} - 1200$$

a. What is the cost of drinking water that is 100% impure?

b. What is the cost of drinking water that is 50% impure?

47. *First-Class Postage* The postage charged for first-class mail is a function of its weight. The U.S. Postal Service uses the following table to describe the rates for 2005.

Weight Increment x(oz)	First-Class (¢) Postage $P(x)$
First ounce or fraction of an ounce	37
Each additional ounce or fraction	23

(Source: pe.usps.gov/text)

a. Convert this table to a piecewise-defined function that represents first-class postage for letters weighing up to 4 ounces, using x as the weight in ounces and P the postage in cents.

b. Find $P(1.2)$ and explain what it means.

c. Give the domain of P as it is defined above.

d. Find $P(2)$ and $P(2.01)$.

e. Find the postage for a 2-ounce letter and for a 2.01-ounce letter.

48. *Postal Rates* The table below gives the local postal rates as a function of the weight of the printed matter that is mailed. Write a step function that gives the postage P as a function of the weight in pounds x for $1.5 \le x < 5$.

Weight (lb)	Postal Rate ($)
$1.5 \le x < 2$	1.14
$2 \le x < 3$	1.16
$3 \le x < 4$	1.20
$4 \le x < 5$	1.24

(Source: 1999 *World Almanac*)

49. *Income Tax* The 2000 U.S. federal income tax owed by a married couple filing jointly can be found from Schedule Y-1.

Schedule Y-1—Use if your filing status is **Married filing jointly** or **Qualifying widow(er)**

If the amount on Form 1040, line 39, is: *Over—*	*But not over—*	Enter on Form 1040, line 40	*of the amount over—*
$0	$43,850	----- 15%	$0
43,850	105,950	$6,577.50 + 28%	43,850
105,950	161,450	23,965.50 + 31%	105,950
161,450	288,350	41,170.50 + 36%	161,450
288,350	-----	86,854.50 + 39.6%	288,350

(Source: Internal Revenue Service, 2000, Form 1040)

a. Write the piecewise-defined function T with input x that models the federal tax dollars owed as a function of x, the taxable income dollars earned, with $0 < x \le 105{,}950$.
b. Use the function to find $T(42{,}000)$.
c. Find the tax owed on a taxable income of $55,000.
d. A friend tells Jack Waddell not to earn any money over $43,850 because it would raise his tax rate to 28% on all of his taxable income. Test this statement by finding the tax on $43,850 and $43,850 + $1. What do you conclude?

50. *Federal Support for Education* The federal on-budget funds for all educational programs (in millions of constant 2000 dollars) between 1965 and 2000 can be modeled by the function

$$P(t) = \begin{cases} 1.965t - 5.65 & \text{when } 5 \le t \le 20 \\ 0.095t^2 - 2.925t + 54.429 & \text{when } 20 < t \le 40 \end{cases}$$

where t is the number of years after 1960.

a. Graph the function P for $5 \le t \le 40$. Describe how the funding for educational programs varied between 1965 and 2000.
b. What was the amount of funding for educational programs in 1980?
c. How much federal funding was allotted for educational programs in 1998?
(Source: U.S. Department of Education, National Center for Educational Statistics)

51. *Average Cost* The monthly average cost of producing 32-inch plasma television sets is given by $\overline{C}(x) = \dfrac{100{,}000}{x} + 150$ dollars, where x is the number of units produced.
a. Graph this function for $x > 0$.
b. Will the average cost function decrease or increase as the number of units produced increases?

52. *Average Cost* The monthly average cost of producing x sofas is $\overline{C}(x) = \dfrac{50{,}000}{x} + 120$ dollars.
a. Graph this function for $x > 0$.
b. Will the average cost function decrease or increase as the number of units produced increases?

53. *Concentration of Body Substances* The concentration C of a substance in the body depends on the quantity of substance Q and the volume V through which it is distributed. For a static substance, the concentration is given by

$$C = \frac{Q}{V}$$

a. For $Q = 1000$ milliliters (mL) of a substance, graph the concentration as a function of the volume on the interval from $V = 1000$ mL to $V = 5000$ mL.
b. For a fixed quantity of a substance, does the concentration of the substance in the body increase or decrease as the volume through which it is distributed increases?

54. *Pollution* The daily cost C (in dollars) of removing pollution from the smokestack of a coal-fired electric power plant is related to the percent of pollution p being removed according to the equation $100C - Cp = 10{,}500$.
a. Solve this equation for C to write the daily cost as a function of p, the percent of pollution removed.
b. What is the daily cost of removing 50% of the pollution?
c. Why would this company probably resist removing 99% of the pollution?

2.4 Quadratic and Power Models

Key Concepts

- Modeling quadratic functions
- Modeling power functions
- Comparison of models; first and second differences
- Modeling with power functions
- Comparison of quadratic and power models

Section Preview Cocaine Addicts

Table 2.9 shows that the number of registered cocaine addicts rocketed in the United Kingdom for the years 1985–1995. The scatter plot of these data, shown in Figure 2.34, shows that a line would not fit these data points well, so a linear function is not a good model for this data. The use of graphing utilities permits us to model nonlinear data with other types of functions by using steps similar to those that we used to model data with linear functions. In this section, we model sets of data with quadratic and power functions.

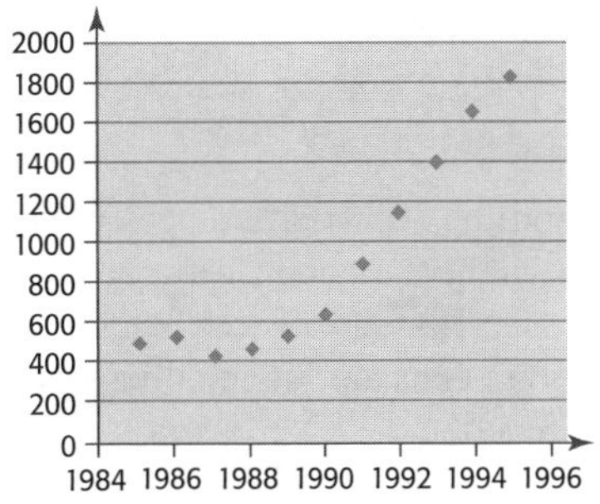

FIGURE 2.34

TABLE 2.9 Cocaine Use in the United Kingdom

Year	Number of Registered Cocaine Addicts	Year	Number of Registered Cocaine Addicts
1985	490	1991	882
1986	520	1992	1131
1987	431	1993	1375
1988	462	1994	1636
1989	527	1995	1809
1990	633		

(Source: www.ukcia.org/lib/stats/ukstats.htm, 4/23/2001)

Modeling with Quadratic Functions

If the graph of a set of data has a pattern that approximates the shape of a parabola or part of a parabola, a quadratic function may be appropriate to model the data. The scatter plot in Figure 2.34 has this shape.

EXAMPLE 1 Cocaine Use in the United Kingdom

Table 2.9 gives the number of registered cocaine addicts in the United Kingdom for the years 1985 to 1995.

a. Create a quadratic function that models the data, using the number of years after 1985 as the input x.

b. Graph the aligned data and the quadratic function on the same axes. Does this model seem like a reasonable fit?

c. Use the model to estimate the number of registered addicts in 1991. Is this estimate close to the actual number?

d. Use the model to estimate the number in 2000. Discuss the reliability of this estimate.

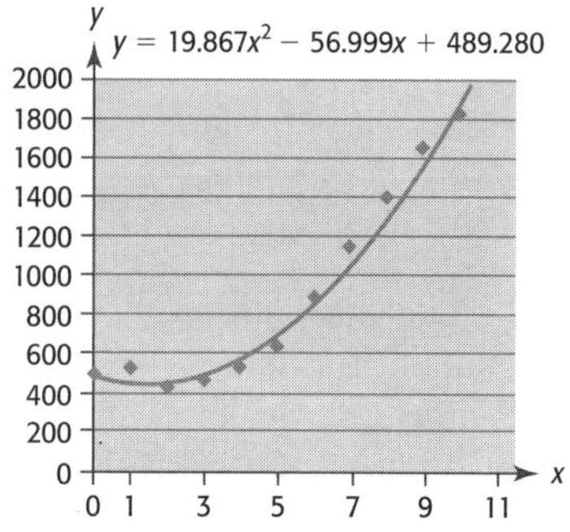

FIGURE 2.35

Solution

a. Figure 2.34 shows the scatter plot of the data. Enter the aligned inputs 0 through 10 in a graphing utility and enter the corresponding output values from the second column of the table. Using quadratic regression in a graphing utility gives a quadratic function that models the data. The function, with coefficients rounded to three decimal places, is

$$y = 19.867x^2 - 56.999x + 489.280$$

where x is the number of years from 1985. Note that using a rounded model in calculations may result in slight calculation errors.

b. The scatter plot of the aligned data and the graph of the function are shown in Figure 2.35. The function appears to be a good fit to the data.

c. Evaluating the function at $x = 6$ gives the number of registered drug addicts as 862 in 1991. This is a reasonable estimate of 882, the actual number.

d. Evaluating the function at $x = 15$ gives the number of registered drug addicts as 4104 in 2000. This estimate is an extrapolation that is 5 years after the last data, and many factors may have changed the pattern that predicts this number. Thus, the number is not very reliable. Finding and including data points for more recent years would give a new model that would be a better predictor for 2000. ■

Comparison of Linear and Quadratic Models

Recall that when the changes in inputs are constant and the (first) differences of the outputs are constant or nearly constant, a linear model will give a good fit for the data. In a similar manner, we can compare the differences of the first differences, which are called the **second differences**. If the second differences are constant for equally spaced inputs, the data can be modeled exactly by a quadratic function.

Consider the data in Table 2.10, which gives the measured height y of a toy rocket x seconds after it has been shot into the air from the ground.

TABLE 2.10 **Height of a Rocket**

Time x (seconds)	Height (meters)
1	68.6
2	117.6
3	147
4	156.8
5	147
6	117.6
7	68.6

Table 2.11 gives the first differences and second differences for the equally spaced inputs for the rocket height data in Table 2.10.

TABLE 2.11

Outputs	68.6		117.6		147		156.8		147		117.6		68.6
First Differences		49		29.4		9.8		−9.8		−29.4		−49	
Second Differences			**−19.6**		**−19.6**		**−19.6**		**−19.6**		**−19.6**		

The first differences are not constant, but each of the second differences is -19.6, which indicates that the data can be fit exactly by a quadratic formula. We can find that the quadratic model for the height of the toy rocket is $y = 78.4x - 9.8x^2$ meters, where x is the time in seconds.

EXAMPLE 2 Aid to Dependent Children

Table 2.12 gives the number of families (in thousands) who were recipients under the Federal Aid to Families with Dependent Children program for the years 1990 to 1996.

a. Find the first and second differences for the data to justify that a quadratic function will be a better model for the data than a linear function.

b. Align the data so that the input is the number of years from 1990 and create a quadratic function that models the data.

c. Graph the aligned data and the quadratic function on the same axes.

d. What does the model give as the maximum number of families who were recipients of federal aid during the period 1990–1996?

TABLE 2.12

Year	1990	1991	1992	1993	1994	1995	1996
Recipient Families (thousands)	4218	4708	4936	5050	4979	4641	4166

(Source: U.S. Census Bureau, *Statistical Abstract of the United States*)

Solution

a. The inputs are equally spaced, and the first and second differences for the data are shown in Table 2.13. The second differences are closer to being constant than the first differences, so a quadratic model will be a better fit for the data.

TABLE 2.13

Outputs	4218		4708		4936		5050		4979		4641		4166
First Differences		490		228		114		−71		−338		−475	
Second Differences			**−262**		**−114**		**−185**		**−267**		**−137**		

b. Enter the aligned inputs 0, 1, 2, 3, 4, 5, and 6 in a graphing utility and enter the corresponding output values from the second row of Table 2.12. Using quadratic regression in a graphing utility gives a quadratic function that models the data. The function, with the coefficients rounded to four decimal places, is

$$y = -95.5357x^2 + 564.3929x + 4219.9286$$

c. The scatter plot of the aligned data and the graph of the function are shown in Figure 2.36(a). We see that a quadratic model fits the data well even though the second differences are not constant.

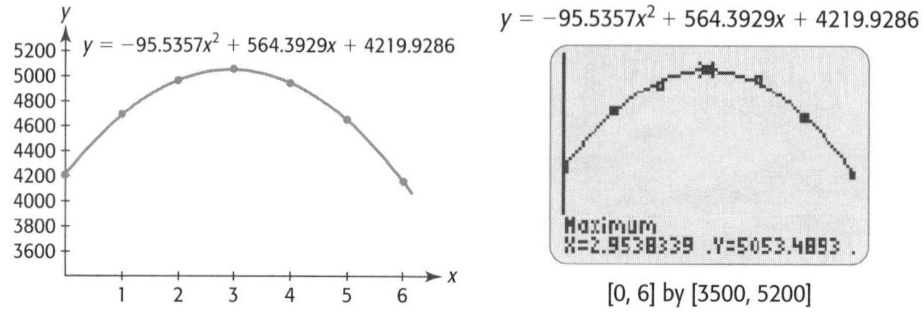

(a) (b)

FIGURE 2.36

d. The maximum value of the rounded function occurs at the vertex of the parabola shown in Figure 2.36(a). The x-coordinate of this vertex is $x = \dfrac{-564.3929}{2(-95.5357)} = 2.9538 \approx 3$, and the y-coordinate is $y \approx 5053$. Thus, the model gives 5053 thousand, or 5,053,000 families as the maximum number of families who received federal aid under this program in $1990 + 3 = 1993$. This can be verified by finding the maximum point on the graph with technology (Figure 2.36(b)). ■

Spreadsheet Solution

We can use graphing calculators, software programs, and spreadsheets to find the quadratic function that is the best fit for data. Table 2.14 shows the Excel spreadsheet for the aligned data of Example 2. Selecting the cells containing the data, using Chart Wizard to get the scatter plot of the data, and selecting Add Trendline and picking Polynomial with order 2 gives the equation of the quadratic function that is the best fit for the data, along with the scatter plot and the graph of the best-fitting parabola. The equation and graph are shown in Figure 2.37 on the next page.

TABLE 2.14

	A	B
1	Year x	Recipient Families (thousands) y
2	0	4218
3	1	4708
4	2	4936
5	3	5050
6	4	4979
7	5	4641
8	6	4166

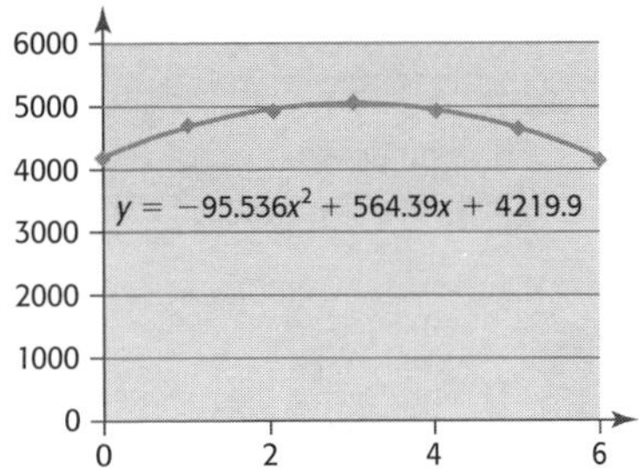

FIGURE 2.37

Modeling with Power Functions

We can model some experimental data by observing patterns. For example, by observing how the area of each of the (square) faces of a cube is found and that a cube has six faces, we can deduce that the surface area of a cube that is x units on each edge is

$$S = 6x^2 \text{ square units}$$

We can also measure and record the surface areas for cubes of different sizes to investigate the relationship between the edge length and the surface areas for cubes. Table 2.15 contains selected measures of edges and the resulting surface areas.

TABLE 2.15

Edge length x (units)	Surface Area of Cube (square units)
1	6
2	24
3	54
4	96
5	150

We can enter the lengths from the table as the independent (x) variable and the corresponding surface areas as the dependent (y) variable in a graphing utility, and then have the utility create the **power function** that is the best model for the data (Figure 2.38). This model also has the equation $y = 6x^2$ square units.

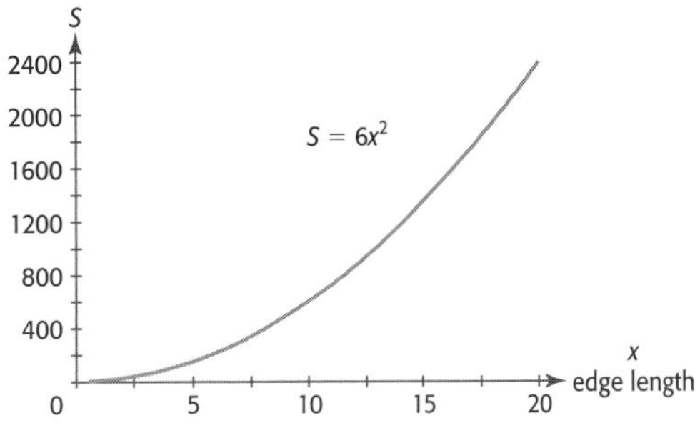

FIGURE 2.38

EXAMPLE 3 Auto Noise

The noise level of a Vauxhall VX220 increases as the speed of the car increases. Table 2.16 gives the noise, in decibels (db), at different speeds.

TABLE 2.16

Speed (mph)	Noise Level (db)
10	50
30	68
50	75
70	79
100	84

(Source: *Auto Car Magazine*, November 2001, p. 63)

a. Fit a power function model to the data.

b. Graph the data points and the model on the same axes.

c. Use the result from part (a) to estimate the noise level at 80 mph.

$y = 30.414x^{0.226}$

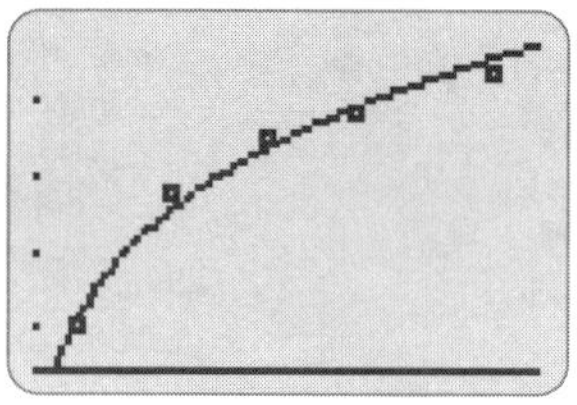

[0, 110] by [40, 100]

FIGURE 2.39

Solution

a. Entering the data in a graphing utility and creating the power regression model with the utility gives the equation. The model is

$$y = 30.414x^{0.226}$$

where x is in miles per hour and y is in decibels.

b. The graphs of the data points and the model are shown in Figure 2.39.

c. Evaluating the model at $x = 80$ gives $y = 81.88$, so the noise level at 80 mph is 81.88 decibels. ■

EXAMPLE 4 Cohabiting Households

The data in Table 2.17 on the next page gives the number of cohabiting (without marriage) households (in thousands) for selected years from 1960 to 1998. The scatter plot of these data, with x representing the number of years from 1950 and y representing thousands of households, is shown in Figure 2.40.

a. Find the power function that models these data.

b. Graph the data and the model on the same axes.

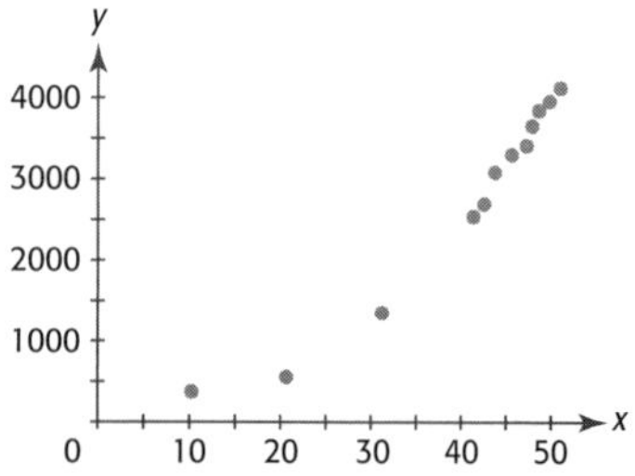

FIGURE 2.40

Solution

Using power regression in a graphing utility gives the power function that models these data. The power function that is the best fit is

$$y = 6.999x^{1.634}$$

TABLE 2.17 Cohabiting Households

Year	Cohabiting Households (thousands)	Year	Cohabiting Households (thousands)
1960	439	1993	3510
1970	523	1994	3661
1980	1589	1995	3668
1985	1983	1996	3958
1990	2856	1997	4130
1991	3039	1998	4236
1992	3308		

(Source: Index of Leading Cultural Indicators)

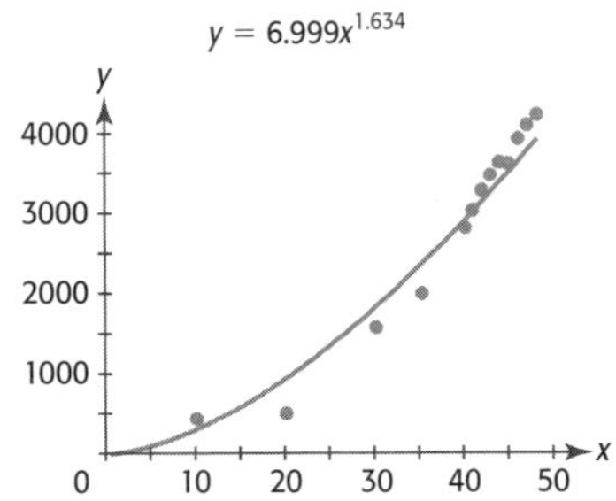

FIGURE 2.41

where y is in thousands and x is the number of years from 1950. The graphs of the data and the power function that models it are shown in Figure 2.41. ■

Spreadsheet Solution

Power models can also be found with Excel. The Excel spreadsheet in Table 2.18 shows the household assets in mutual funds, measured in billions of dollars, for 1995 to 1999, listed as years after 1990. Selecting the cells containing the data, using Chart Wizard to get the scatter plot of the data, selecting Add Trendline, and picking Power gives the equation of the power function that is the best fit for the data, along with the scatter plot and the graph of the best-fitting power function. The equation and graph of the function are shown in Figure 2.42.

TABLE 2.18

	A	B
1	Year x	Assets in mutual funds y ($ dollars)
2	5	1265
3	6	1586
4	7	2057
5	8	2501
6	9	3104

(Source: "Federal Reserve," *Time*, April 3, 2000)

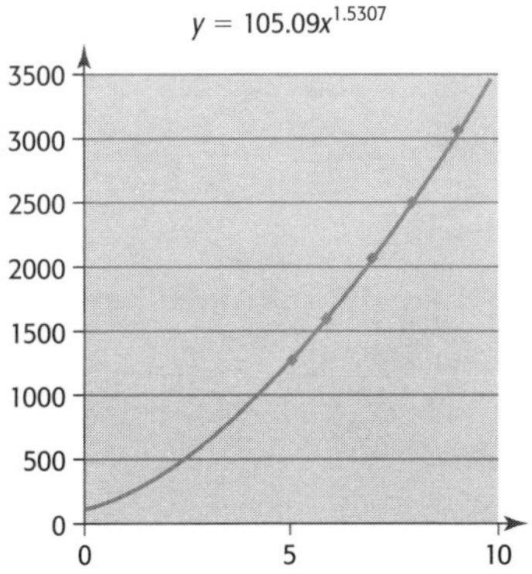

FIGURE 2.42

Comparison of Power and Quadratic Models

We found a power function that is a good fit for the data in Table 2.18, but a linear function or a quadratic function may also be a good fit for this data. A quadratic function may fit data points even if there is no obvious "turning point" in the graph of the data points. If the data points appear to rise (or fall) more rapidly than a line, then a quadratic model or a power model may fit the data well. In some cases it may be necessary to find both models to determine which is the better fit for the data.

EXAMPLE 5 Voting

Table 2.19 shows the percent of registered voters who voted in presidential elections for the years 1960–2000.

TABLE 2.19 Percent Voting in Presidential Elections

Year	Percent	Year	Percent
1960	63.1	1984	53.1
1964	61.9	1988	50.1
1968	60.8	1992	55.1
1972	55.2	1996	49.1
1976	53.6	2000	51.0
1980	52.6		

(Source: Federal Election Commission)

a. Graph the data points, representing the years from 1950 as x and the percent as y.

b. Find the quadratic model that is the best fit for the data.

c. Find the power model that is the best fit for the data.

d. Discuss the use of the two models to predict the percent voting after 2000.

e. Which type of function would give the better fit if a point were added giving the percent voting as 58.1 in 2004?

Solution

a. The scatter plot of the data is shown in Figure 2.43(a). It has no obvious "turning point," so either a quadratic or a power model may possibly be a good fit.

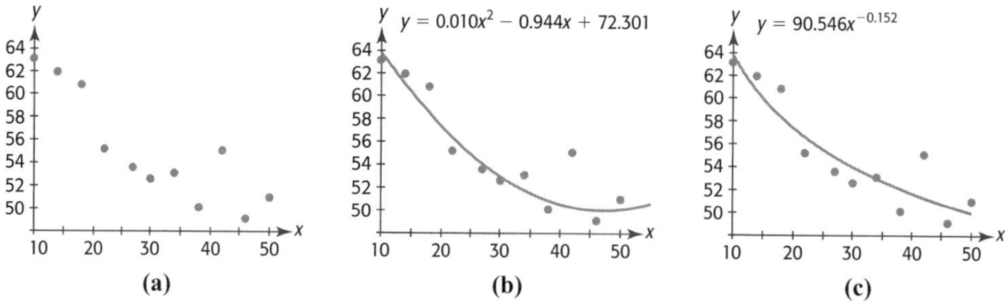

FIGURE 2.43

b. The quadratic model that is the best fit for the data is

$$y = 0.010x^2 - 0.944x + 72.301$$

The graphs of this model and the data points are shown in Figure 2.43(b).

c. The power model that is the best fit for the data is

$$y = 90.546x^{-0.152}$$

The graphs of this model and the data points are shown in Figure 2.43(c).

d. The graphs of the two models appear to be very similar until $x = 50$, which represents 2000. The quadratic model looks like it is beginning to rise after 1996 ($x = 46$) and could be used for prediction if the percent voting is expected to increase. The graph of the power model could be used for prediction if the percent voting is expected to continue to fall.

e. Adding the point (54, 58.1) indicates a significant increase in the voting percent, so the quadratic model is better in this case. ■

Statistical measurements exist that can be used to measure the goodness of fit of models to the data, but care must be used in interpreting these measures. For example, the statistical measurement for goodness of fit for quadratic functions is different from that for linear functions.

2.4 SKILLS CHECK

1. Find the quadratic function that models the data in the table below.

x	−2	−1	0	1	2	3	4
y	16	5	0	1	8	21	40

x	5	6	7	8	9	10
y	65	96	133	176	225	280

2. The following table has the inputs, x, and the outputs for three functions, f, g, and h. Use second differences to determine which function is exactly quadratic, which is approximately quadratic, and which is not quadratic.

x	$f(x)$	$g(x)$	$h(x)$
0	0	2	0
2	399	0.8	110
4	1601	1.2	161
6	3600	3.2	195
8	6402	6.8	230
10	9998	12	290

3. As you can verify, the following data points give constant second differences, but the points do not fit on the graph of a quadratic function. How can this be?

x	1	2	4	8	16	32	64
y	1	6	15	28	45	66	91

4. **a.** Make a scatter plot of the data in the table below.
b. Does it appear that a linear model or a power model is the best fit for the data?

x	y
1	4
2	9
3	11
4	21
5	32
6	45

5. Find the quadratic function that is the best fit for $f(x)$ defined by the table in Exercise 2.

6. Find the quadratic function that is the best fit for $g(x)$ defined by the table in Exercise 2.

7. **a.** Find a power function that models the data in the table in Exercise 4.
   b. Find a linear function that models the data.
   c. Visually determine which model is the better fit for the data.

8. **a.** Make a scatter plot of the data in the table below.
   b. Does it appear that a linear model or a power model is the better fit for the data?

x	y
3	4.7
5	8.6
7	13
9	17

9. **a.** Find a power function that models the data in the table in Exercise 8.
   b. Find a linear function that models the data.
   c. Visually determine if each model is a good fit.

10. Find the quadratic function that models the data in the table below.

x	−2	−1	0	1	2	3
y	15	5	2	1	3	10

x	4	5	6	7	8
y	20	35	55	75	176

11. Find the power function that models the data in the table below.

x	1	2	3	4	5	6	7	8
y	3	4.5	5.8	7	8	9	10	10.5

12. Is a power function or a quadratic function the better model for the data below?

x	0.5	1	2	3	4	5	6
y	1	7	17	32	49	70	90

2.4 EXERCISES

Calculate numerical results with the unrounded models, unless otherwise instructed.

13. *Income* The median annual incomes of male householders in the United States with children under 18 years of age for selected years between 1969 and 1996 are shown in the table below.

Year	Median Annual Income ($)	Year	Median Annual Income ($)
1969	33,749	1984	36,002
1972	36,323	1987	34,747
1975	33,549	1990	33,769
1978	37,575	1993	29,320
1981	33,337	1996	31,020

(Source: U.S. Bureau of the Census)

   a. Find the quadratic function that models the median income as a function of x, the number of years after 1960.
   b. Use the model from part (a) to estimate the median annual income of male householders with children under 18 years of age in 1997 and in 2002.
   c. Do you feel that these estimates are reliable? Explain.

14. *Aircraft Accidents* The table on the next page shows data recorded on December 31 of the indicated years for the number of U.S. air carrier accidents for all military services.
   a. Plot the data in a scatter plot, with x equal to the number of years after 1970.
   b. Fit a quadratic function to these data.
   c. Use the model from part (b) to estimate the number of aircraft accidents in 1992 and in 2001.

Do you feel that these estimates are reliable? Explain.

Year	1975	1980	1985	1990
Accidents	37	19	21	24

Year	1995	1996	1997
Accidents	36	38	49

(Source: National Traffic Safety Council)

15. *Cell Phones* The following table gives the number of millions of U.S. cellular telephone subscribers.
  a. Create a scatter plot for the data with x equal to the number of years from 1985. Does it appear that the data could be modeled with a quadratic function?
  b. Find the quadratic function that is the best fit for these data, with x equal to the number of years from 1985 and y equal to the number of subscribers in millions.
  c. Use the model to estimate the number in 2005.
  d. What part of the U.S. population does this estimate equal?

Year	Subscribers (millions)	Year	Subscribers (millions)
1985	0.340	1994	24.134
1986	0.682	1995	33.786
1987	1.231	1996	44.043
1988	2.069	1997	55.312
1989	3.509	1998	69.209
1990	5.283	1999	86.047
1991	7.557	2000	109.478
1992	11.033	2001	128.375
1993	16.009	2002	140.767

(Source: Semiannual CTIA Wireless Industry Survey)

16. *Unemployment* Total unemployment in South Carolina for the years 1982–1992 is given by the data in the table at the top of the next column.
  a. Create a scatter plot for the data, with x equal to the number of years after 1980.
  b. Does it appear that a quadratic model will fit the data? If so, find the best-fitting quadratic model.
  c. Does the y-intercept of the function in part (b) have meaning in the context of this problem? If so, interpret the value.

Year	Total Unemployment (thousands)	Year	Total Unemployment (thousands)
1982	162	1988	76
1983	148	1989	80
1984	105	1990	81
1985	107	1991	108
1986	99	1992	111
1987	91		

(Source: South Carolina Employment Security Commission, September 1993)

17. *Foreign-Born Population* The following table gives the percent of U.S. population that is foreign born.
  a. Create a scatter plot for the data, with x equal to the number of years from 1900 and y equal to the percent.
  b. Does it appear that the data could be modeled with a quadratic function?
  c. Find the best-fitting quadratic function for the data.
  d. Use the function to estimate the percent in 2005.

Year	Foreign Born (%)	Year	Foreign Born (%)
1900	13.6	1960	5.4
1910	14.7	1970	4.7
1920	13.2	1980	6.2
1930	11.6	1990	8.0
1940	8.8	2000	10.4
1950	6.9	2002	11.5

(Source: U.S. Bureau of the Census)

18. *U.S. Population* The data in the table below gives the U.S. population for selected years from 1790 to 2000.

a. Create a quadratic function that models this data, with y equal to the population in millions and x equal to the number of years from 1700.

b. During what year after 1700 does this model indicate that the population was 140.2 million?

Year	Population
1790	3,929,214
1870	38,558,371
1930	123,202,624
1990	248,790,925
2000	281,421,906

(Source: *Newsweek*, January 8, 2001)

19. *Educational Funding* Data that gives the amount of federal on-budget funds for research programs at universities and related institutions appear in the table below.

Year	1965	1970	1975	1980	1985
Federal Funds ($ billions)	1.82	2.28	3.42	5.80	8.84

Year	1990	1995	1998	1999	2000
Federal Funds ($ billions)	12.61	15.68	18.48	20.24	21.02

(Source: U.S. Department of Education, National Center for Educational Statistics)

a. Using an input equal to the number of years after 1960, find a quadratic function that models the data.

b. Using an input equal to the number of years after 1965, find a quadratic function that models the data.

c. Use the model in part (a) to find when (between 1965 and 2000) federal funds for research first exceeded $10 billion.

d. If you had used the model in part (b) instead of the one in part (a) to answer part (c), how would your results have differed?

20. *Medicare Trust Fund Balance* The year 1994 marked the 30th anniversary of Medicare. A 1994 pamphlet by Representative Lindsey O. Graham gave projections for the Medicare Trust Fund Balance as shown in the table below.

a. Representative Graham stated, "The fact of the matter is that Medicare is going broke." Use the data in the table to find when he predicted that this would happen.

b. Find a quadratic model to fit the data, with x equal to the number of years after 1990. When does this model predict that the Medicare Trust Fund balance will be 0?

c. The pamphlet reported the Medicare trustees as saying "the present financing schedule for the hospital insurance program is sufficient to ensure the payment of benefits only over the next 7 years." Does the function in part (b) confirm or refute the value 7 in this statement?

d. Find and interpret the vertex of the quadratic function in part (b).

Year	Medicare Trust Fund Balance ($ billions)	Year	Medicare Trust Fund Balance ($ billions)
1993	128	1999	98
1994	133	2000	72
1995	136	2001	37
1996	135	2002	−7
1997	129	2003	−61
1998	117	2004	−126

21. *Unemployment* The unemployment as a percent of the labor force in South Carolina for the years 1982–1992 is given by the table on the next page.

a. Create a scatter plot for the data, with x equal to the number of years from 1980.

b. Does it appear that a quadratic model will fit the data? If so, find the best-fitting quadratic model.

c. Does the y-intercept of the function in part (b) have meaning in the context of this problem? If so, interpret the value.

Year	Total Unemployment as a Percent of Labor Force	Year	Total Unemployment as a Percent of Labor Force
1982	10.8	1988	4.6
1983	10.1	1989	4.7
1984	7.1	1990	4.7
1985	6.9	1991	6.2
1986	5.6	1992	6.2
1987	5.6		

(Source: South Carolina Employment Security Commission, September 1993)

22. *Shrimp-Baiting Permits* The table below gives the number of shrimp-baiting permits in South Carolina in each of the years from 1994 through 1998.

a. Create a quadratic model for the data, using x as the number of years from 1990.

b. Use the model to predict the number of permits in 1999.

c. Because of a busy hurricane season, the number of permits issued in 1999 was 15,894. How does this compare with the prediction from the model, and what does it tell us about the dangers of extrapolating from models?

Year	1994	1995	1996	1997	1998
Number of Permits	13,366	13,919	14,150	15,488	17,467

(Source: South Carolina Department of Natural Resources)

23. *Volume* The measured volume of a pyramid with each edge of the base equal to x units and with its altitude (height) equal to x units is given in the table at the top of the next column.

a. Determine if the second differences of the outputs are constant.

b. If the answer is yes, find the quadratic model that is the best fit for the data. Otherwise, find the power function that is the best fit.

Edge Length x (units)	Volume of Pyramid (cubic units)
1	1/3
2	8/3
3	9
4	64/3
5	125/3
6	72

24. *Age and Income* The median annual income of men for certain average ages is given in the table below.

a. Find the quadratic function that models the annual income as a function of the average age.

b. Does the model appear to be a good fit?

c. Find and interpret the vertex of the graph of the *rounded* function.

Average Age (yr)	Median Income ($)	Average Age (yr)	Median Income ($)
19.5	6960	49.5	36,526
29.5	25,179	59.5	29,526
39.5	32,167	69.5	16,684

(Source: *Statistical Abstract of the United States, 1998*)

25. *World Population* One projection of the world population by the United Nations for selected years (a low projection scenario) is given in the table below.

Year	Projected Population (millions)	Year	Projected Population (millions)
1995	5666	2075	6402
2000	6028	2100	5153
2025	7275	2125	4074
2050	7343	2150	3236

(Source: *World Population Prospects: 1998 Revision*, United Nations, 1999)

a. Find a quadratic function that fits these data, using the number of years after 1990 as the input.

b. Find the positive x-intercept of this graph, to the nearest year.

c. When can we be certain that this model no longer applies?

26. *Personal Savings* The table below gives Americans' personal-savings rate for selected years from 1960 to 1998.

a. Find a quadratic function that models the personal-savings rate as a function of the number of years from 1960.

b. Find and interpret the vertex of the graph of the *rounded* model from part (a).

c. Find the positive x-intercept of this graph, to the nearest tenth.

d. When can we be certain that this model no longer applies?

Year	Personal-Savings Rate (%)	Year	Personal-Savings Rate (%)
1960	6.6	1985	6.9
1965	7.8	1990	5.1
1970	8.5	1995	3.4
1975	9.3	1998	0.5
1980	8.5		

(Source: Index of Leading Cultural Indicators)

27. *Mortgages* The balance owed y on a \$50,000 mortgage after x monthly payments is shown in the table below. Graph the data points with each of the equations below to determine, which of them is the better model for the data, if x is the number of months that payments have been made.

a. $y = 338{,}111.278x^{-0.676}$

b. $y = 4700\sqrt{110 - x}$

Monthly Payments	Balance Owed ($)
12	47,243
24	44,136
48	36,693
72	27,241
96	15,239
108	8074

28. *International Visitors* The number of international visitors to the U.S. for selected years 1986–1997 is given in the table at the top of the next column.

Year	U.S. Visitors (millions)	Year	U.S. Visitors (millions)
1986	26	1992	47.3
1987	29.5	1994	45.5
1988	34.1	1995	44
1989	36.6	1996	46.3
1990	39.5	1997	48.9
1991	43		

(Source: *1998 World Almanac*)

a. Using an input equal to the number of years after 1980, graph the aligned data points and both of the given equations to determine which equation is the better model for the aligned data.

i. $y = 12\sqrt{x}$

ii. $y = -.2327x^2 + 7.237x - 9.167$

b. If you had to pick one of these models to predict the number of international visitors in the year 2002, which model would be the more reasonable choice?

29. *Trust in the Government* The percent of people who say they trust the government in Washington always or most of the time has decreased over the years.

Year	Trust Government Always or Most of Time (%)	Year	Trust Government Always or Most of Time (%)
1964	76	1982	33
1966	65	1984	44
1968	61	1986	38
1970	54	1988	41
1972	53	1990	28
1974	36	1992	29
1976	34	1994	21
1978	29	1996	27
1980	25		

(Source: Pew Research Center)

a. Find a power function that models the data in the table using an input equal to the number of years from 1960.

b. According to the model, what percent of people said they trust the government always or most of the time in 2000?

30. *Box-Office Revenues* The data in the table below give the box-office revenues, in billions of dollars, for movies released in selected years between 1980 and 1998.
   a. Find the power function that best fits the revenues as a function of the number of years from 1970.
   b. What does the unrounded model estimate as the revenue in 2005?
   c. Discuss the reliability of this estimation.

Year	Revenue ($ billions)	Year	Revenue ($ billions)
1980	2.75	1994	5.39
1985	3.75	1995	5.49
1990	5.02	1996	5.91
1991	4.80	1997	6.37
1992	4.87	1998	6.95
1993	5.15		

(Source: Index of Leading Cultural Indicators)

31. *Classroom Size* The data in the table below give the numbers of students per teacher for selected years between 1960 and 1998.

Year	Students per Teacher	Year	Students per Teacher
1960	25.8	1992	17.4
1965	24.7	1993	17.4
1970	22.3	1994	17.3
1975	20.4	1995	17.3
1980	18.7	1996	17.1
1985	17.9	1997	17.0
1990	17.2	1998	17.2
1991	17.3		

(Source: U.S. Department of Education)

   a. Find the power function that is the best fit for the data, using as input the number of years after 1950.
   b. According to the unrounded model, how many students per teacher were there in 2000?
   c. Is this function increasing or decreasing during this time period?
   d. What does the model predict will happen to the number of students per teacher as time goes on?

32. *Classroom Teachers* The percent of all full-time school staff that are classroom teachers in the United States has decreased over the years since 1960, as seen in the table below.

Year	Classroom Teachers (%)	Year	Classroom Teachers (%)
1960	64.8	1992	52.2
1970	60.0	1993	52.1
1980	52.4	1994	52.0
1990	53.4	1995	52.0
1991	53.3	1996	52.1

(Source: U.S. Department of Education)

   a. Find a power function that models the data, using an input equal to the number of years from 1950.
   b. What does the unrounded model predict as the percent of all full-time school staff that are classroom teachers in 2008?
   c. Do we have enough data to be confident that the 2008 prediction is accurate?
   d. Use graphical or numerical methods to predict when the model output will fall below 50%.

33. *Cigarette Advertising* The data in the table on the next page give the total U.S. cigarette advertising and promotional expenditures, in millions of dollars, for the years 1975–1993.
   a. Using an input equal to the number of years after 1970, find a power function that models the data.
   b. Use the model to predict the expenditure for cigarette advertising and promotion in 2000.
   c. Use the model to estimate the year when the expenditure was $3.962 billion.
   d. Cigarette smoking has decreased during this period. Has advertising expenditure increased or decreased?

Year	Cigarette Advertising ($ millions)	Year	Cigarette Advertising ($ millions)
1975	491.3	1985	2476.4
1976	639.1	1986	2382.4
1977	779.5	1987	2580.5
1978	875.0	1988	3274.9
1979	1083.4	1989	3617.0
1980	1242.3	1990	3992.0
1981	1547.7	1991	4650.1
1982	1800.4	1992	5231.9
1983	1901.5	1993	6035.4
1984	2095.2		

(Source: Federal Trade Commission Report to Congress for 1990–1996)

34. *Travel and Tourism Spending* The global spending on travel and tourism (in billions of dollars) for the years 1988–1997 is given in the table below.

a. Write the equation of a power function that models the data, letting your input represent the number of years after 1980.

b. Use the model to estimate the global spending for 1999.

c. When did the global spending reach $300 billion, according to this model?

Year	Spending ($ billions)	Year	Spending ($ billions)
1988	204.7	1993	322.3
1989	221.0	1994	353.5
1990	269.2	1995	403.6
1991	277.6	1996	438.8
1992	315.5	1997	447.7

(Source: *1998 World Almanac*)

35. *Insurance Rates* The following table gives the monthly insurance rates for a $100,000 life insurance policy for smokers 35–50 years of age.

a. Create a scatter plot for the data.

b. Does it appear that a quadratic function can be used to model the data? If so, find the best-fitting quadratic model.

c. Find the power model that is the best fit for the data.

d. Compare the two models by graphing each model on the same axes with the data points. Which model appears to be the better fit?

Age (yr)	Monthly Insurance Rate ($)	Age (yr)	Monthly Insurance Rate ($)
35	17.32	43	23.71
36	17.67	44	25.11
37	18.02	45	26.60
38	18.46	46	28.00
39	19.07	47	29.40
40	19.95	48	30.80
41	21.00	49	32.55
42	22.22	50	34.47

(Source: American General Life Insurance Company, November 1999)

36. *U.S. Gross Domestic Product* The graph in the figure on the next page gives the U.S. gross domestic product (in billions of dollars) for selected years from 1940 through 1995.

a. Create a scatter plot for the data, with x equal to the number of years from 1930 and y equal to the gross domestic product in billions of dollars.

b. Does it appear that a quadratic function can be used to model these data? If so, find the best-fitting quadratic model.

c. Find the power model that is the best fit for the data.

d. Compare the two models by graphing each model on the same axes with the data points. Which model appears to be the better fit?

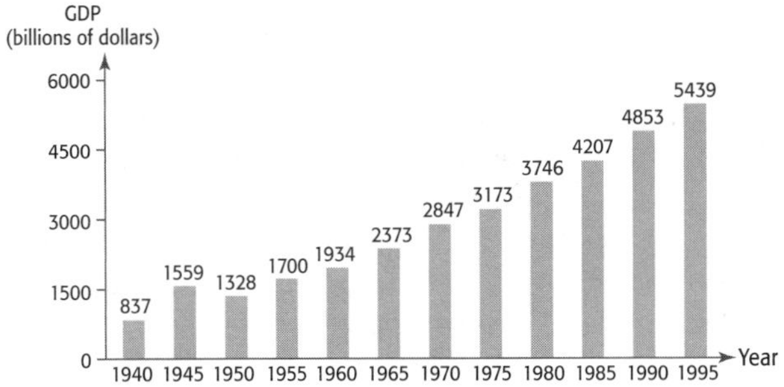

(Source: U.S. Department of Management and Budget)

37. *Hospital Discharges* The numbers of discharges per 1000 people in nonfederal short-stay hospitals for the indicated year from 1980 to 1996 are given in the table below.

a. Align the data as the number of years after 1970 and find both quadratic and power models for the data in the table. Discuss the fit of each of the models.

b. Even though you know the dangers of extrapolation, use the data in the table to make a quick estimate of the discharge rate in 2005. Discuss how your estimate for the year 2005 compares with the 1996 value.

c. Use both of the functions in part (a) to estimate the number of discharges per 1000 people in 2005. Compare the values with what you expected from the discussion in part (b).

Year	Discharges per 1000 People
1980	158.5
1985	137.7
1988	117.6
1990	113.0
1992	110.5
1994	106.5
1996	102.3

(Source: National Hospital Discharge Survey)

38. *Abortions* The total numbers of abortions, in thousands, in the United States for selected years from 1973 through 1996 are given in the table below.

a. Use the data to find a quadratic function that models the number of abortions as a function of years from 1970.

b. Does the function reported in part (a) yield a maximum or minimum? When is this value obtained, and what does the value represent?

Year	Number of Abortions (thousands)	Year	Number of Abortions (thousands)
1973	744.6	1992	1528.9
1975	1034.2	1993	1500.0
1980	1553.9	1994	1431.0
1985	1588.6	1995	1363.7
1990	1608.6	1996	1365.7
1991	1556.5		

(Source: Alan Guttmacher Institute)

CHAPTER 2 *Summary*

In this chapter, we presented the building blocks for function construction. We began with an in-depth discussion of quadratic functions, including realistic applications that involve the vertex and x-intercepts of parabolas and solving quadratic equations. We then studied a library of functions, including power, root, reciprocal, absolute value, and piecewise-defined functions, and transformations of these functions. Real-world data are provided throughout the chapter, and we learned how to fit power and quadratic functions to some of these data. We finished the chapter by learning how to combine two or more functions to create new functions, how to find the inverse of a function, and how to solve additional equations and inequalities.

Key Concepts and Formulas

2.1 Quadratic Functions; Parabolas

Quadratic function Also called a *second-degree polynomial function*, this function can be written in the form $f(x) = ax^2 + bx + c$ where $a \neq 0$.

Parabola A parabola is the graph of a quadratic function.

Vertex The turning point on a parabola is the vertex.

Maximum point If a quadratic function has $a < 0$, the vertex of the parabola is the maximum point, and the parabola opens downward.

Minimum point If a quadratic function has $a > 0$, the vertex of the parabola is the minimum point, and the parabola opens upward.

Axis of symmetry The axis of symmetry is the vertical line through the vertex of the parabola.

Forms of quadratic functions

$y = a(x - h)^2 + k$	$y = ax^2 + bx + c$
Vertex at (h, k)	x-coordinate of vertex at $x = -\dfrac{b}{2a}$
Axis of symmetry is the line $x = h$.	Axis of symmetry is the line $x = -\dfrac{b}{2a}$
Parabola opens up if a is positive and down if a is negative.	Parabola opens up if a is positive and down if a is negative.

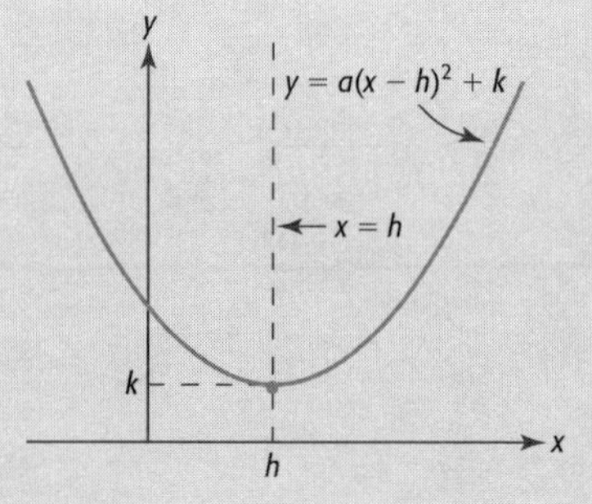

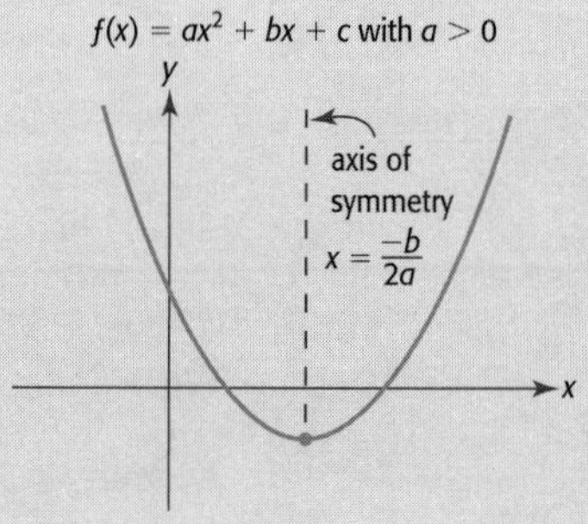

x-Intercepts of the graph of a quadratic function The x-intercepts occur where the graph of the function crosses or touches the x-axis. The parabola will cross the x-axis at zero, one, or two points.

2.2 *Solving Quadratic Equations*

Solving quadratic equations An equation that can be written in the form $ax^2 + bx + c = 0$, $a \neq 0$, is called a *quadratic equation*.

Zero product property For real numbers a and b, the product $ab = 0$ if and only if either $a = 0$ or $b = 0$ or both a and b are zero.

Solving by factoring To solve a quadratic equation by factoring, write the equation in a form with 0 on one side. Then factor the nonzero side of the equation, if possible, and use the zero product property to convert the equation into two linear equations that are easily solved.

Solving and checking graphically Solutions or decimal approximations of solutions to quadratic equations can be found by using TRACE, ZERO, or INTERSECT with a graphing utility.

Factor Theorem The factorization of a polynomial function $f(x)$ and the solutions to $f(x) = 0$ can be verified by graphing $y = f(x)$ and observing where the graph crosses the x-axis. If $x = a$ is a solution to $f(x) = 0$, then $x - a$ is a factor of f.

Solving using the square root method When a quadratic equation has the simplified form $x^2 = C$, the solutions are $x = \pm\sqrt{C}$. This method can also be used to solve equations of the form $(ax + b)^2 = C$.

Completing the square A quadratic equation can be solved by converting one side to a perfect binomial square and taking the square root of both sides to solve the equation.

Solving using the quadratic formula The solutions of the quadratic equation $ax^2 + bx + c = 0$, with $a \neq 0$, are given by the formula

$$x = \frac{-b \pm \sqrt{b^2 - 4ac}}{2a}$$

Solutions, zeros, x-intercepts, and factors If a is a real number, the following three statements are equivalent:

2.3 *A Library of Functions*

Power functions A power function is a function of the form $y = ax^b$, with $b > 0$, where a and b are real numbers.

Cubing function A special power function, $y = x^3$, is called the cubing function.

Root functions A root function is a function of the form $y = ax^{1/n}$, or $y = a\sqrt[n]{x}$, where n is an integer, $n \geq 2$.

Reciprocal function This function is formed by the quotient of a constant function and the identity function, $f(x) = \dfrac{1}{x}$

Piecewise-defined functions This is a function that is created by combining two or more functions. Piecewise-defined functions can be graphed by graphing their pieces on the same axes. A familiar example of a piecewise-defined function is the *absolute value function*.

Absolute value function

$$|x| = \begin{cases} x & \text{if } x \geq 0 \\ -x & \text{if } x < 0 \end{cases}$$

2.4 Quadratic and Power Models

Quadratic modeling The use of graphing utilities permits us to fit quadratic functions to data by using technology.

Second differences If the second differences of data are constant for equally spaced inputs, a quadratic function is the exact fit for the data.

Power modeling The use of graphing utilities permits us to fit power functions to nonlinear data by using technology.

Chapter 2 Skills Check

1. Find the coordinates of the vertex of the graph of $f(x) = 3x^2 - 6x - 24$.

2. Graph $f(x) = 3x^2 - 6x - 24$.

3. Use graphical and algebraic methods to find the x-intercepts of the graph of $f(x) = 3x^2 - 6x - 24$.

4. Find the solutions to $f(x) = 0$ if $f(x) = 3x^2 - 6x - 24$.

In Exercises 5 and 6, use factoring to solve the equations.

5. $x^2 - 5x + 4 = 0$
6. $6x^2 + x - 2 = 0$

Use a graphing utility as an aid in factoring to solve the equations in Exercises 7 and 8.

7. $5x^2 - x - 4 = 0$
8. $3x^2 + 4x - 4 = 0$

In Exercises 9 and 10, use the quadratic formula to solve the equations.

9. $x^2 - 4x + 3 = 0$
10. $4x^2 + 4x - 3 = 0$

11. **a.** Graph the functions $f(x) = \sqrt{x}$ and $g(x) = \sqrt{x+2} - 3$.
    b. How are the graphs related?

12. What is the domain of the function $g(x) = \sqrt{x+2} - 3$?

13. Determine if the function $y = -3x^2$ is increasing or decreasing
    a. For $x < 0$. **b.** For $x > 0$.

14. For each of the functions, determine if the function is concave up or concave down.
    a. $y = x^{3/2}$ **b.** $y = x^{1/2}$

15. Find the inverse of $f(x) = 3x - 2$.

16. Find the inverse of $g(x) = \sqrt[3]{x} - 1$.

17. Graph $f(x) = (x+1)^2$ and its inverse $f^{-1}(x)$ on the domain $[-1, 10]$.

18. Is the function $f(x) = \dfrac{x}{x-1}$ a one-to-one function?

19. Find a power function that models the data below.

x	1	2	3	4	5	6
y	4	9	11	21	32	45

20. Find a quadratic function that models the data below.

x	1	3	4	6	8
y	2	8	15	37	63

For Exercises 21–23, match each graph with the correct equation.

a. $y = |x| + 2$

b. $y = x^3$

c. $y = \dfrac{3}{x-1} + 1$

d. $y = \dfrac{-3}{x+1} + 1$

e. $y = 2x^3$

f. $y = |x + 2|$

21.

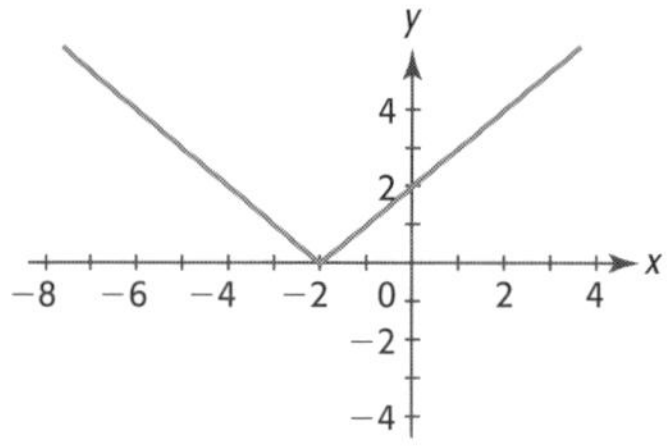

22.

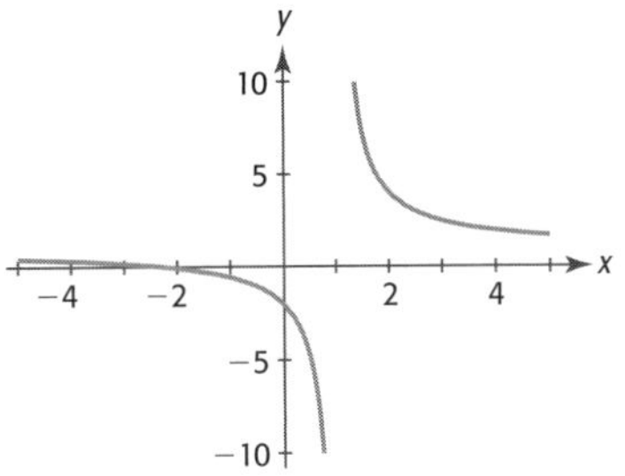

23.

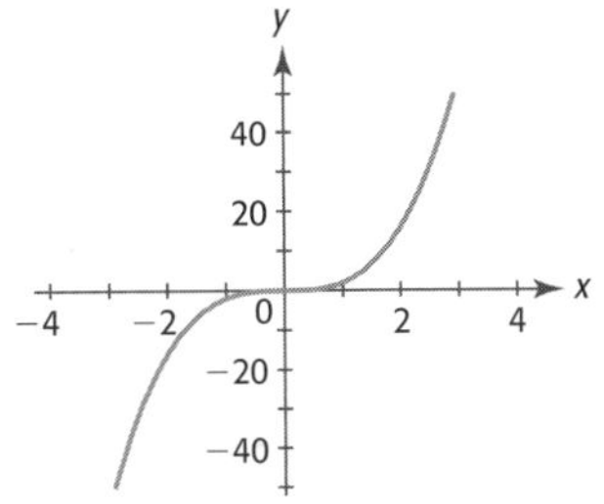

24. Graph $f(x) = \begin{cases} 4 - x & \text{if } x \leq 3 \\ x^2 - 5 & \text{if } x > 3 \end{cases}$.

25. Solve the inequality $x^2 - 7x \leq 18$.

26. Solve the inequality $2x^2 + 5x \geq 3$.

In Exercises 27–30, find the exact solutions to $f(x) = 0$ in the complex numbers.

27. $z^2 - 4z + 6 = 0$

28. $w^2 - 4w + 5 = 0$

29. $4x^2 - 5x + 3 = 0$

30. $4x^2 + 2x + 1 = 0$

31. Solve $(x - 4)^3 < 4096$.

32. Solve $(x + 2)^2 \geq 512$.

33. Solve $\sqrt{4x^2 + 1} = 2x + 2$.

34. Solve $\sqrt{3x^2 - 8} + x = 0$.

35. Solve $|3x - 6| = 24$.

36. Solve $|2x + 3| = 13$.

37. Solve $|2x - 4| \leq 8$.

38. Solve $|4x - 3| \geq 15$.

Chapter 2 Review

39. *Maximizing Profit* The monthly profit from producing and selling x units of a product is given by the function $P(x) = -0.01x^2 + 62x - 12{,}000$.
   a. Producing and selling how many units will result in the maximum profit for this product?
   b. What is the maximum possible profit for the product?

40. *Profit* The revenue from sales of x units of a product is given by $R(x) = 200x - 0.01x^2$, and the cost of producing and selling the product can be described by $C(x) = 38x + .01x^2 + 16{,}000$.
   a. Producing and selling how many units will give maximum profit?
   b. What is the maximum possible profit for the product?

41. *Height of a Ball* If a ball is thrown into the air at 64 feet per second from a height of 192 feet, its height (in feet) is given by $S = 192 + 64t - 16t^2$, where t is in seconds.
   a. In how many seconds will the ball reach its maximum height?
   b. What is the maximum possible height for the ball?

42. *Height of a Ball* If a ball is thrown into the air at 29.4 meters per second from a height of 60 meters, its height (in meters) is given by $S = 60 + 29.4t - 9.8t^2$, where t is in seconds.
   a. In how many seconds will the ball reach its maximum height?
   b. What is the maximum possible height for the ball?

43. *Ballistics* Ballistic experts are able to identify the weapon that fired a certain bullet by studying the markings on the bullet. If a test is conducted by firing the bullet into a bale of paper, the distance that the bullet will travel is given by $s = 64 - (4 - 10t)^3$ inches, for $0 \leq t \leq 0.4$, t in seconds.
   a. Graph this function for $0 \leq t \leq 0.4$.
   b. How far does the bullet travel during this time period?

44. *Projectile* A toy rocket is fired into the air from the top of a building, with its height given by $S = -16(t - 4)^2 + 380$ feet, with t in seconds.
   a. In how many seconds will the rocket reach its maximum height?
   b. What is the maximum possible height?

c. Describe the transformations needed to obtain the graph of $S(t)$ from the graph of $y = t^2$.

45. *Airplane Crashes* The number of all airplane crashes (in thousands) is given by the equation $y = 0.0057x^2 - 0.197x + 3.613$, where x is the number of years after 1980.
 a. During what year does the model indicate that the number of crashes was at a minimum?
 b. Use a graphing utility to determine the years during which the model indicates that the annual number of crashes is below 3000.

46. *Unemployment* For the years 1983–1992, the unemployment percent in South Carolina can be modeled by $y = 0.144x^2 - 2.482x + 15.467$, where x is the number of years from 1980.
 a. Use numerical methods to find the year when the unemployment percent is a minimum.
 b. What is the unemployment percent for the year in part (a)?
 c. During what year after 1988 does the model indicate that the unemployment was 5.047%?
 (Source: *South Carolina Statistical Abstract, 1994*)

47. *Break Even* The profit for a product is given by $P = -3600 + 150x - x^2$, where x is the number of units produced and sold. How many units will give break even (that is, return a profit of 0)?

48. *Falling Ball* If a ball is dropped from the top of a 400-foot-high building, its height S in feet is given by $S = 400 - 16t^2$, where t is in seconds. In how many seconds will it hit the ground?

49. *Profit* The profit from producing and selling x units of a product is given by the function $P(x) = -0.3x^2 + 1230x - 120{,}000$ dollars. Producing and selling how many units will result in a profit of \$324,000 for this product?

50. *Tobacco Sales* The domestic sales of tobacco (in millions of kilograms) in Canada can be described by $y = -0.084x^2 + 1.124x + 4.028$ dollars, where x is the number of years after 1986. During what years does this model indicate that the sales will equal 6.745 million kilograms?
 (Source: Canadian Council on Smoking and Health)

51. *Homicide Rate* The homicide rate (the number of homicides per 100,000 people) is given by the function $r = -0.0128x^2 + 0.586x + 2.986$, where x is the number of years after 1960. Use a graphing utility to find during what years this model indicates that the rate will be above 8 per 100,000 people.
 (Source: Bureau of Justice Statistics)

52. *Hotel Room Supply* The percentage change p (from the previous year) in the hotel room supply is given by $p = 0.025x^2 - 0.645x + 4.005$, where x is the number of years after 1998. If this model is accurate, use a graphing utility to find during what years between 1998 and 2010 the percentage change will fall below 3%.
 (Source: Hotel and Motel Management, September 6, 1999)

53. *Firefighter Injuries* The number of on-duty injuries to firefighters during the years 1988–1998 is given by $y = 161{,}488.4931x^{-0.4506}$, with $x = 0$ in 1980.
 a. What does this model indicate the number of injuries to be in 1998?
 b. Graph the function and use the graph to determine in what year the model indicates that the number of injuries was 57,219.

54. *Taxes* The tax burden has grown rapidly in the United States during the 20th century. Using data from 1910 to 1999, the per capita (per person) tax paid can be modeled by the power function $S = 11.23t^{1.44}$, where t is the number of years from 1900.
 a. Does this model indicate that the function is increasing?
 b. Graph the function on the interval from $t = 0$ to $t = 100$.
 c. Is the graph concave up or concave down?
 d. What does the model estimate the per capita tax paid to be in 1995?

55. *Federal Support for Education* The federal on-budget funds for all educational agencies between 1965 and 2000 can be modeled by the function

$$f(t) = \begin{cases} 0.011t^3 - 0.696t^2 + 14.380t - 29.370 & \text{when } 5 \le t < 35 \\ 0.931t^2 - 67.374t + 1295.785 & \text{when } 35 \le t \le 40 \end{cases}$$

 where t is the number of years after 1960 and $f(t)$ is in millions of constant 2000 dollars.
 a. Graph the function f for $5 \le t \le 40$.
 b. What was the amount of funding given to educational agencies in 1989?
 c. How much federal funding was given to educational agencies in 1997?
 (Source: U.S. Department of Education, National Center for Educational Statistics)

56. *Average Cost* Suppose the total cost function for a product is determined to be $C(x) = 30x + 3150$, where x is the number of units produced and sold.
 a. Write the average cost function, which is the quotient of two functions.
 b. Graph this function for $0 < x \le 20$.

57. *Supply and Demand* The price per unit of a product is \$$p$ and the number of units of the product is denoted by q. Suppose the supply function for a product is given by $p = \frac{180 + q}{6}$, and the demand for the product is given by $p = \frac{300}{q} - 20$.

a. Which of these functions is a shifted reciprocal function?

b. Graph the supply and demand functions on the same axes.

58. *Supply and Demand* The supply function for a commodity is given by $p = 58 + \frac{q}{2}$, and the demand function for this commodity is given by $p = \frac{2555}{q + 5}$.

a. Which of these functions is a shifted reciprocal function?

b. Graph the supply and demand functions on the same axes.

59. *Indiana Population* The population of Indiana (in thousands) between 1980 and 1997 can be described by

$$P(x) = \begin{cases} 2.320x^2 - 389x + 21762 & \text{where } 80 \le x \le 90 \\ 46.0x + 1411.667 & \text{where } x > 90 \end{cases}$$

and x is the number of years after 1900.

a. What was the Indiana population in 1990?

b. Use P to predict when the population of Indiana reached 6 million people. What assumptions are made if this prediction is to be considered valid? (Source: *Statistical Abstract of the United States, 1998*)

60. *Marijuana Use* The percent of persons 12 years of age and over in the United States who said that they had used marijuana at least once within the month prior to being asked during selected years is described by the function

$$M(x) = \frac{-1}{12}\left(x - \frac{789}{10}\right)^2 + \frac{15{,}541}{1200}$$

when $79 \le x \le 88$, where x is the number of years after 1900.

a. Describe the graph of the function M in the context of the problem.

b. Find and interpret $M(79)$, $M(80)$, and $M(88)$.

c. Using the information in parts (a) and (b), sketch a graph of M.

d. The survey on which M is based was taken once during each of the selected years. State the domain of the related discrete function.

e. At what value of x does the maximum of the function M occur? If you put this result in the context of this problem, do you feel that it might be valid? (Source: National Household Survey on Drug Abuse, Series H6 and H7)

61. *U.S. Foreign Trade* The *trade balance* is defined as the value of exports minus the value of imports. Between 1880 and 1980, the value of exports from the United States exceeded the value of imports into the United States, resulting in a positive trade balance. However, this has not been the case since the early 1980s. The value of U.S. imports between 1990 and 1999 is modeled by $I(x) = 61487.24x + 435{,}606.84$ million dollars, and the value of U.S. exports between 1990 and 1999 can be modeled by the function $E(x) = -191.73x^2 + 39{,}882.69x + 377{,}849.85$ million dollars. For both functions, x is the number of years after 1990.

a. Use the functions I and E to write a model for the U.S. trade balance between 1990 and 1999.

b. Estimate the year 2000 trade deficit. (Source: Office of Trade and Economic Analysis, U.S. Department of Commerce)

62. *Southwest Airlines* The number of employees of Southwest Airlines between 1990 and 1998 can be modeled by the function $E(t) = 0.017t^2 + 2.164t + 8.061$ thousand employees, where t is the number of years after 1990. Southwest Airlines' revenue during the same time period is given by the model $R(E) = 0.165E - 0.226$ billion dollars when there are E thousand employees of the company.

a. Find and interpret the meaning of the function $R(E(t))$.

b. Use the result of part *a* to find $R(E(3))$. Interpret this result.

c. How many employees worked for Southwest Airlines in 1997?

d. What was the 1997 revenue for this airline? (Source: Hoover's Online Capsules)

63. *Prison Sentences* The mean time in prison y for certain crimes can be found as a function of the mean sentence length x, using $f(x) = 0.554x - 2.886$, where x and y are measured in months.

a. Find the inverse of this function.

b. Interpret the inverse function from part (a). (Source: Index of Leading Cultural Indicators)

64. *Dairy Cows* The number of cows and heifers kept for milk production in the United States between 1990 and 1997 can be approximated by the function $C(x) = -0.093x + 9.929$ million head, where x is the number of years after 1990.

a. Find a formula for the inverse of the function C.
b. Interpret your answer to part (a).
c. If the inverse function is called C^{-1}, find $C^{-1}(C(8))$.
(Source: *Statistical Abstract of the United States, 1998*)

65. *Internet Usage* The worldwide Internet usage from 1997 through 2002 (projected) is shown in the table on the next page.
a. Find the quadratic function that models this data, with x equal to the number of years after 1990.
b. Graph the data and the function that models the data, for $6 \leq x \leq 12$.
c. Does this model appear to be a good fit for the data?
d. Use the result of part (a) to estimate when the Internet usage is 550 billion.

Year	Internet Usage (billions)	Year	Internet Usage (billions)
1997	75	2000	255
1998	105	2001	340
1999	175	2002	490

(Source: *CyberAtlas*, 1999)

66. *Personal Income* The income received by persons from all sources minus their personal contributions for Social Security insurance is called *personal income*. The table below lists the personal income, in billions of dollars, of persons living in the United States for the indicated years.

Year	Personal Income ($ billions)
1990	4802.4
1991	4981.6
1992	5277.2
1993	5519.2
1994	5791.8
1995	6150.8
1996	6495.2
1997	6873.9

(Source: *Statistical Abstract of the United States, 1998*)

a. Using an input equal to the number of years after 1990, find a quadratic function that models this data.
b. Use your unrounded model to estimate when the personal income was $5500 billion.
c. If the pattern indicated by the model remains valid, when will the personal income be double its 1990 value?

67. *Resident Population* The U.S. resident population from 15 to 19 years of age and the percent of this resident population that are males are given for selected years in the table below. Fill in the blank column of the table and write the quadratic equation (to two decimal places) that models the 15- to 19-year-old male population as a function of the years from 1980. Include the description of this population and its units of measure.

Year	Age 15–19 Population (thousands)	Male Population (%)	Age 15–19 Male Population (thousands)
1980	21,168	48.6	
1985	18,727	48.7	
1990	17,890	48.7	
1995	18,152	48.9	
1997	19,068	49.0	

(Source: U.S. Bureau of the Census, *Current Population Reports*)

68. *Insurance Premiums* The table below gives the monthly premiums required for a $250,000 term-life insurance policy on a 35-year-old female nonsmoker for different guaranteed term periods.
a. Find a quadratic function that models the monthly premium as a function of the length of term for a 35-year-old female nonsmoking policyholder.
b. Assuming that the domain of the function contains integer values between 10 years and 30 years, what term in years could a 35-year-old nonsmoking female purchase for $130 a month?

Term Period (years)	Monthly Premium for 35-Year-Old Female (dollars)
10	103
15	125
20	145
25	183
30	205

(Source: Quotesmith.com, December 1999)

69. *Computer Usage* The number of students per computer in U.S. public schools from the 1983–1984 through the 1997–1998 school years is shown in the table on the next page.

School Year	Students per Computer
1983–1984	125
1985–1986	50
1987–1988	32
1989–1990	22
1991–1992	18
1993–1994	14
1995–1996	10
1997–1998	6.1

(Source: *CyberAtlas*, 1999)

a. Align the input data as the number of years after the beginning of the 1980–1981 school year and find a power function f to fit the data. Graph the data and the function $y = f(t)$ that models the data. Does this model appear to be a good fit?

b. If we assume that the function in part (a) is valid for all school years after the 1997–1998 one, will the function in part (a) ever indicate that there will be one student per computer? Explain.

c. Another function that might be used to model this data is

$$C(t) = \frac{380}{t + 0.3} - 15 \text{ students per computer}$$

where t is the number of years after the beginning of the 1980–1981 school year. What is the basic function that can be transformed to obtain C? Describe the transformations.

d. Do you feel that the function $y = f(t)$ in part (a) or $y = C(t)$ better fits the data? Why?

e. There were 5.7 students per computer in 1998–1999. Which function, $y = f(t)$ or $y = C(t)$, comes closer to estimating the actual value? Does this result agree with your thoughts in part (d)?

70. *Hawaii Population* The data in the table below give the 1990 through 1997 population of Hawaii.

Year	Population (thousands)	Year	Population (thousands)
1990	1108	1994	1173
1991	1131	1995	1179
1992	1150	1996	1183
1993	1160	1997	1187

(Source: *Statistical Abstract of the United States, 1998*)

a. Align the data with x = the number of years after 1985 and find a power function to fit the aligned data.

b. If the pattern indicated by the model remains valid, estimate in what year Hawaii's population will rise to 1.3 million people.

71. *Hospital Utilization* The average length of stay in noninstitutional, short-stay hospitals (exclusive of federal hospitals and not counting newborn infants) for selected years is given in the table below. Align the years to be the number of years after 1980.

Year	Average Stay (days)	Year	Average Stay (days)
1980	7.3	1993	6.0
1985	6.5	1994	5.7
1990	6.4	1995	5.4
1991	6.4	1996	5.2
1992	6.2		

a. Fit a quadratic model to the 1980–1991 aligned data.

b. Fit a linear model to the 1992–1996 aligned data.

c. Combine the results of parts (a) and (b) to form a piecewise model for the 1980–1996 data. The function should be defined for all years between 1980 and 1991.

d. Use the piecewise-defined model to answer the following:

i. Find and interpret the output in 1987.

ii. When was the average stay 6.1 days?

iii. What was the average stay in 1991?

72. *Consumer Price Index* Prices as measured by the U.S. Consumer Price Index (CPI) have risen steadily since World War II. The data in the table gives the CPI for selected years between 1975 and 1995. The CPI in this table has 1967 as a reference year; that is, what cost \$1 in 1967 cost about \$1.61 in 1975 and \$4.57 in 1995.

Year	CPI
1975	161.2
1980	248.8
1985	322.2
1990	391.4
1995	456.5

(Source: Congressional Budget Office, *Investor's Business Daily*, July 28, 1999)

Align the input data as the number of years after 1975 and report any models with two decimal places.

a. Find a linear model for the data. Discuss the fit to the data.

b. Find a quadratic model for the data. How good is the fit?

c. The CPI for 2000 was predicted to be 511.5. Evaluate the unrounded linear and quadratic functions at the input corresponding to 2000. Which does a better job of predicting this value?

d. Use the model you chose in part (c) to estimate when, before 2010, the CPI is 550.

73. *Profit* The monthly profit from producing and selling x units of a product is given by

$$P(x) = -0.01x^2 + 62x - 12{,}000$$

Producing and selling how many units will result in profit for this product ($P(x) > 0$)?

74. *Unemployment* For the years 1983–1992, the unemployment rate (as a percent) in South Carolina can be modeled by $y = 0.144x^2 - 2.482x + 15.467$, where y is the percent and x is the number of years from 1980. Use graphical or numerical methods to find the years during which the model indicates that the unemployment rate is above 5.5%.
(Source: *South Carolina Statistical Abstract*)

GROUP ACTIVITIES/EXTENDED APPLICATIONS

I. Cost, Revenue, and Profit

The following table gives the weekly revenue and cost, respectively, for a selected number of units of production and sale of a product by the Quest Manufacturing Company.

Number of Units	Revenue ($)	Number of Units	Cost ($)
100	6,800	100	32,900
300	20,400	300	39,300
500	34,000	500	46,500
900	61,200	900	63,300
1,400	95,200	1,400	88,800
1,800	122,400	1,800	112,800
2,500	170,000	2,500	162,500

Provide the information requested and answer the questions.

1. Use technology to determine the equations that model revenue and cost functions for this product, using x as the number of units produced and sold.
2. a. Combine the revenue and cost functions with the correct operation to create the profit function for this product.
   b. Use the profit function to complete the following table:

x (units)	Profit, $P(x)$ ($)
0	
100	
600	
1600	
2000	
2500	

3. Find the number of units of this product that must be produced and sold to break even.
4. Find the maximum possible profit and the number of units that gives the maximum profit.
5. a. Use operations with functions to create the average cost function for the product.
   b. Complete the following table.

x (units)	Average Cost, $\overline{C}(x)$ ($/unit)
1	
100	
300	
1400	
2000	
2500	

   c. Graph this function using the viewing window [0, 2500] by [0, 400].
6. Graph the average cost function using the viewing window [0, 4000] by [0, 100]. Determine the number of units that should be produced to minimize the average cost and the minimum average cost.
7. Compare the number of units that produced the minimum average cost with the number of units that produced the maximum profit. Are they the same number of units? Discuss which of these values is more important to the manufacturer and why.

2. Modeling

Graphs displaying linear and nonlinear growth are frequently displayed in periodicals such as *Newsweek* and *Time*, in newspapers such as *USA Today* and the *Wall Street Journal*, and on numerous Web sites on the Internet. Tables of data can also be found in these sources, especially in federal and state government Web sites, such as www.census.gov.

Your mission is to find a company sales, stock price, biological growth, or sociological trend over a period of years that is nonlinear, to determine which type of nonlinear function is the best fit for the data, and to find a nonlinear function that is a model for this data.

A quadratic model will be a good fit for the data:

1. If the data are presented as a graph that resembles a parabola or part of a parabola.
2. If the data are presented in a table and the plot of the data points lie near some parabola or part of a parabola.
3. If the data are presented in a table and the second differences of the outputs are nearly constant for equally spaced inputs.

A power model may be a good fit if the shape of the graph or graph of the data points resembles part of a parabola, but a parabola is not a good fit for the data.

After creating the model, you should test the goodness of fit of the model to the data and discuss uses that you could make of the model.

Your completed project should include

a. A complete citation of the source of the data that you are using.
b. An original copy or photocopy of the data being used.
c. A scatter plot of the data.
d. The equation that you have created.
e. A graph containing the scatter plot and the modeled equation.
f. A statement about how the model could be used to make estimations or predictions about the trend you are observing.

Some helpful hints:

1. If you decide to use a relation determined by a graph that you have found, read the graph very carefully to determine the data points or contact the source of the data to get the data from which the graph was drawn.
2. Align the independent variable by letting x represent the number of years from some convenient year and then enter the data into a graphing utility and create a scatter plot.
3. Use your graphing utility to create the equation of the function that is the best fit for the data. Graph this equation and the data points on the same axes to see if the equation is reasonable.

CHAPTER 3

Exponential and Logarithmic Functions

The intensities of earthquakes like the one in San Francisco in 1989 are measured with the Richter scale, which uses logarithmic functions. Logarithmic functions are also used to measure loudness (in decibels) and stellar magnitude and in calculating pH values. Exponential functions are used to compute the growth of investments, the growth of populations, and in sales decay and radioactive decay.

SECTIONS	TOPICS	APPLICATIONS
3.1 Exponential Functions	Graphing exponential functions; exponential growth and decay; transformations of graphs; spreadsheets; the number *e*	Growth of paramecia, Head Start program growth, sales decay, carbon-14 dating
3.2 Logarithms, Logarithmic Functions, and Equations	Graphing and evaluating logarithmic functions; exponential and logarithmic forms; common and natural logarithms; solving logarithmic equations	Investments, Richter scale, doubling time for investments, mothers in the workforce
3.3 Exponential and Logarithmic Equations; Properties of Logarithms	Solving exponential equations; logarithmic properties; change of base; exponential inequalities	Carbon-14 dating, measuring earthquake intensity, sales decay
3.4 Exponential and Logarithmic Models	Modeling with exponential functions; spreadsheets; constant percent change; logarithmic models; comparison of models	Insurance premiums, sales decay, carbon-14 dating, smokers, inflation, mothers in the workforce, international visitors to the United States
3.5 Exponential Functions and Investing	Compound interest; continuous compounding; the number *e*; present value; investment models	Future value of an investment, present value of an investment, mutual fund growth
3.6 Annuities; Loan Repayment	Future value of an annuity; present value of an annuity; amortization	Future value of an ordinary annuity, present value of an ordinary annuity, home mortgage, loan repayment

3.1 Exponential Functions

Key Concepts

- Exponential functions
- Transformations of graphs of exponential functions
- Horizontal asymptotes
- Exponential growth function: $y = a^x \quad (a > 1)$
- Exponential decay function: $y = a^{-x} \quad (a > 1)$
- The number e

Section Preview Paramecia

The primitive single-cell animal called a paramecium reproduces by splitting into two pieces (called binary fission) so that the population doubles each time there is a split. If we assume that the population begins with 1 paramecium and doubles each hour, then there will be 2 paramecia after 1 hour, 4 paramecia after 2 hours, 8 after 3 hours, and so on. If we let y represent the number of paramecia in the population after x hours have passed, the points (x, y) that satisfy this function for the first 9 hours of growth are described by the data in Table 3.1 and the plot in Figure 3.1(a). We can show that each of these points and the data points for $x = 10, 11,$ and so forth lie on the graph of the function $y = 2^x$, with $x \geq 0$ (Figure 3.1(b)).

Functions like $y = 2^x$, which have a constant base raised to a variable power, are called **exponential functions**. We discuss exponential functions and their applications in this section.

TABLE 3.1

x (hours)	0	1	2	3	4	5	6	7	8	9
y (paramecia)	1	2	4	8	16	32	64	128	256	512

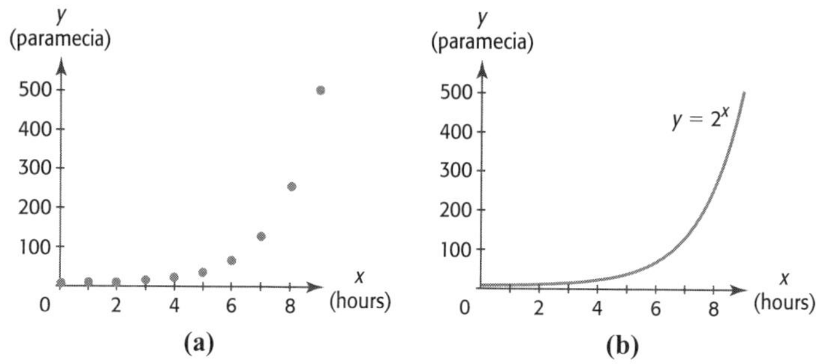

FIGURE 3.1

Exponential Functions

Note that not every point on the graph of $y = 2^x$ in Figure 3.1(b) describes a number of paramecia. For example, the point $(7.1, 2^{7.1})$ is on this graph, but $2^{7.1} \approx 137.187$ does not represent a number of paramecia because it is not a whole number. However, every point describing the number of paramecia is on the graph of $y = 2^x$, so the function $y = 2^x$ with $x \geq 0$ is a continuous model for this discrete growth function.

The function that models the growth of paramecia was restricted to $x \geq 0$ because of the physical setting, but if the domain of the function $y = 2^x$ is not restricted, it is the set of real numbers. Some values satisfying this function are shown in Table 3.2 on the next page, and the graph of the function is shown in Figure 3.2. Note that the range of this function is the set of all positive real numbers because there is no power of 2 that results in a value of 0 or a negative number.

TABLE 3.2

x	y
-3	$2^{-3} = 0.125$
-1.5	$2^{-1.5} \approx 0.35$
-0.5	$2^{-0.5} \approx 0.71$
0	$2^{0} = 1$
0.5	$2^{0.5} \approx 1.41$
1.5	$2^{1.5} \approx 2.83$
3	$2^{3} = 8$
5	$2^{5} = 32$

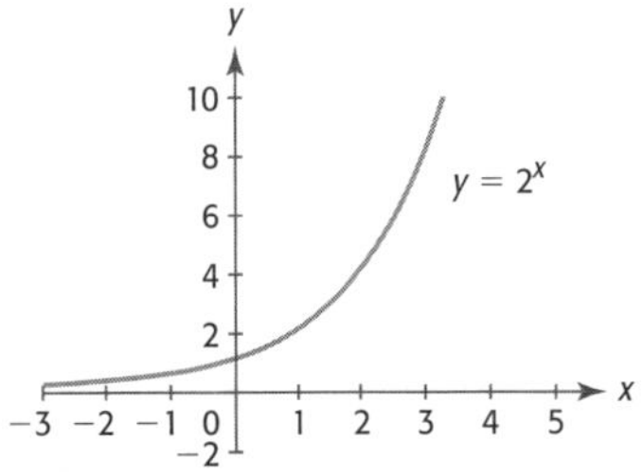

FIGURE 3.2

The graph in Figure 3.2 approaches but never touches the x-axis as x becomes more negative (approaching $-\infty$), so the x-axis is a **horizontal asymptote** for the graph. As previously mentioned, the function $y = 2^x$ is an example of a special class of functions called **exponential functions**. In general, we define an exponential function as follows.

> **EXPONENTIAL FUNCTION**
>
> If b is a positive real number, $b \neq 1$, then the function $f(x) = b^x$ is an exponential function. The constant b is called the *base* of the function and the variable x is the *exponent*.

If we graph another exponential function $y = b^x$ that has a base b greater than 1, the graph will have the same basic shape as the graph of $y = 2^x$. (See the graphs of $y = 1.5^x$ and $y = 12^x$ in Figure 3.3.)

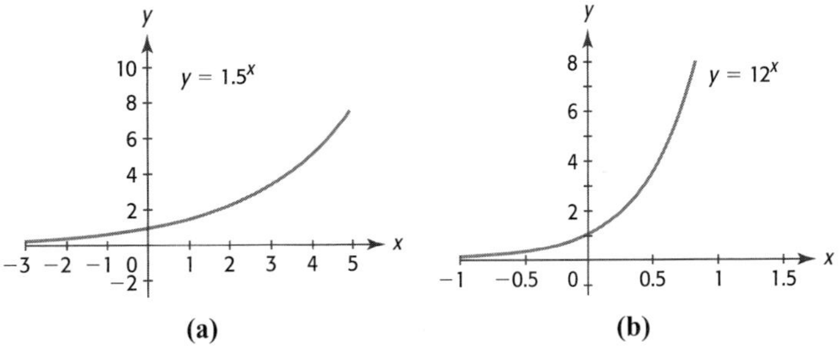

FIGURE 3.3

Exponential Growth

Exponential functions can be used to model growth in many different applications. Whenever the base b of $y = a(b^x)$ is greater than 1 and $a > 0$, the exponential function is increasing and can be used to model growth.

EXPONENTIAL GROWTH FUNCTION

Equation:	$y = a(b^x)$, $b > 1$, $a > 0$
x-intercept:	none
y-intercept:	$(0, a)$
Domain:	all real numbers
Range:	all real numbers $y > 0$
Horizontal asymptote:	x-axis (the line $y = 0$)
Shape:	increasing on domain and concave up

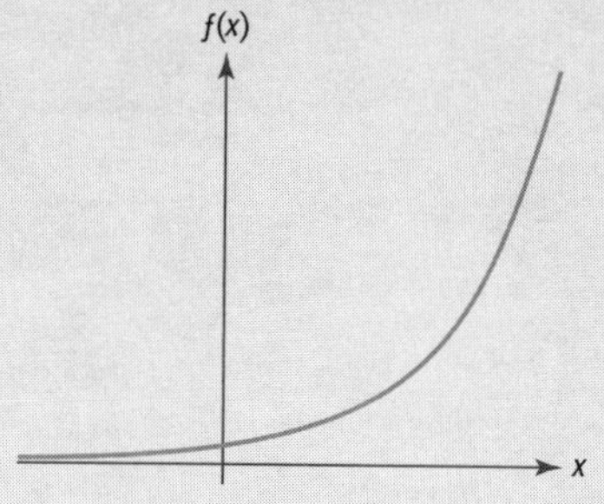

EXAMPLE 1 Head Start Program Growth

Federal funds for the Head Start Program in the U.S. Department of Health and Human Services increased dramatically between 1975 and 2000. The on-budget (federal) funds that were allocated for Head Start during this time period can be described by the model

$$f(x) = 400(1.107)^x \text{ million dollars}$$

where x is the number of years after 1975.

a. Graph this function. Find and interpret the vertical-axis intercept.

b. According to the model, how much was allocated for Head Start in 2000? (Source: *Federal Support for Education, Fiscal Years 1980–2000*, National Center for Educational Statistics, U.S. Department of Education, September 2000.)

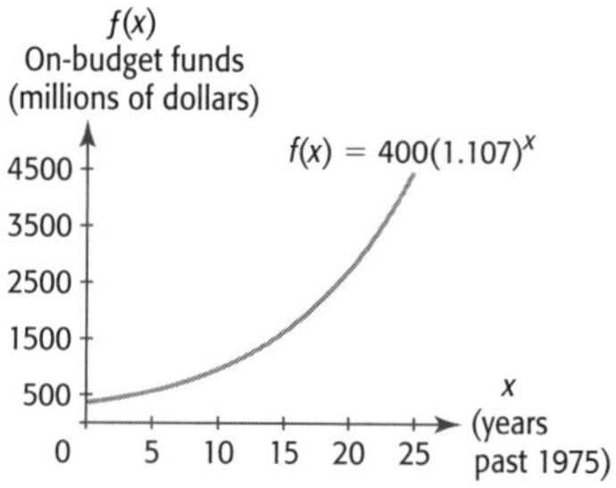

FIGURE 3.4(a)

Solution

a. The graph is shown in Figure 3.4(a). The vertical-axis intercept is found by evaluating the function at $x = 0$, obtaining

$$f(0) = 400(1.107)^0 = 400$$

Because our data is aligned so that $x = 0$ at the end of 1975, one possible interpretation is this: $400 million in federal funds were allocated for the Head Start Program in 1975.

b. The year 2000 is 25 years after 1975, so we find the allocation for 2000 by evaluating $f(25)$:

$$f(25) = 400(1.107)^{25} \approx 5078.68 \text{ million dollars}$$

so federal funds allocated for Head Start in 2000 amounted to about $5.079 billion. Note that we could find the value of the function at $x = 25$ graphically or numerically by using the table feature of a graphing utility. ■

Spreadsheet Solution

As with linear and quadratic functions, computer software of several types can be used to create exponential graphs, and spreadsheets like Excel can also be used. Figure 3.4(b) shows the output of $y = 400(1.107)^x$ at several values of x and the graph of this function on Excel.

	A	B
1	0	400
2	5	664.9639
3	10	1105.443
4	15	1837.699
5	20	3055.008
6	25	5078.676

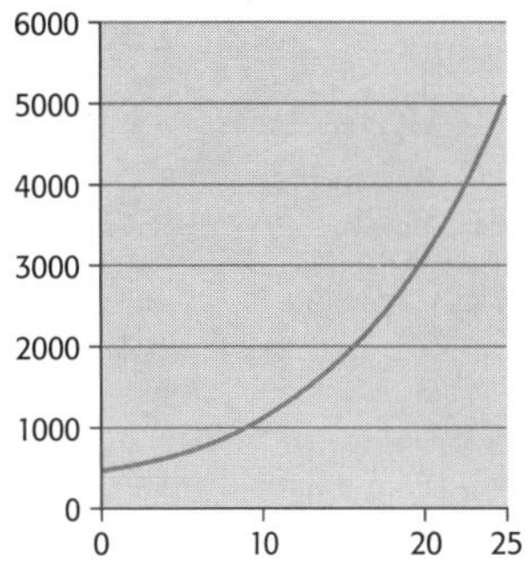

FIGURE 3.4(b)

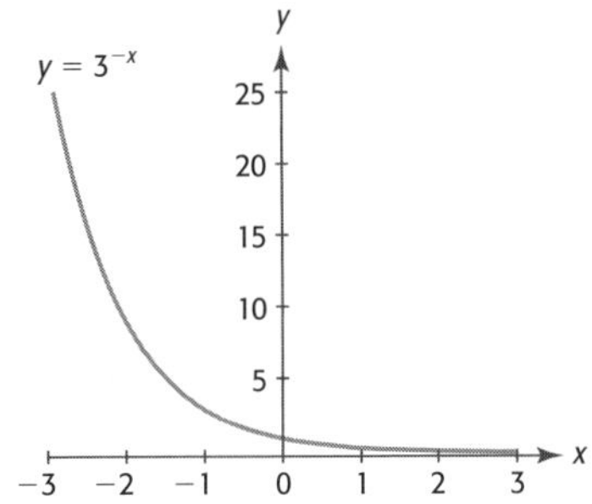

FIGURE 3.5

Exponential Decay

Replacing x with $-x$ in $y = f(x)$ gives a function whose graph is a reflection of the graph of $y = f(x)$ about the y-axis. For example, the graph of $y = 3^{-x}$ is the reflection of the graph of $y = 3^x$, and the function $y = 3^{-x}$ decreases rather than increases (see Figure 3.5).

We can also write the function $y = 3^{-x}$ in the form $y = \left(\frac{1}{3}\right)^x$ because

$$3^{-x} = (3^{-1})^x = \left(\frac{1}{3}\right)^x$$

In general, a function of the form $f(x) = a(b^{-x})$ with $b > 1$ can also be written in the form $f(x) = a(c^x)$ with $0 < c < 1$ and $c = \frac{1}{b}$. Functions with equations of this form can be used to model **exponential decay**.

EXPONENTIAL DECAY FUNCTION

Equation:	$y = a(b^{-x}), b > 1, a > 0$ or $y = a(c^x), 0 < c < 1, a > 0$
x-intercept:	none
y-intercept:	$(0, a)$
Domain:	all real numbers
Range:	$y > 0$
Horizontal asymptote:	x-axis (the line $y = 0$)
Shape:	decreasing on domain

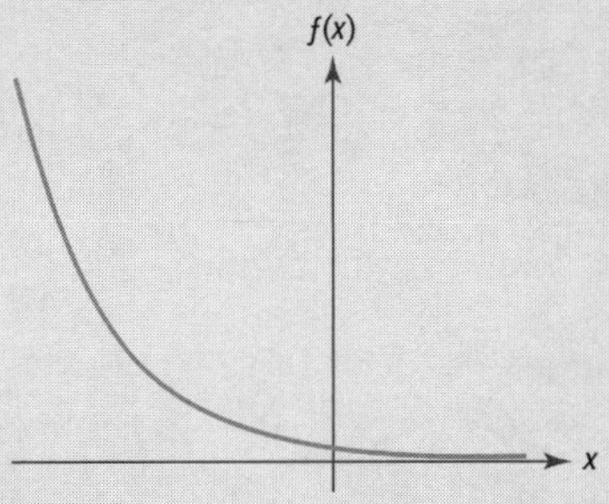

EXAMPLE 2 Sales Decay

It pays to advertise, and it is frequently true that weekly sales will drop rapidly for certain products after an advertising campaign ends. This decline in sales is called *sales decay*. Suppose that the decay in the sales of a product is given by

$$S = 1000(2^{-0.5x}) \text{ dollars}$$

where x is the number of weeks after the end of a sales campaign. Use this function to answer the following.

a. What is the level of sales when the advertising campaign ends?

b. What is the level of sales 1 week after the end of the campaign?

c. Use a graph of the function to estimate the week in which sales equal $500.

d. According to this model, will sales ever fall to zero?

Solution

a. The campaign ends when $x = 0$, so $S = 1000(2^{-0.5(0)}) = 1000(2^0) = 1000(1) =$ $1000.

b. At 1 week after the end of the campaign, $x = 1$. Thus, $S = 1000(2^{-0.5(1)}) = \$707.11$.

c. The graph of this sales decay function is shown in Figure 3.6(a). One way to find the x-value for which $S = 500$ is to graph $y_1 = 1000(2^{-0.5x})$ and $y_2 = 500$ and find the point of intersection of the two graphs. See Figure 3.6(b), which shows that $y = 500$ when $x = 2$.

Thus, sales fall to half their original amount after 2 weeks.

d. The graph of this sales decay function approaches the positive x-axis as x gets large, but it never reaches the x-axis. Thus, sales will never reach a value of $0.

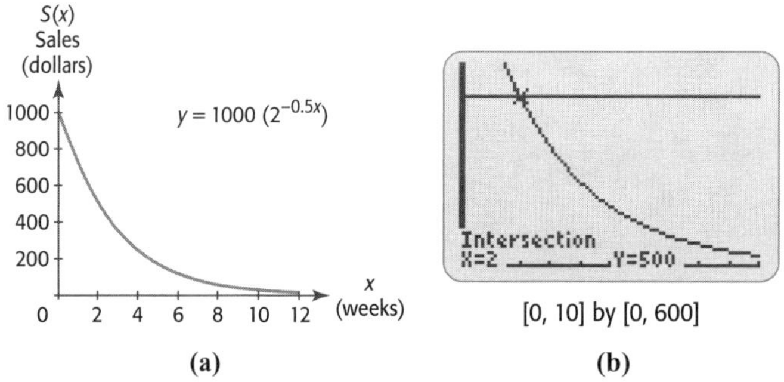

FIGURE 3.6

The equations of growth and decay functions are related as follows.

GROWTH AND DECAY

For initial amount $y_0 > 0$, the equation $y = y_0 b^{kx}$ for $b > 1$ defines a growth function if $k > 0$ and a decay function if $k < 0$.

The Number e

Many real applications involve exponential functions with the base e. The number e is an irrational number with decimal approximation 2.718281828 (to nine decimal places).

$$e \approx 2.718281828$$

The exponential function with base e occurs frequently in biology and in finance. We discuss the derivation of e and how it is used in finance in Section 3.5. Because e is close in value to 3, the graph of the exponential function $y = e^x$ has the same basic shape as $y = 3^x$. (See Figure 3.10, which compares the graphs of $y = e^x$, $y = 3^x$, and $y = 2^x$.)

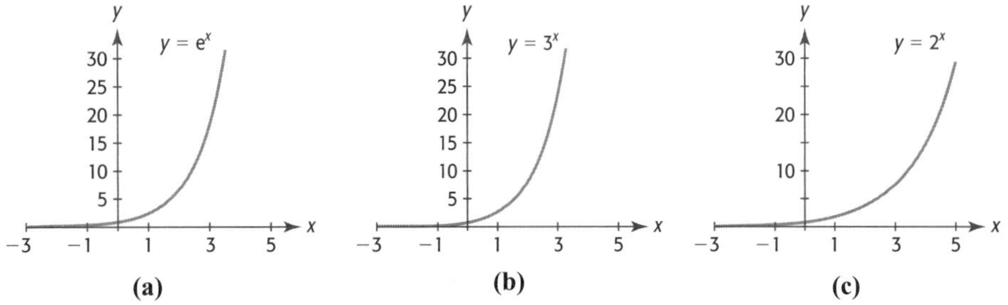

FIGURE 3.7

The growth in the value of an investment is another important application that can be modeled by an exponential function. As we prove in Section 3.5, the future value S of an investment of P dollars for t years at interest rate r, compounded continuously, is given by

$$S = Pe^{rt} \text{ dollars}$$

EXAMPLE 3 Growth of an Investment

If \$10,000 is invested for 15 years at 10%, compounded continuously, what is the future value of the investment?

Solution

The future value of this investment is

$$S = Pe^{rt} = 10{,}000e^{0.10(15)} = 10{,}000e^{1.5} = 44{,}816.89 \text{ dollars}$$ ■

EXAMPLE 4 Comparing Exponential Functions

Compare the functions $f(x) = 5e^{2x}$ and $g(x) = 5e^{-2x}$ algebraically and graphically.

Solution

To compare these functions algebraically, we see that the base in each function is e and that the function g can be created by replacing x with $-x$ in $f(x) = 5e^{2x}$. Thus, the graph of $g(x) = 5e^{-2x}$ is a reflection of the graph of $f(x) = 5e^{2x}$ about the y-axis. The function $f(x) = 5e^{2x}$ is an exponential growth function, and the function $g(x) = 5e^{-2x}$ is an exponential decay function.

The graphs of $y = f(x)$ in Figure 3.8(a) and $y = g(x)$ in Figure 3.8(b) confirm the results of the algebraic investigation. In addition, both graphs have (0, 5) as the y-intercept and the x-axis as a horizontal asymptote.

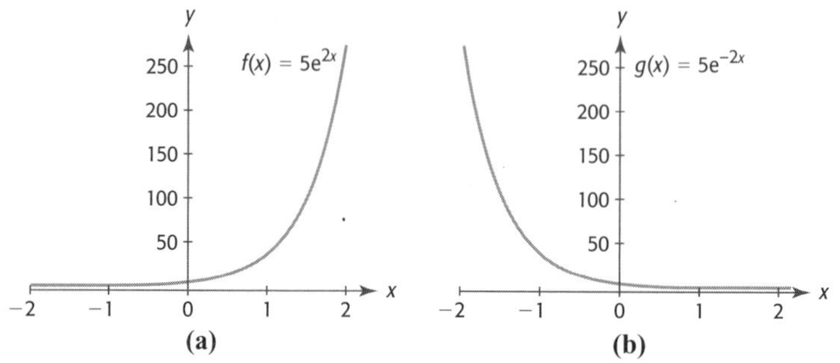

FIGURE 3.8

Note in Example 5 that the exponent of $f(x) = 5e^{2x}$ is $2x$. To see the effect of the 2 in this exponent, we can use a property of exponents* to rewrite $f(x) = 5e^{2x}$ as $f(x) = 5(e^x)^2$, so multiplying the exponent by 2 has the effect of squaring the exponential e^x. In general,

$$y = e^{rx}$$

will increase more rapidly than $y = e^x$ if $r > 1$, will increase more slowly than $y = e^x$ if $0 < r < 1$, and will decrease if $r < 0$.

An exponential decay function can be used to model the amount of a radioactive material that remains after a period of time. Carbon-14 decays over time, with the amount remaining after t years given by

$$f(t) = y_0 e^{-0.00012378t}$$

where y_0 is the original amount. This formula is used by archeologists to date fossils and artifacts from ancient civilizations.

EXAMPLE 5 Carbon-14 Dating

a. If the original amount of carbon-14 present in an artifact is 100 grams, how much remains after 2000 years?

b. What percent of the original amount of carbon-14 remains after 4500 years?

Solution

a. The initial amount y_0 is 100 grams, and the time t is 2000 years, so we substitute these values in $f(t) = y_0 e^{-0.00012378t}$ and calculate:

$$f(2000) = 100e^{-0.00012378(2000)} \approx 78.07$$

Thus the amount of carbon-14 present after 2000 years is approximately 78 grams.

*See the Algebra Toolbox in Appendix C.

b. If y_0 is the original amount, the amount that will remain after 4500 years is

$$f(4500) = y_0 e^{-0.00012378(4500)} = 0.57y_0$$

Thus, approximately 57% of the original amount will remain after 4500 years. (Note that you do not need to know the original amount of carbon-14 in order to answer this question.) ■

3.1 SKILLS CHECK

1. Which of the following functions are exponential functions?
   a. $y = 3x + 3$ **b.** $y = x^3$ **c.** $y = 3^x$
   d. $y = x^2 - 3x$ **e.** $y = e^{x^2-3x}$ **f.** $y = x^e$
2. Determine if each of the following functions is a growth exponential or a decay exponential.
   a. $y = 2^{0.1x}$ **b.** $y = 3^{-1.4x}$
   c. $y = 4e^{-5x}$ **d.** $y = 0.8^{3x}$
3. **a.** Graph the function $f(x) = e^x$ on the window [0, 5] by [−2, 100].
   b. Find $f(1)$, $f(-1)$, and $f(4)$, rounded to three decimal places.
   c. What is the horizontal asymptote of the graph?
   d. What is the y-intercept?
4. **a.** Graph the function $f(x) = 5^x$ on the window [−5, 5] by [−60, 100].
   b. Find $f(1)$, $f(3)$, and $f(-2)$.
   c. What is the horizontal asymptote of the graph?
   d. What is the y-intercept?
5. Graph the function $y = 3^x$ on [0, 5] by [−2, 100].
6. Graph $y = 5e^{-x}$ on [−4, 4] by [−1, 20].

In Exercises 7–12, graph each function.

7. $y = 5^{-x}$
8. $y = 3^{-2x}$
9. $y = 3^x + 5$
10. $y = 2(1.5)^{-x}$
11. $y = 3^{(x-2)} - 4$
12. $y = 3^{x-1} + 2$

In Exercises 13–18, use your knowledge of transformations to compare the graph of the function with the graph of $f(x) = 4^x$. Then graph each function for $-4 \le x \le 4$.

13. $y = 4^x + 2$
14. $y = 4^{x-1}$
15. $y = 4^{-x}$
16. $y = -4^x$
17. $y = 3(4^x)$
18. $y = 3 \cdot 4^{(x-2)} - 3$
19. How does the graph of Exercise 18 compare with the graph of Exercise 17?
20. Which of the functions in Exercises 13–18 are increasing functions, and which are decreasing functions?
21. **a.** Graph $f(x) = 12e^{-0.2x}$ for $-5 \le x \le 15$.
    b. Find $f(10)$ and $f(-10)$.
    c. Does this function represent growth or decay?
22. If $y = 200(2^{-0.01x})$,
    a. find the value of y (to two decimal places) when $x = 20$.
    b. Use graphical or numerical methods to determine the value of x that gives $y = 100$.

3.1 EXERCISES

23. *Sales Decay* At the end of an advertising campaign, weekly sales declined according to the equation $y = 12{,}000(2^{-0.08x})$ dollars, where x is the number of weeks after the end of the campaign.
    a. Determine the sales at the end of the ad campaign.
    b. Determine the sales 6 weeks after the end of the campaign.
    c. Does this model indicate that sales will eventually reach \$0?
24. *Sales Decay* At the end of an advertising campaign, weekly sales declined according to the equation $y = 10{,}000(3^{-0.05x})$ dollars, where x is the number of weeks after the campaign ended.
    a. Determine the sales at the end of the ad campaign.
    b. Determine the sales 8 weeks after the end of the campaign.
    c. How do we know, by inspecting the equation, that this function is decreasing?

25. *Continuous Compounding* If \$8000 is invested for t years at 8% interest compounded continuously, the future value is given by $S = 8000e^{0.08t}$ dollars.
   a. Graph this function for $0 \le t \le 15$.
   b. Use the graph to estimate when the future value will be \$20,000.

26. *Continuous Compounding* If \$35,000 is invested for t years at 9% interest compounded continuously, the future value is given by $S = 35{,}000e^{0.09t}$ dollars.
   a. Graph this function for $0 \le t \le 8$.
   b. Use the graph to estimate when the future value will be \$60,000.

27. *Radioactive Decay* The amount of radioactive isotope thorium-234 present at time t is given by $A(t) = 500e^{-0.02828t}$ grams, where t is the time in years that the isotope decays. The initial amount present is 500 grams.
   a. How many grams remain after 10 years?
   b. Graph this function for $0 \le t \le 100$.
   c. If the half-life is the time it takes for half of the initial amount to decay, use graphical methods to estimate the half-life of this isotope.

28. *Radioactive Decay* A breeder reactor converts stable uranium-238 into the isotope plutonium-239. The decay of this isotope is given by $A(t) = 100e^{-0.00002876t}$, where $A(t)$ is the amount of the isotope at time t (in years) and 100 grams is the original amount.
   a. How many grams remain after 100 years?
   b. Graph this function for $0 \le t \le 50{,}000$.
   c. The half-life is the time it takes for half of the initial amount to decay; use graphical methods to estimate the half-life of this isotope.

29. *Sales Decay* At the end of an advertising campaign, weekly sales at an electronics store declined according to the equation $y = 2000(2^{-0.1x})$ dollars, where x is the number of weeks after the end of the campaign.
   a. Graph this function for $0 \le x \le 60$.
   b. Use the graph to find the weekly sales 10 weeks after the campaign ended.
   c. Comment on "It pays to advertise" for this store.

30. *Sales Decay* At the end of an advertising campaign, weekly retail sales of a product declined according to the equation $y = 40{,}000(3^{-0.1x})$ dollars, where x is the number of weeks after the campaign ended.
   a. Graph this function for $0 \le x \le 50$.
   b. Find the weekly sales 10 weeks after the campaign ended.
   c. Should the retailers consider another advertising campaign even if it costs \$5000?

31. *Purchasing Power* The purchasing power (real value of money) decreases if inflation is present in the economy. For example, the purchasing power of \$$R$ after t years of 5% inflation is given by the model

$$P = R(0.95^t) \text{ dollars}$$

   a. What will be the purchasing power of \$40,000 after 20 years of 5% inflation?
   b. How does this impact people planning to retire at age 50?
   (Source: *Viewpoints*, VALIC, 1993)

32. *Purchasing Power* If a retired couple has a fixed income of \$60,000 per year, the purchasing power (real value of money) after t years of 5% inflation is given by the equation $P = 60{,}000(0.95^t)$ dollars.
   a. What is the purchasing power of their income after 4 years?
   b. Test numerical values in this function to determine the years when their purchasing power will be less than \$30,000.

33. *Real Estate Inflation* During a 5-year period of constant inflation, the value of a \$100,000 property will increase according to the equation $v = 100{,}000e^{0.05t}$.
   a. What will be the value of this property in 4 years?
   b. Use a table or graph to estimate when this property will double in value.

34. *Inflation* An antique table increases in value according to the function $v(x) = 850(1.04^x)$ dollars, where x is the number of years after 1990.
   a. How much was the table worth in 1990?
   b. If the pattern indicated by the function remains valid, what was the value of the table in 2005?
   c. Use a table or graph to estimate the year when this table will reach double its 1990 value.

35. *Population* The population in a certain city was 53,000 in 2000, and its future size is predicted to be $P(t) = 53{,}000e^{0.015t}$ people, where t is the number of years after 2000.
   a. Does this model indicate that the population is increasing or decreasing?
   b. Use this function to estimate the population of the city in 2005.
   c. Use this function to predict the population of the city in 2010.
   d. What is the average rate of growth between 2000 and 2010?

36. *Population* The population in a certain city was 800,000 in 2003, and its future size is predicted to be $P = 800{,}000e^{-0.020t}$ people, where t is the number of years after 2003.
   a. Does this model indicate that the population is increasing or decreasing?
   b. Use this model to predict the population of the city in 2010.
   c. Use this model to predict the population of the city in 2020.
   d. What is the average rate of change in population between 2003 and 2010?

37. *Carbon-14 Dating* An exponential decay function can be used to model the number of grams of a radioactive material that remain after a period of time. Carbon-14 decays over time, with the amount remaining after t years given by $y = 100e^{-0.00012378t}$ if 100 grams is the original amount.
   a. How much remains after 1000 years?
   b. Use graphical methods to estimate the number of years until 10 grams of carbon-14 remain.

38. *Drugs in the Bloodstream* If a drug is injected into the bloodstream, the percent of the maximum dosage that is present at time t is given by

$$y = 100(1 - e^{-0.35(10-t)})$$

where t is in hours, with $0 \le t \le 10$.
   a. What percent of the drug is present after 2 hours?
   b. Graph this function.
   c. When is the drug totally gone from the bloodstream?

39. *Normal Curve* The "curve" on which many students like to be graded is the bell-shaped normal curve. The equation $y = \frac{1}{\sqrt{2\pi}}e^{-(x-50)^2/2}$ describes the normal curve for a standardized test, where x is the test score before curving.
   a. Graph this function for x between 47 and 53 and for y between 0 and 0.5.
   b. The average score for the test is the score that gives the largest output y. Use the graph to find the average score.

40. *IQ Measure* The frequency of IQ measures x follows a bell-shaped normal curve with equation $y = \frac{1}{\sqrt{2\pi}}e^{-(x-100)^2/20}$. Graph this equation for x between 80 and 120 and for y between 0 and 0.4.

3.2 Logarithms, Logarithmic Functions, and Equations*

Key Concepts

- Logarithmic functions
  Base
  Logarithm
  Exponent
- Logarithmic versus exponential form
- Common logarithms
  Richter scale
- Natural logarithms
  Doubling time
- Logarithmic equations

Section Preview Investing

Suppose \$2500 is invested in an account earning 10% annual interest compounded continuously. How long will it take for the amount to grow to \$5000 (that is, to double its value)? To find the time t in which this investment will double, we can solve

$$5000 = 2500e^{0.10t} \quad \text{or} \quad 2 = e^{0.10t}$$

We can solve this equation graphically as we did in the last section (Figure 3.9). In this section, we learn how to rewrite this equation in a new form, called the **logarithmic form** of the equation, to find the value of t analytically. We also introduce logarithmic functions, discuss the relationship between exponential and logarithmic functions, apply logarithmic functions in real situations, and solve logarithmic equations.

* With contributions by Sandee House.

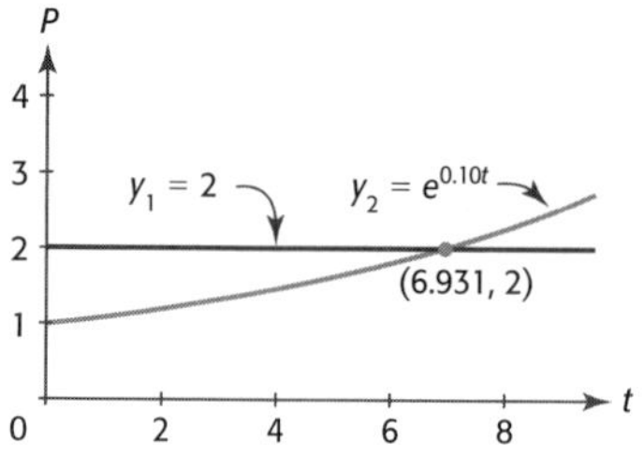

FIGURE 3.9

Before we discuss logarithms, we must first review a few basic terms related to exponents.

In the equation $b^x = A$, b is call the *base,* x is the *exponent,* and A is the *answer* that results when b is raised to the x power.

For example, in $5^2 = 25$, 5 is the base, 2 is the exponent, and 25 is the result of raising 5 to the exponent 2.

If we have an equation such as $5^x = 25$, we can solve the equation by rewriting the right side of the equation with the same base as the left side and then using the property of exponential equations which states the following:

For $A > 0$ and $A \neq 1$, if $A^x = A^y$, then x *must* equal y.

In other words, if the bases are equivalent, the exponents must also be equivalent.

Exponential Equations

EXAMPLE 1 Solving an Exponential Equation using Equivalent Bases

Solve. $5^{x-3} = 25.$

$5^{x-3} = 5^2$ Rewrite 25 as 5^2.

$x - 3 = 2$ Since the bases are equivalent, the exponents are equal.

So, $x = 5$

In some cases, however, it will not be possible for us to determine the base so we must have an alternative method for solving the equation.

Such is the case with $5^x = 100$. Since $5^2 = 25$ and $5^3 = 125$, the value of x lies somewhere between 2 and 3, and can be estimated with the help of a calculator. If this is the case, we will need to find what is known as a *logarithm.*

Definition of Logarithm

A logarithm is an *exponent* to which a base must be raised in order to produce a certain result. For example, we can say the base 5 logarithm of 25 is 2, since 2 is the exponent that base 5 must be raised to in order to get a result of 25. The base 10 logarithm of 1000 is 3, since $10^3 = 1000$.

These logarithm statements can be written in a shorter form.

The base 5 logarithm of 25 is 2 can be written $\log_5 25 = 2$.

The base 10 logarithm of 1000 is 3 can be written $\log 1000 = 3$.

Note that when the base is 10, it not necessary to write the base as it is implied to be 10. This type of logarithm is called a common logarithm and is discussed in more detail later in the chapter. The [log] key on your calculator can be used for base 10 logarithms only.

Equivalent Forms

The equations $b^x = A$ and $x = \log_b A$ are equivalent because in both equations b is the base, x is the exponent, and A is the answer (also referred to as the argument) that results when b is raised to the x power.

LOGARITHMIC FORM OF AN EXPONENTIAL EQUATION

If $x > 0, b > 0, b \neq 1$, then

the exponential equation $b^x = A$ can be written in logarithmic form: $x = \log_b A$.

EXAMPLE 2 Exponential and Logarithmic Forms

a. Write the equation $y = \log_3 x$ in exponential form.

b. Write $x = 3^y$ in logarithmic form.

Solution

a. The base of $y = \log_3 x$ is 3, and y, which equals the logarithm, is the exponent. Thus, the exponential form is $x = 3^y$.

b. To write $x = 3^y$ in logarithmic form, note that the base of the logarithm will be 3, the same as the base of the exponential expression. Also note that y is the exponent in $x = 3^y$, so y is the value of the logarithm. That is, the exponential form of $x = 3^y$ is $y = \log_3 x$. ■

Table 3.3 shows some logarithmic equations and their equivalent exponential forms.

TABLE 3.3

Logarithmic Form	Exponential Form
$\log_2 x = y$	$2^y = x$
$\log_{10}100 = 2$	$10^2 = 100$
$\log_a 1 = 0\ (a > 0)$	$a^0 = 1$
$\log_a a = 1\ (a > 0)$	$a^1 = a$

We can sometimes more easily evaluate logarithmic functions for different inputs by changing the function from logarithmic form to exponential form.

EXAMPLE 3 Evaluating Logarithms

a. Write $y = \log_2 x$ in exponential form.

b. Use the exponential form from part (a) to find y when $x = 8$.

c. Evaluate $\log_2 8$.

Solution

a. To rewrite $y = \log_2 x$ in exponential form, note that the base of the logarithm, 2, becomes the base of the exponential expression, and y equals the exponent. That is, the exponential form is

$$2^y = x$$

b. To find y when x is 8, we solve $2^y = 8$. We can easily see that $y = 3$ because $2^3 = 8$.

c. To evaluate $\log_2 8$, we write $\log_2 8 = y$ in its equivalent exponential form $2^y = 8$. From part (b), $y = 3$. Thus, $\log_2 8 = 3$. ■

Recall that exponential and logarithmic functions are inverse functions. This relationship is shown in the following example.

Solving Exponential Equations using Logarithms

Using the logarithmic form will allow us to solve an equation such as $5^x = 100$. Since the unknown is the exponent, we are looking for the base 5 logarithm of 100. The equivalent form of this equation is $x = \log_5 100$, therefore the value of x can be written $\log_5 100$.

Recall that log key on the calculator works for base 10 logarithms only, so to use the calculator to find an estimate for the value of $\log_5 100$, we will also need the *change of base formula.*

CHANGE OF BASE FORMULA

If $x > 0, a > 0, b > 0, a \neq 1, b \neq 1$, then

$$\log_b A = \frac{\log A}{\log b}$$

These are common logarithms and can be evaluated on the calculator.

EXAMPLE 4 Solving an Exponential Equation using Logarithms

Solve. $5^x = 100.$

Solution: $5^x = 100$ can be written $\log_5 = 100.$

$$5^x = 100$$
$$x = \log_5 100$$
$$x = \frac{\log 100}{\log 2} \quad \text{Change of base formula.}$$
$$x \approx 2.861353116$$

Note: When entering in the calculator, be sure to close the parentheses after the argument of the logarithm.

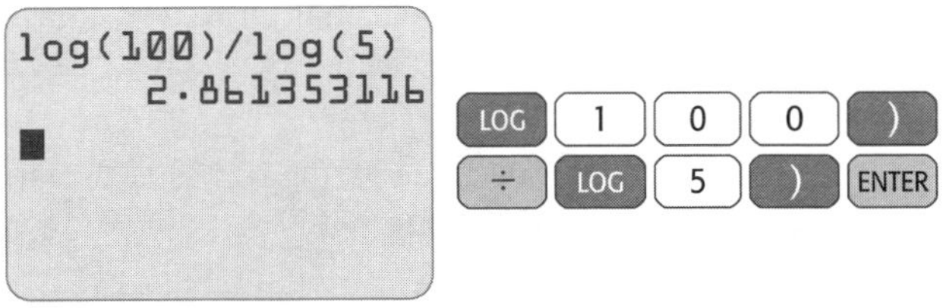

LOGARITHMIC FUNCTION

Equation:	$y = \log_b x$ $(b > 0, b \neq 1)$
x-intercept:	$(1, 0)$
y-intercept:	none
Domain:	$x > 0$
Range:	all real numbers
Vertical asymptote:	y-axis (the line $x = 0$)

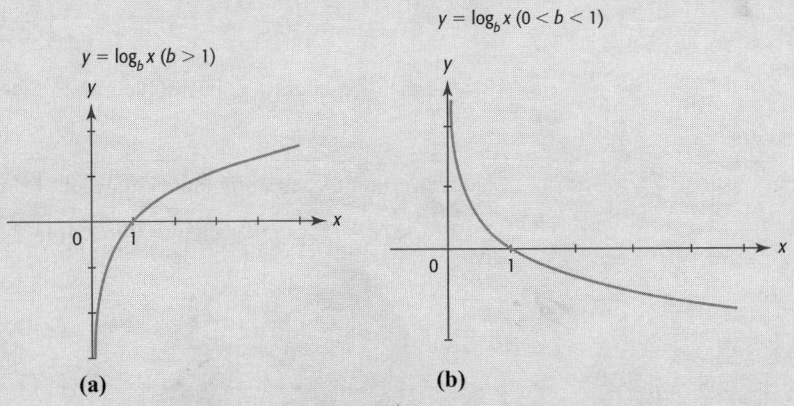

FIGURE 3.10

Figure 3.10 shows graphs of the logarithmic function $y = \log_b x$ for (a) base $b > 1$ and (b) a base $0 < b < 1$.

EXAMPLE 5 Evaluating Base 10 Logarithms

Using each of the indicated methods, find $f(10{,}000)$ if $f(x) = \log x$.

a. Write the equation $y = \log x$ in exponential form and find y when $x = 10{,}000$.

b. Use technology to evaluate log 10,000.

FIGURE 3.11

Solution

a. The exponential form of $y = \log x = \log_{10} x$ is $10^y = x$. Substituting 10,000 for x in this equation gives $10^y = 10{,}000$. Because $10^4 = 10{,}000$, we see that $y = 4$. Thus, $f(10{,}000) = 4$.

b. The value of $\log 10{,}000$ is shown to be 4 on the calculator screen in Figure 3.11, so $f(10{,}000) = 4$. ■

Graphing utilities and spreadsheets can also be used to graph $y = \log x$. Figure 3.12(a) shows the graph of $y = \log x$ on a graphing calculator. The graph of a logarithm with any base can be created with a spreadsheet such as Excel by entering the formula = log(x, base). The graph in Figure 3.12(b) results from the Excel command " = log(x, 10)."

Natural Logarithms

The logarithmic function with base e is $y = \log_e x$. It is defined by the exponential equation $x = e^y$. Because logarithms with base e are used frequently, especially in science, they are called **natural logarithms**. By convention, we use the special notation $\ln x$ to denote $\log_e x$. Most graphing utilities and spreadsheets have the key or command ln x to use in evaluating or graphing functions involving natural logarithms.

We have previously used graphical methods to find the time it would take for an investment to double. We can now develop a formula to find the *doubling time* when the interest is compounded continuously.

If P dollars are invested for t years at an annual interest rate r compounded continuously (at every instant), then the investment will grow to a future value S given by the exponential function

$$S = Pe^{rt}$$

and the investment will be doubled when $S = 2P$, giving

$$2P = Pe^{rt} \quad \text{or} \quad 2 = e^{rt}$$

Converting the equation to its equivalent logarithmic form gives $\log_e 2 = rt$ and solving this equation for t gives the **doubling time**, $t = \dfrac{\ln 2}{r}$.

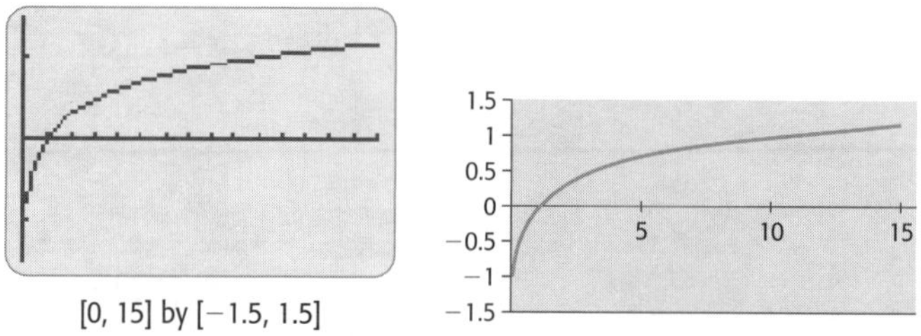

[0, 15] by [−1.5, 1.5]
(a)

(b)

FIGURE 3.12

DOUBLING TIME

The time it takes for an investment to grow to double its value is

$$t = \frac{\ln 2}{r}$$

if the interest rate is r, compounded continuously.

EXAMPLE 6 Time for an Investment to Double

Suppose \$2500 is invested in an account earning 6% annual interest compounded continuously. How long will it take for the amount to grow to \$5000?

Solution

We seek the time it takes to double the investment, and the doubling time is given by

$$t = \frac{\ln 2}{.06} \approx 11.5525$$

Thus, it will take approximately 11.6 years for an investment of \$2500 to grow to \$5000 if it is invested at an annual rate of 6% compounded continuously. Note that this is the same time it would take any investment at 6%, compounded continuously, to double. We can confirm this solution graphically or numerically (Figure 3.13).

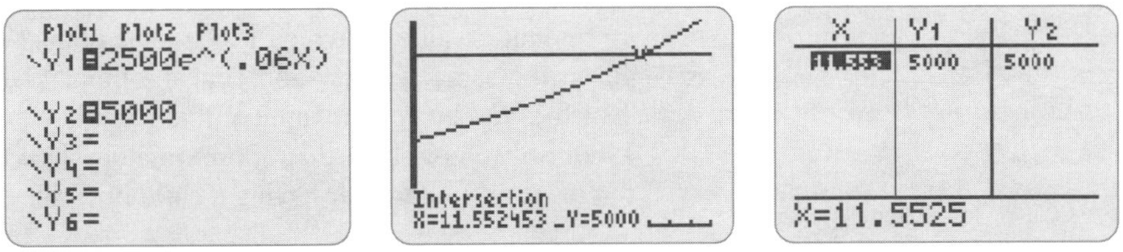

[0, 15] by [0, 6000]

(a) (b) (c)

FIGURE 3.13

EXAMPLE 7 Mothers in the Workforce

Between the years 1976 and 1998, the percent of moms who returned to the workforce within 1 year after having a child is given by $w(x) = 0.815 + 17.062 \ln x$, where x is the number of years after 1970.

(Source: Associated Press)

a. Graph this function.

b. What does this model estimate the percent of mothers returning to the workforce to be for 1994?

c. Use the graph drawn in part (a) to estimate the year in which the percent reached 40.

Solution

a. We are given that the model is valid between 1976 and 1998, so an appropriate domain is $6 \leq x \leq 28$. This logarithmic function increases as the input becomes larger, so a reasonable range includes the interval $w(6) \leq y \leq w(28)$, or $31.386 \leq y \leq 57.669$. We choose to use the viewing window [0, 28] by [0, 60], but there are many other appropriate windows. The graph of $w(x) = 0.815 + 17.062 \ln x$ appears in Figure 3.14(a).

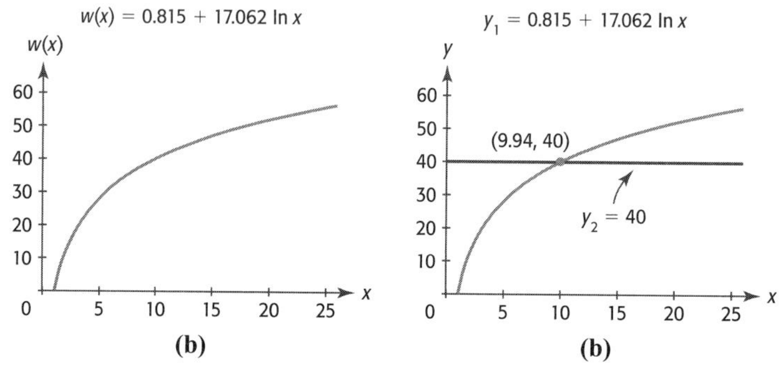

FIGURE 3.14

b. The year 1994 is 24 years after 1970, so we find

$$w(24) = 0.815 + 17.062 \ln 24 \approx 55.04$$

So the model estimates that in 1994 about 55% of moms returned to the workforce within 1 year after having a child.

c. To find the x-value for which $y = 40$, we graph $y_1 = 0.815 + 17.062 \ln x$ and $y_2 = 40$ on the same window and find the point of intersection of the graphs (Figure 3.14(b)). We see that the input of this point is $x \approx 10$, which corresponds to the year 1980. So we conclude that approximately 40% of mothers returned to the workforce within 1 year after having a child in 1980. ■

Solution of Logarithmic Equations

Some logarithmic equations can be solved analytically by using the properties that relate exponential and logarithmic forms of an equation.

EXAMPLE 8 Solving a Logarithmic Equation

a. Solve $6 + 3 \ln x = 12$ by writing the equation in exponential form.

b. Solve the equation graphically.

Solution

a. We first solve the equation for $\ln x$:

$$\begin{aligned} 6 + 3 \ln x &= 12 \\ 3 \ln x &= 6 \\ \ln x &= 2 \end{aligned}$$

Writing $\log_e x = 2$ in exponential form gives

$$x = e^2$$

b. We solve $6 + 3 \ln x = 12$ graphically by the intersection method. Entering $y_1 = 6 + 3 \ln x$ and $y_2 = 12$, we graph using the window $[-2, 10]$ by $[-10, 15]$ and find the point of intersection to be about (7.38906, 12) (Figure 3.15). Thus, the solution to the equation $6 + 3 \ln x = 12$ is $x \approx 7.38906$. Because $e^2 \approx 7.38906$, the solutions agree.

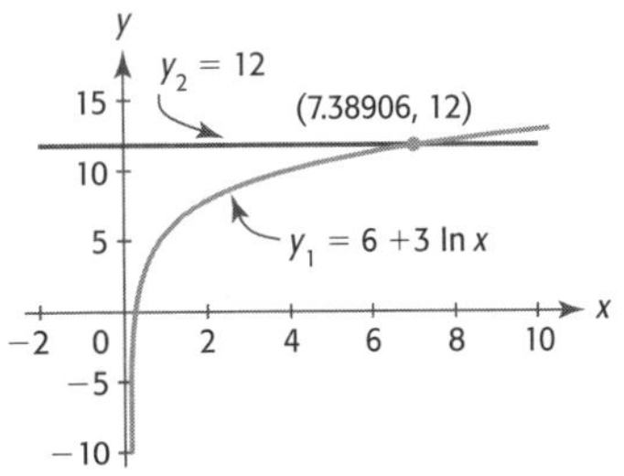

FIGURE 3.15

Some exponential and logarithmic equations are difficult or impossible to solve analytically, and finding approximate solutions to real data problems is frequently easier with graphical methods.

3.2 SKILLS CHECK

In Exercises 1–4, write the logarithmic equations in exponential form.

1. $y = \log_3 x$ **2.** $2y = \log_5 x$

3. $y = \ln(2x)$ **4.** $y = \log(-x)$

In Exercises 5–8, write the exponential equations in logarithmic form.

5. $x = 4^y$ **6.** $m = 3^p$

7. $32 = 2^5$ **8.** $9^{2x} = y$

In Exercises 9–14, evaluate the logarithms, if possible. Round each answer to three decimal places.

9. $\log 7$ **10.** $\log 456$

11. $\ln 86$ **12.** $\log(-12)$

13. $\log 63{,}980$ **14.** $\ln 10$

In Exercises 15–28, find the value of the logarithms without *using a calculator.*

15. $\log_2 32$ **16.** $\log_9 81$

17. $\log_3 27$ **18.** $\log_4 64$

19. $\log_5 625$ **20.** $\log_2 64$

21. $\log_9 27$ **22.** $\log_4 2$

23. $\ln(e^3)$ **24.** $\log 100$

25. $\log_3 \frac{1}{27}$ **26.** $\ln 1$

27. $\ln e$ **28.** $\log 0.0001$

In Exercises 29 and 30, graph the functions.

29. $y = \log_3 x$ **30.** $y = \log_5 x$

In Exercises 31–34, graph the functions with technology.

31. $y = 2 \ln x$

32. $y = 4 \log x$

33. $y = \log(x + 1) + 2$

34. $y = \ln(x - 2) - 3$

35. **a.** Write the inverse of $y = 4^x$ in logarithmic form.
b. Graph $y = 4^x$ and its inverse and discuss the symmetry of their graphs.

36. **a.** Write the inverse of $y = 3^x$ in logarithmic form.
b. Graph $y = 3^x$ and its inverse and discuss the symmetry of their graphs.

37. Write $\log_a a = x$ in exponential form and find x to evaluate $\log_a a$ for any $a > 0, a \neq 1$.

38. Write $\log_a 1 = x$ in exponential form and find x to evaluate $\log_a 1$ for any $a > 0, a \neq 1$.

In Exercises 39–42, solve the equation.

39. $\log_2 x = 3$

40. $\log_4 x = -2$

41. $5 + 2 \ln x = 8$

42. $4 + 3 \log x = 10$

3.2 EXERCISES

43. *Life Span* On the basis of data for the years 1910 through 1998, the expected life span of people in the United States can be described by the function $f(x) = 12.734 \ln x + 17.875$ years, where x is the number of years from 1900 to the person's birth year.
a. What does this model estimate the life span to be for people born in 1925? In 1996? (Give each answer to the nearest year.)
b. Explain why these numbers are so different.
(Source: National Center for Health Statistics)

44. *Japan's Population* The population of Japan for the years 1984–1999 is approximated by the logarithmic function $y = 114.198 + 4.175 \ln x$ million people, with x equal to the number of years after 1980. According to the model, what is the estimated population in 1986? in 2000?
(Source: www.jinjapan.org/stat/)

45. *Poverty Threshold* A single person is considered as living in poverty if his or her income falls below the federal government's official poverty level, which is adjusted every year for inflation. The function that models the poverty threshold for the years 1987–1999 is $f(x) = 282.1666 + 2771.0125 \ln x$ dollars, where x is the number of years after 1980.
a. What does this model give as the poverty threshold in 1990? In 1999?
b. Is this function increasing or decreasing?
c. What factors have led to this change in the threshold?
(Source: U.S. Bureau of the Census)

46. *Out-of-Wedlock Births* For the years 1960–1997, the percent of births among blacks that were out of wedlock is given by the model $y = -61.0399 + 34.326 \ln x$, where x is the number of years after 1950.
a. What does this model give as the percent in 1960? In 1995?
b. Is the percent of out-of-wedlock births increasing or decreasing?
(Source: National Center for Health Statistics)

47. *Doubling Time* If \$4000 is invested in an account earning 10% annual interest compounded continuously, then the number of years that it takes for the amount to grow to \$8000 is $n = \frac{\ln 2}{0.10}$. Find the number of years.

48. *Doubling Time* If \$5400 is invested in an account earning 7% annual interest compounded continuously, then the number of years that it takes for the amount to grow to \$10,800 is $n = \frac{\ln 2}{0.07}$. Find the number of years.

49. *Doubling Time* The number of quarters needed to double an investment when a lump sum is invested at 8% compounded quarterly is given by $n = \frac{\log 2}{0.0086}$. In how many years will the investment double?

50. *Doubling Time* The number of periods needed to double an investment when a lump sum is invested at 12% compounded semiannually is $n = \frac{\log 2}{0.0253}$. How many years pass before the investment doubles in value?

51. *Cost* Suppose the weekly cost for the production of x units of a product is given by $C(x) = 3452 + 50 \ln(x + 1)$ dollars. Use graphical methods to estimate the number of units produced if the total cost is \$3556.

52. *Supply* Suppose the daily supply function for a product is $p = 31 + \ln(x + 2)$, where p is in dollars and x is the number of units supplied. Use graphical methods to estimate the number of units that will be supplied if the price is \$35.70.

53. *Life Span* Based on data for the years 1910–1998, the expected life span of people in the United States can be described by the function $f(x) = 12.734 \ln x + 17.875$, where x is the number of years from 1900 to the person's birth year.
(Source: National Center for Health Statistics)

a. Replace $f(x)$ with y and solve the equation for $\ln x$.
b. Write the exponential form of the result of part (a).
c. Use the exponential function from part (b) to estimate the birth year for which the expected life span is 75 years.
d. Use graphical methods to determine the birth year for which the expected life span is 75 years. Does this agree with the solution in part (c)?

54. *Mothers in Workforce* Between the years 1976 and 1998, the percent of moms who returned to the workforce within 1 year after they had a child is given by $w(x) = 1.11 + 16.94 \ln x$, where x is the number of years past 1970.

a. Replace $w(x)$ with y and solve the equation for $\ln x$.
b. Write the exponential form of the result of part (a).
c. Use the exponential function from part (b) to estimate the year in which the percent is 50.
d. Use graphical methods to determine the birth year for which the percent is 50. Does this agree with the solution in part (c)?
(Source: Associated Press)

Use the following information to answer the questions in Exercises 55–57. To simplify the measurement of the acidity or basicity of a solution, the pH (hydrogen potential) measure was developed. The pH is given by the formula $\text{pH} = -\log[H^+]$, *where* $[H^+]$ *is the concentration of hydrogen ions in moles per liter in the solution.*

55. *pH Levels* Find the pH value of beer for which $[H^+] = 0.0000631$.

56. *pH Levels* The pH value of eggs is 7.79. Find the hydrogen-ion concentration for eggs.

57. *pH Levels* The most common solutions have a pH range between 1 and 14. Use an exponential form of the pH formula to find the values of $[H^+]$ associated with these pH levels.

3.3 Exponential and Logarithmic Equations; Properties of Logarithms

Key Concepts

- Solving an exponential equation by writing it in logarithmic form
- Properties of logarithms
- Special cases of logarithm properties
- Change of base formula
- Solving an exponential equation by using properties of logarithms
- Exponential inequalities
- Solving logarithmic equations

Section Preview Carbon-14 Dating

Carbon-14 dating is a process by which scientists can tell the age of many fossils. To find the age of a fossil that originally contained 1000 grams of carbon-14 and now contains 1 gram, we solve the equation

$$1 = 1000e^{-0.00012378t}$$

An approximate solution of this equation can be found graphically, but sometimes a window that shows the solution is hard to find. If this is the case, it may be easier to solve an exponential equation like this analytically. In this section, we consider two analytical methods of solving exponential equations. The first method involves converting the equation to logarithmic form and then solving the logarithmic equation for the variable. The second method involves taking the logarithm of both sides of the equation and using the properties of logarithms to solve for the variable.

Solving Exponential Equations Using Logarithmic Forms

When we wish to solve an equation for a variable that is contained in an exponent, we can remove the variable from the exponent by converting the equation to its logarithmic form. The steps in this solution method follow.

SOLVING EXPONENTIAL EQUATIONS USING LOGARITHMIC FORMS

To solve an exponential equation using logarithmic form:

1. Rewrite the equation with the term containing the exponent by itself on one side.
2. Divide both sides by the coefficient of the term containing the exponent.
3. Change the new equation to logarithmic form.
4. Solve for the variable.

This method is illustrated in the following example.

EXAMPLE 1 Solving a Base 10 Exponential Equation

a. Solve the equation $3000 = 150(10^{4t})$ for t analytically by converting it to logarithmic form.

b. Solve the equation graphically to confirm the solution.

Solution

a. 1. The term containing the exponent is by itself on one side of the equation.
2. We divide both sides of the equation by 150.

$$3000 = 150(10^{4t})$$
$$20 = 10^{4t}$$

3. We rewrite this equation in logarithmic form.

$$4t = \log_{10} 20 \quad \text{or} \quad 4t = \log 20$$

4. Solving for t gives the solution:

$$t = \frac{\log 20}{4} \approx 0.32526$$

b. We solve the equation graphically by using the intersection method. Entering $y_1 = 3000$ and $y_2 = 150(10^{4t})$, we graph using the window [0, 0.4] by [0, 3500] and find the point of intersection to be about (0.32526, 3000) (Figure 3.16). Thus, the solution to the equation $3000 = 150(10^{4t})$ is $t \approx 0.32526$, as we found analytically in part (a).

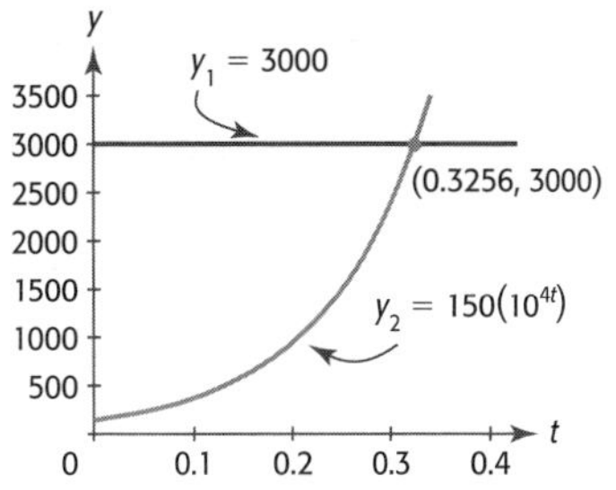

FIGURE 3.16

EXAMPLE 2 Carbon-14 Dating

Radioactive carbon-14 decays according to the equation

$$y = y_0 e^{-0.00012378t}$$

where y_0 is the original amount and y is the amount of carbon-14 at time t years. To find the age of a fossil if the original amount of carbon-14 was 1000 grams and the present amount is 1 gram, we solve the equation

$$1 = 1000e^{-0.00012378t}$$

a. Find the age of the fossil analytically by converting the equation to logarithmic form.

b. Find the age of the fossil using graphical methods.

Solution

a. To write the equation so that the coefficient of the exponential term is 1, we divide both sides by 1000, obtaining

$$0.001 = e^{-0.00012378t}$$

Writing this equation in logarithmic form gives the equation

$$-0.00012378t = \log_e 0.001 \quad \text{or} \quad -0.00012378t = \ln 0.001$$

FIGURE 3.17

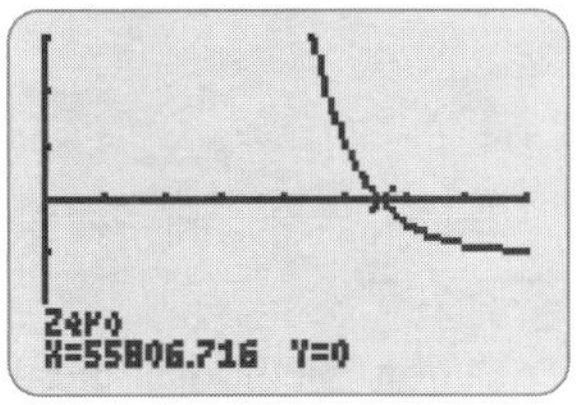

FIGURE 3.18

Solving the equation for t gives

$$t = \frac{\ln 0.001}{-0.00012378} \approx 55{,}806.7$$

Thus, the age of the fossil is calculated to be about 55,807 years.

We can confirm that this is the solution to $1 = 1000e^{-0.00012378t}$ by using a table of values. The result of entering 55,806.7 for x is shown in Figure 3.17.

b. We can solve the equation graphically by using the x-intercept method. Entering

$$y_1 = 1000e^{-0.00012378t} - 1$$

and graphing using the window [0, 80000] by [−3, 3], we find the x-intercept to be 55806.716 (see Figure 3.18.) Thus, as we found in part (a), the age of the fossil is about 55,807 years.

While reading this solution from a graph is easy, actually finding a window that contains the intersection is not. It may be easier to solve the equation analytically, as we did in part (a), and then use graphical or numerical methods to confirm the solution. ■

Logarithmic Properties

We have learned that logarithms are exponents and that a logarithmic function with base a is the inverse of an exponential function with base a . Thus, the properties of logarithms can be derived from the properties of exponents. These properties are useful in solving equations involving exponents and logarithms.

The basic properties of logarithms are easy to derive from the definition of a logarithm.

BASIC PROPERTIES OF LOGARITHMS

For $b > 0, b \neq 1$

1. $\log_b b = 1$
2. $\log_b 1 = 0$
3. $\log_b b^x = x$
4. $b^{\log_b x} = x$

These properties are easily justified by rewriting the logarithmic form as an exponential form or vice versa.

LOGARITHMIC VERSUS EXPONENTIAL FORM OF LOGARITHMIC PROPERTIES

Logarithmic Form	Exponential Form
$\log_b b = 1$	$b^1 = b$
$\log_b 1 = 0$	$b^0 = 1$
$\log_b b^x = x$	$b^x = b^x$
$\log_b x = \log_b x$	$b^{\log_b x} = x$

We can write these properties using base 10 and base e to obtain the following special cases.

LOGARITHMIC PROPERTIES USING BASE 10 AND BASE *e*

Property	Base 10	Base e
1.	$\log 10 = 1$	$\ln e = 1$
2.	$\log 1 = 0$	$\ln 1 = 0$
3.	$\log 10^x = x$	$\ln e^x = x$
4.	$10^{\log x} = x$	$e^{\ln x} = x$

EXAMPLE 3 Logarithmic Properties

Use the basic properties of logarithms to simplify the following:

a. $\log_5 5^{10}$ **b.** $\log_4 4$ **c.** $\log_4 1$ **d.** $\log 10^7$ **e.** $\ln e^3$ **f.** $\log\left(\frac{1}{10^3}\right)$

Solution

a. $\log_5 5^{10} = 10$ **b.** $\log_4 4 = 1$ **c.** $\log_4 1 = 0$

d. $\log 10^7 = 7$ **e.** $\ln e^3 = 3$ **f.** $\log\left(\frac{1}{10^3}\right) = \log 10^{-3} = -3$ ■

Other logarithmic properties are helpful in simplifying logarithmic expressions, which in turn help us solve some logarithmic equations. These properties are also directly related to the properties of exponents.

Change of Base

Most calculators and other graphing utilities can be used to evaluate logarithms and graph logarithmic functions if the base is 10 or e. Graphing logarithmic functions with other bases usually requires converting to exponential form to determine outputs and then graphing the functions by hand. However, we can use a special formula called the **change of base formula** to rewrite logarithms so that the base is 10 or e.

CHANGE OF BASE FORMULA

If $b > 0, b \neq 1, a > 0, a \neq 1$, and $x > 0$, then

$$\log_a x = \frac{\log_b x}{\log_b a}$$

In particular, for base 10 and base e,

$$\log_a x = \frac{\log x}{\log a} \quad \text{and} \quad \log_a x = \frac{\ln x}{\ln a}$$

EXAMPLE 4 Applying the Change of Base Formula

a. Evaluate $\log_8 124$.

Graph the functions in parts (b) and (c) by changing each logarithm to a common logarithm and then by changing the logarithm to a natural logarithm.

b. $y = \log_3 x$

c. $y = \log_2(-3x)$

Solution

a. The change of base formula (to base 10) can be used to evaluate this logarithm:

$$\log_8 24 = \frac{\log 124}{\log 8} = 2.318 \text{ approximately}$$

b. Changing to base 10 gives $\log_3 x = \dfrac{\log x}{\log 3}$, so we graph $y = \dfrac{\log x}{\log 3}$ (Figure 3.19(a)). Changing to base e gives $\log_3 x = \dfrac{\ln x}{\ln 3}$, so we graph $y = \dfrac{\ln x}{\ln 3}$ (Figure 3.19(b)). Note that the graphs appear to be identical (as they should be because they both represent $y = \log_3 x$).

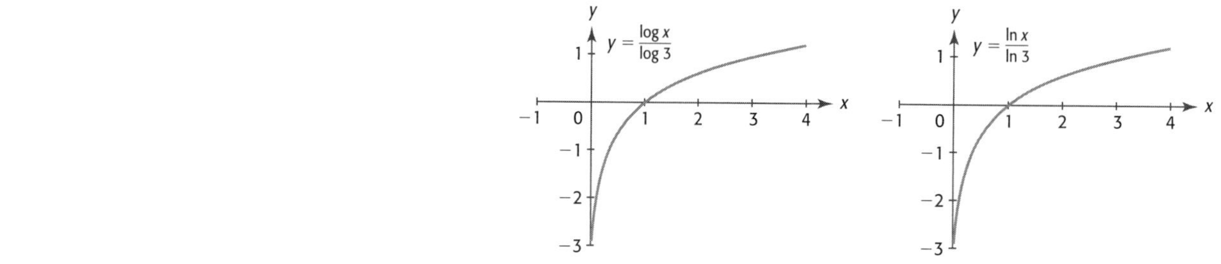

(a) (b)

FIGURE 3.19

c. Changing to base 10 gives $\log_2(-3x) = \dfrac{\log(-3x)}{\log 2}$, so we graph $y = \dfrac{\log(-3x)}{\log 2}$ (Figure 3.20(a)). Changing to base e gives $\log_2(-3x) = \dfrac{\ln(-3x)}{\ln 2}$, so we graph $y = \dfrac{\ln(-3x)}{\ln 2}$ (Figure 3.20(b)). Note that the graphs are identical. Also notice that because logarithms are defined only for positive inputs, x must be negative so that $-3x$ is positive. Thus, the domain of $y = \log_2(-3x)$ is the interval $(-\infty, 0)$.

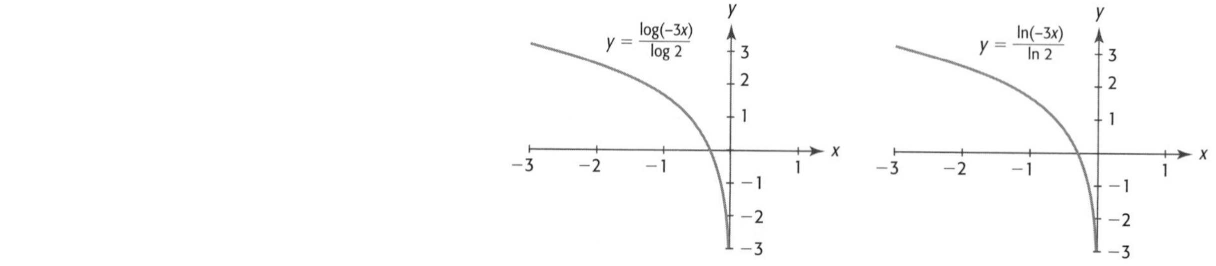

(a) (b)

FIGURE 3.20 ■

The change of base formula is also useful in solving exponential equations whose base is neither 10 nor e.

EXAMPLE 5 Investment

If \$10,000 is invested for t years at 10% compounded annually, the future value is given by

$$S = 10{,}000(1.10^t)$$

In how many years will the investment grow to \$45,950?

Solution

We seek to solve the equation

$$45{,}950 = 10{,}000(1.10^t)$$

Dividing both sides by 10,000 gives

$$4.5950 = 1.10^t$$

Writing the equation in logarithmic form solves the equation for t:

$$t = \log_{1.10} 4.5950$$

Using the change of base formula gives the number of years:

$$t = \log_{1.10} 4.5950 = \frac{\log 4.5950}{\log 1.10} = 16 \text{ years}$$

■

Solving Exponential Equations Using Logarithmic Properties

Now that we have the properties of logarithms at our disposal, we consider a second method for solving exponential equations, which may be easier to use. This method of solving an exponential equation involves taking the logarithm of both sides of the equation and then using properties of logarithms to write the equation in a form that we can solve.

SOLVING EXPONENTIAL EQUATIONS USING LOGARITHMIC PROPERTIES

To solve an exponential equation using logarithmic properties:

1. Rewrite the equation with a base raised to a power on one side.
2. Take the logarithm, base e or 10, of both sides of the equation.
3. Use a logarithmic property to remove the variable from the exponent.
4. Solve for the variable.

This method is illustrated in the following example.

EXAMPLE 6 Solution of Exponential Equations

Solve the following exponential equations:

a. $4096 = 8^{2x}$ **b.** $500 = 10e^{0.1x}$

Solution

a. 1. This equation has the base 8 raised to a variable power on one side.
2. Taking the logarithm, base 10, of both sides of the equation $4096 = 8^{2x}$ gives

$$\log 4096 = \log 8^{2x}$$

3. Using Logarithmic Property 7 removes the variable x from the exponent:

$$\log 4096 = 2x \log 8$$

4. Solving for x gives the solution.

$$\frac{\log 4096}{2 \log 8} = x$$

$$x = 2$$

b. To get the equation in a form with a base raised to a power on one side of the equation, we divide both sides by 10.

$$500 = 10e^{0.1x}$$

$$50 = e^{0.1x}$$

Taking the logarithm, base e, of both sides of this equation gives

$$\ln 50 = \ln e^{0.1x}$$

By Logarithmic Property 3 for base e, $\ln e^{0.1x} = 0.1x$, so

$$\ln 50 = 0.1x$$

$$\frac{\ln 50}{0.1} = x$$

$$x \approx 39.120$$

■

EXAMPLE 7 Sales Decay

After the end of an advertising campaign, the daily sales of Genapet fell rapidly, with daily sales given by $S = 3200e^{-0.08x}$ dollars, where x is the number of days from the end of the campaign. For how many days after the campaign ended were sales at least \$1980?

Solution

To solve this problem, we find the solution to the inequality $3200e^{-0.08x} > 1980$. We begin our solution by solving the related equation.

$$3200e^{-0.08x} = 1980$$

$$e^{-0.08x} = .61875 \quad \text{Divide both sides by 3200.}$$

$$\ln e^{-0.08x} = \ln .61875 \quad \text{Take the logarithm, base } e\text{, of both sides.}$$

$$-0.08x \approx -.4801 \quad \text{Use Logarithmic Property 3.}$$

$$x \approx 6$$

We now investigate the inequality graphically.

To solve this inequality graphically, we graph $y_1 = 3200e^{-0.08x}$ and $y_2 = 1980$ on the same axes, with nonnegative x-values since x represents the number of days. The graph shows that $y_1 = 3200e^{-0.08x}$ is above $y_2 = 1980$ for $x \leq 6$ (Figure 3.21). Thus, the daily revenue is at least \$1980 for each of the first 6 days after the end of the sales campaign.

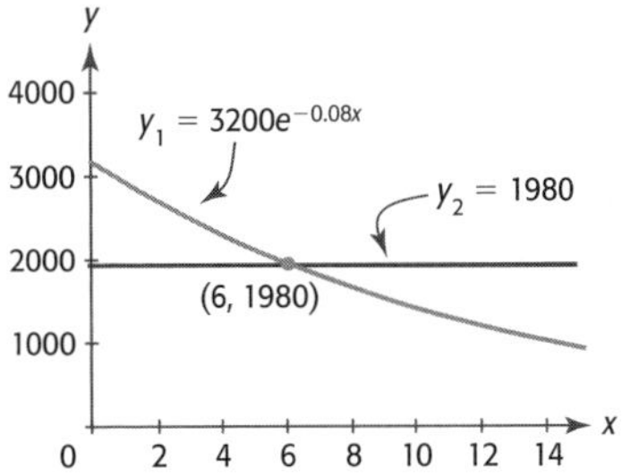

FIGURE 3.21

■

3.3 SKILLS CHECK

In Exercises 1–10, solve the equations analytically and check graphically. Round to 3 decimal places.

1. $1600 = 10^x$ **2.** $4600 = 10^x$

3. $2500 = e^x$ **4.** $54.6 = e^x$

5. $8900 = e^{5x}$ **6.** $2400 = 10^{8x}$

7. $4000 = 200e^{8x}$ **8.** $5200 = 13e^{12x}$

9. $8000 = 500(10^x)$ **10.** $9000 = 400(10^x)$

In Exercises 11–14, use the properties of logarithms to evaluate the expressions.

11. $\log 10^{14}$ **12.** $\ln e^5$

13. $10^{\log_{10} 12}$ **14.** $6^{\log_6 25}$

In Exercises 15–18, use $\log_a(20) = 1.4406$ *and* $\log_a(5) = 0.7740$ *to evaluate each expression.*

15. $\log_a(100)$ **16.** $\log_a(4)$

17. $\log_a 5^3$ **18.** $\log_a \sqrt{20}$

In Exercises 19–22, rewrite each expression as the sum, difference, or product of logarithms, and simplify if possible.

19. $\ln \dfrac{3x - 2}{x + 1}$ **20.** $\log[x^3(3x - 4)^5]$

21. $\log_3 \dfrac{\sqrt[4]{4x + 1}}{4x^2}$ **22.** $\log_3 \dfrac{\sqrt[3]{3x - 2}}{5x^2}$

In Exercises 23–26, rewrite each expression as a single logarithm.

23. $3 \log_2 x + \log_2 y$ **24.** $\log x - \dfrac{1}{3} \log y$

25. $4 \ln(2a) - \ln b$ **26.** $6 \ln(5y) + 2 \ln x$

In Exercises 27–30, use a change of base formula to evaluate each logarithm. Give your answers rounded to four decimal places.

27. $\log_6(18)$ **28.** $\log_7(215)$

29. $\log_8(\sqrt{2})$ **30.** $\log_4(\sqrt[3]{10})$

In Exercises 31–38, solve the equations.

31. $8^x = 1024$

32. $9^x = 2187$

33. $2(5^{3x}) = 31{,}250$

34. $2(6^{2x}) = 2592$

35. $5^{x-2} = 11.18$

36. $3^{x-4} = 140.3$

37. $18{,}000 = 30(2^{12x})$

38. $5880 = 21(2^{3x})$

In Exercises 39–42, solve the logarithmic equations.

39. $5 + \ln(8x) = 23 - 2\ln x$

40. $3\ln x + 8 = \ln(3x) + 12.18$

41. $2\log x - 2 = \log(x - 25)$

42. $\ln(x - 6) + 54 = \ln x - 1000$

In Exercises 43–46, solve the inequalities.

43. $3^x < 243$

44. $7^x \geq 2401$

45. $5(2^x) \geq 2560$

46. $15(4^x) \leq 15{,}360$

3.3 EXERCISES

47. *Supply* The supply function for a certain size boat is given by $p = 340(2^q)$ boats, where p dollars is the price per boat and q is the quantity of boats supplied at that price. What quantity will be supplied if the price is \$10,880 per boat?

48. *Demand* The demand function for a dining room table is given by $p = 4000(3^{-q})$ dollars per table, where p is the price and q is the quantity, in thousands of tables, demanded at that price. What quantity will be demanded if the price per table is \$256.60?

49. *Sales Decay* After a television advertising campaign ended, the weekly sales of Korbel champagne fell rapidly. Weekly sales in a city were given by $S = 25{,}000e^{-0.072x}$ dollars, where x is the number of weeks after the campaign ended.
a. Write the logarithmic form of this function.
b. Use the logarithmic form of this function to find the number of weeks after the end of the campaign before weekly sales fell to \$16,230.

50. *Sales Decay* After a television advertising campaign ended, sales of Genapet fell rapidly, with daily sales given by $S = 3200e^{-0.08x}$ dollars, where x is the number of days after the campaign ended.
a. Write the logarithmic form of this function.
b. Use the logarithmic form of this function to find the number of days after the end of the campaign that passed before daily sales fell to \$2145.

51. *Sales Decay* After the end of an advertising campaign, the daily sales of Genapet fell rapidly, with daily sales given by $S = 3200e^{-0.08x}$ dollars, where x is the number of days from the end of the campaign.
a. What were daily sales when the campaign ended?
b. How many days passed after the campaign ended before daily sales were below half of what they were at the end of the campaign?

52. *Sales Decay* After the end of a television advertising campaign, the weekly sales of Korbel champagne fell rapidly, with weekly sales given by $S = 25{,}000e^{-0.072x}$ dollars, where x is the number of weeks from the end of the campaign.
a. What were weekly sales when the campaign ended?
b. How many weeks passed after the campaign ended before weekly sales were below half of what they were at the end of the campaign?

53. *Snapple Beverage Revenues* Prior to the November 1994 \$1.7 billion takeover proposal by Quaker Oats, Snapple Beverage Corporation's revenues were given by the function $B(t) = 1.337e^{0.718t}$ million dollars, where t is the number of years after 1985.
a. According to the model, what was Snapple's 1995 revenue?
b. If the revenue continued to increase as described by this model, when did it reach \$3599 million? (Source: *U.S. News and World Report*, November 14, 1994)

54. *Population Growth* The population in a certain city was 53,000 in 2000, and its future size is predicted to be $P = 53{,}000e^{0.015t}$ people, where t is the number of years after 2000. Determine when the population will reach 60,000.

55. *Purchasing Power* The purchasing power (real value of money) decreases if inflation is present in the economy. For example, the purchasing power of \$40,000 after t years of 5% inflation is given by the model

$$P = 40{,}000e^{-0.05t} \text{ dollars}$$

How long will it take for the value of a $40,000 pension to have a purchasing power of $20,000 under 5% inflation?
(Source: *Viewpoints*, VALIC, 1993)

56. *Purchasing Power* If a retired couple has a fixed income of $60,000 per year, the purchasing power (adjusted value of the money) after t years of 5% inflation is given by the equation $P = 60{,}000e^{-0.05t}$. In how many years will the purchasing power of their income be half of their current income?

57. *Real Estate Inflation* During a 5-year period of constant inflation, the value of a $100,000 property increases according to the equation $v = 100{,}000e^{0.03t}$ dollars. In how many years will the value of this building be double its current value?

58. *Real Estate Inflation* During a 10-year period of constant inflation, the value of a $200,000 property is given by the equation $v = 200{,}000e^{0.05t}$ dollars. In how many years will the value of this building be $254,250?

59. *Doubling Time* If $$P$ is invested at an annual interest rate r compounded continuously for t years, the future value of the investment is given by $S = Pe^{rt}$. Find a formula for the number of years it will take to double the initial investment.

60. *Doubling Time* If $$P$ is invested at 10% APR compounded continuously for t years, the future value of the investment is given by $S = Pe^{0.10t}$. Find the number of years (to the nearest whole year) it will take to double the investment.

61. *Doubling Time* The number of quarters needed to double an investment when a lump sum is invested at 8% compounded quarterly is given by $n = \log_{1.02} 2$.
a. Use the change of base formula to find n.
b. In how many years will the investment double?

62. *Doubling Time* The number of periods needed to double an investment when a lump sum is invested at 12% compounded semiannually is given by $n = \log_{1.06} 2$.
a. Use the change of base formula to find n.
b. How many years pass before the investment doubles in value?

63. *Annuities* If $2000 is invested at the end of each year in an annuity that pays 5% compounded annually, the number of years it takes for the future value to amount to $40,000 is given by $t = \log_{1.05} 2$. Use the change of base formula to find the number of years until the future value is $40,000.

64. *Annuities* If $1000 is invested at the end of each year in an annuity that pays 8% compounded annually, the number of years it takes for the future value to amount to $30,000 is given by $t = \log_{1.08} 3.4$. Use the change of base formula to find the number of years until the future value is $30,000.

65. *Investment* At the end of t years, the future value of an investment of $10,000 in an account that pays 8% APR compounded monthly is

$$S = 10{,}000\left(1 + \frac{0.08}{12}\right)^{12t} \text{ dollars}$$

Assuming no withdrawals or additional deposits, how long will it take for the investment to amount to $40,000?

66. *Investment* At the end of t years, the future value of an investment of $25,000 in an account that pays 12% APR compounded quarterly is

$$S = 25{,}000\left(1 + \frac{0.12}{4}\right)^{4t} \text{ dollars}$$

In how many years will the investment amount to $60,000?

67. *Drugs in a Bloodstream* The concentration of a drug in the bloodstream from the time the drug is injected until 8 hours later is given by

$$y = 100(1 - e^{-0.312(8-t)}) \text{ percent}$$

where the drug is administered at time $t = 0$. In how many hours will the drug concentration be 79% of the initial dose?

68. *Drugs in the Bloodstream* If a drug is injected into the bloodstream, the percent of the maximum dosage that is present at time t is given by $y = 100(1 - e^{-0.35(10-t)})$, where t is in hours, with $0 \le t \le 10$. In how many hours will the percent reach 65%?

69. *Recording Media Sales* On the basis of data collected between 1982 and 1997, manufacturer's shipments of vinyl record albums (LPs and EPs) declined according to the function $A(t) = 482.794e^{-0.351t}$ million albums, where t is the number of years after 1980.
a. Graph the function A for $0 \le t \le 20$.
b. If album sales continue according to the pattern indicated in the graph, what sales level are vinyl record albums approaching?
c. According to the function, when did vinyl record album sales fall to 5 million?
d. Why are the sales decreasing?
(Source: *Statistical Abstract of the United States, 1998*)

70. *Sales Decay* After a television advertising campaign ended, the weekly sales of Korbel champagne fell rapidly. Weekly sales in a city were given by $S = 25{,}000e^{-0.072x}$ dollars, where x is the number of weeks after the campaign ended.
 a. Use the logarithmic form of this function to find the number of weeks after the end of the campaign that passed before weekly sales fell below \$16,230.
 b. Check your solution by graphical or numerical methods.

71. *Sales Decay* After a television advertising campaign ended, the weekly sales of Turtledove bars fell rapidly. Weekly sales are given by $S = 600e^{-0.05x}$ thousand dollars, where x is the number of days after the campaign ended.
 a. Use the logarithmic form of this function to find the number of weeks after the end of the campaign before weekly sales fell below \$269.60.
 b. Check your solution by graphical or numerical methods.

72. *Investment* At the end of t years, the future value of an investment of \$10,000 in an account that pays 8% compounded monthly is

$$S = 10{,}000\left(1 + \frac{0.08}{12}\right)^{12t} \text{ dollars}$$

Assuming no withdrawals or additional deposits, for what time period (in years) will the future value of the investment be less than \$40,000?

73. *Investment* At the end of t years, the future value of an investment of \$38,500 in an account that pays 12% compounded quarterly is

$$S = 38{,}500\left(1 + \frac{.08}{12}\right)^{12t} \text{ dollars}$$

For what time period (in years) will the future value of the investment be more than \$100,230?

74. *Drugs in the Bloodstream* The concentration of a drug in the bloodstream from the time the drug is administered until 8 hours later is given by $y = 100(1 - e^{-0.312t})$ percent, where the drug is administered at time $t = 0$. For what time period is the amount of drug present less than 79%?

75. *Doubling Time* The future value of a lump sum P that is invested for n years at 10% compounded annually is $S = P(1.10)^n$. Show that the number of years it would take for this investment to double is $n = \log_{1.10} 2$.

3.4 Exponential and Logarithmic Models

Key Concepts
- Modeling with exponential functions
- Constant percent change
- Comparing quadratic and exponential models
- Logarithmic modeling
- Exponents, logarithms, and linear regression

Section Preview Head Start

We saw in Section 3.1 that the federal funds allocated for the Head Start Program can be modeled by the exponential function

$$f(x) = 400.004(1.107)^x \text{ million dollars}$$

where x is the number of years after 1975, and that the decay of carbon-14 can be modeled by the exponential function

$$y = y_0e^{-0.00012378t}$$

where y_0 is the original amount and t is the number of years.

Many other different sources provide data that can be modeled by exponential growth and decay functions, and technology can be used to find exponential functions that model data. In this section we model real data with exponential functions and determine when this type of model is appropriate. We also create exponential functions that model phenomena characterized by constant percent change, and we model real data with logarithmic functions when appropriate.

Modeling with Exponential Functions

The exponential functions discussed above were used to model real data because the data exhibited rapid growth or decay. When the scatter plot of data shows a very rapid increase or decrease, it is possible that an exponential function can be used to model the data. Consider the following examples.

EXAMPLE 1 Insurance Premiums

The monthly premiums for $250,000 in term-life insurance over a 10-year term period increase with the age of the men purchasing the insurance. The monthly premiums for nonsmoking males are shown in Table 3.4.

TABLE 3.4

Age (years)	Monthly Premium for 10-Year Term Insurance (dollars)	Age (years)	Monthly Premium for 10-Year Term Insurance (dollars)
35	123	60	783
40	148	65	1330
45	225	70	2448
50	338	75	4400
55	500		

(Source: Quotesmith.com)

a. Graph the data in the table with x as age and y in dollars.

b. Create an exponential function that models these premiums as a function of age.

c. Graph the data and the exponential function that models the data on the same axes.

Solution

a. The scatter plot of the data is shown in Figure 3.22(a). It shows that the outputs rise rapidly as the inputs increase.

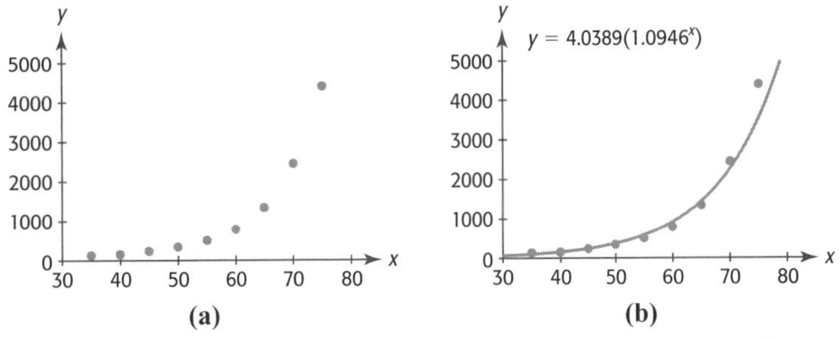

FIGURE 3.22

b. Using technology gives the exponential model, rounded to four decimal places, as

$$y = 4.0389(1.0946^x) \text{ dollars}$$

as the monthly premium where x is the age in years.

c. Figure 3.22(b) shows a scatter plot of the data and a graph of the (unrounded) exponential equation used to model it. The graph of the model appears to be a good, but not perfect, fit for the data. ■

EXAMPLE 2 Smokers

The percent of U.S. males 18 years of age and older who smoked in selected years from 1965 to 1995 is given in Table 3.5.

TABLE 3.5

Year	Percent	Year	Percent
1965	51.6	1990	28
1974	42.9	1991	27.5
1979	37.2	1992	28.2
1983	34.7	1993	27.5
1985	32.1	1994	27.8
1987	31	1995	26.7

(Source: Centers for Disease Control, www.cdc.gov/nchs/datawh/statabl/)

a. Graph the data, with x equal to the number of years after 1960.

b. Find an exponential function that models the data, using as input the number of years after 1960.

c. Graph the data and the exponential function that models the data on the same axes.

d. Use the model to estimate the percent of U.S. male smokers age 18 and over in 2000.

Solution

a. The graph of this data is shown in Figure 3.23(a).

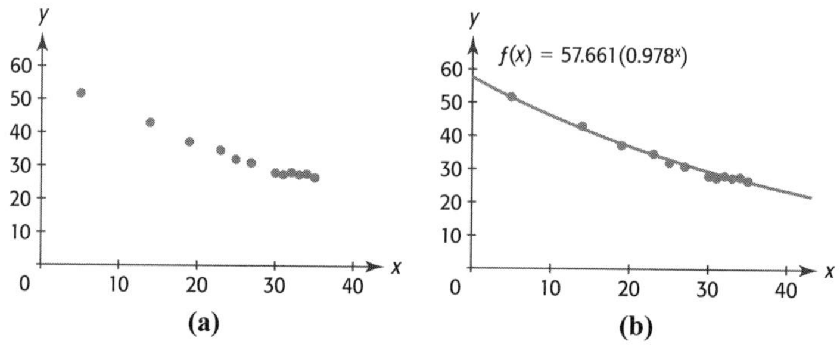

FIGURE 3.23

b. Exponential regression gives the function $f(x) = 57.661(0.978^x)$, with three-decimal place accuracy.

c. The graphs of the (unrounded) model and the data are shown in Figure 3.23(b).

d. Evaluating the function at $x = 40$ gives $f(40) \approx 23.3$, so the estimated percent in 2000 is 23. ■

Spreadsheet Solution

We can use software programs and spreadsheets to find the exponential function that is the best fit for the data. Table 3.6 shows the Excel spreadsheet for the aligned data of Example 2. Selecting the cells containing the data, using Chart Wizard to get the scatter plot of the data, selecting Add Trendline, and picking Exponential give the equation of the exponential function that is the best fit for the data, along with the scatter plot and the graph of the best-fitting function. The equation and graph are shown in Figure 3.24.

TABLE 3.6

	A	B
1	Year x	Percent y
2	5	51.6
3	14	42.9
4	19	37.2
5	23	34.7
6	25	32.1
7	27	31
8	30	28
9	31	27.5
10	32	28.2
11	33	27.5
12	34	27.8
13	35	26.7

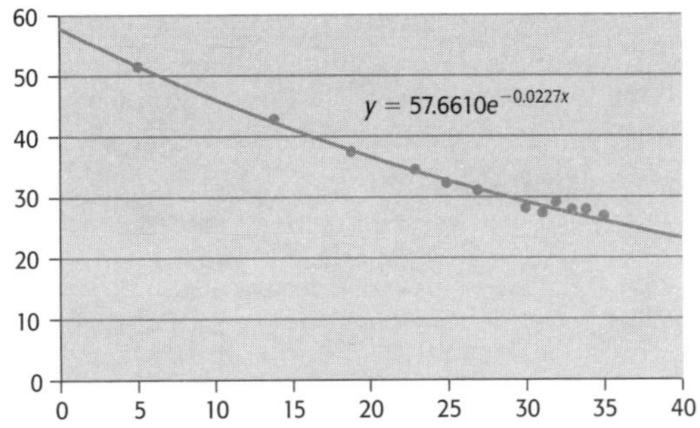

FIGURE 3.24

The function created by Excel,

$$y = 57.661(e^{-0.0227x})$$

is in a different form than the model found in Example 2, which is

$$y = 57.661(0.978^x)$$

However, they are equivalent because $e^{-0.0227} \approx 0.978$ which means that

$$57.661(e^{-0.0227})^x \approx 57.661(0.978)^x$$

Constant Percent Change in Exponential Models

How can we determine if a set of data can be modeled by an exponential function? To investigate this, we look at Table 3.7, which gives the growth of paramecia for each hour up to 9 hours.

TABLE 3.7

x (hours)	0	1	2	3	4	5	6	7	8	9
y (paramecia)	1	2	4	8	16	32	64	128	256	512

If we look at the first differences in the output, we see that they are not constant (Table 3.8). But if we calculate the percent change of the outputs for equally spaced inputs, we see that they are constant. For example, from hour 2 to hour 3, the population grew by 4 units; this represents a 100% increase from the hour-2 population. From hour 3 to hour 4, the population grew by 8, which is a 100% increase over the hour-3 population. In fact, this population increases by 100% each hour. This means that the population 1 hour from now will be the present population plus 100% of the present population.

TABLE 3.8

Outputs	1	2	4	8	16	32	64	128	256	512
First Differences	1	2	4	8	16	32	64	128	256	
Percent Change	100	100	100	100	100	100	100	100	100	

Because the percent change in the outputs is constant for equally spaced inputs in this example, an exponential model ($y = 2^x$) fits the data perfectly.

> **CONSTANT PERCENT CHANGES**
>
> If the percent change of the outputs of a set of data is constant for equally spaced inputs, an exponential function will be a perfect fit for the data.
>
> If the percent change of the outputs is approximately constant for equally spaced inputs, an exponential function will be an approximate fit for the data.

EXAMPLE 3 Sales Decay

Suppose a company develops a product that is released with great expectations and extensive advertising, but sales suffered due to bad word of mouth from dissatisfied customers.

a. Use the monthly sales data shown in Table 3.9 to determine the percent change for each of the months given.

TABLE 3.9

Month	1	2	3	4	5	6	7	8
Sales ($ thousands)	780	608	475	370	289	225	176	137

b. Find the exponential function that models the data.

c. Graph the data and the model on the same axes.

Solution

a. The inputs are equally spaced; the differences of outputs and the percent changes are shown in Table 3.10.

TABLE 3.10

Outputs	780		608		475		370		289		225		176		137
First Differences		−172		−133		−105		−81		−64		−49		−39	
Percent of Change		−22.1		−21.9		−22.1		−21.9		−22.1		−21.8		−22.2	

The percent change of the outputs is approximately −22%. This means that the sales 1 month from now will be approximately 22% less than the sales now.

b. Because the percent change is nearly constant, an exponential function should fit these data well. Technology gives the model

$$y = 999.781(0.780^x)$$

where x is the month and y is the sales in \$1000.

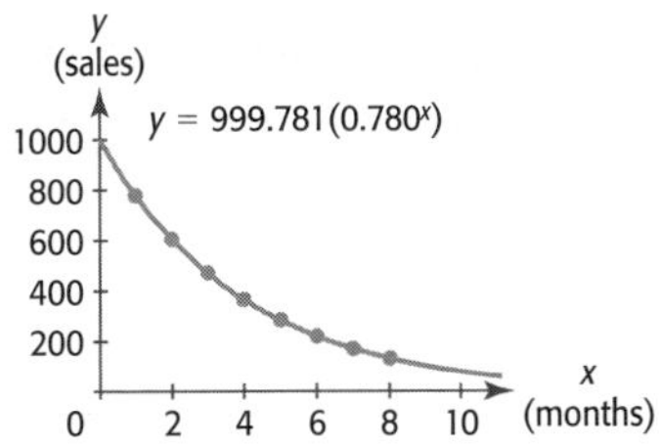

FIGURE 3.25

c. The graphs of the data and the (unrounded) model are shown in Figure 3.25. ■

Exponential Models

Most graphing utilities give the best-fitting exponential model in the form

$$y = a \cdot b^x$$

It can be shown that the constant percent change of this function is

$$(b - 1) \cdot 100\%$$

Thus, if $b > 1$, the function is increasing (growing), and if $0 < b < 1$, the function is decreasing (decaying).

For the function $y = 2^x$, $b = 2$, so the constant percent change is $(2 - 1) \cdot 100\% = 100\%$ (which we found in Table 3.6); for the function $y = 1000(0.78^x)$, the base is $b = 0.78$, and the percent change is $(0.78 - 1) \cdot 100\% = -22\%$.

Because $a \cdot b^0 = a \cdot 1 = a$, the value of $y = a \cdot b^x$ is a when $x = 0$, so a is the y-intercept of the graph of the function. Because $y = a$ when $x = 0$, we say that the **initial value** of the function is a. If the constant percent change (as a decimal) is $r = b - 1$, then $b = 1 + r$, and we can write $y = a \cdot b^x$ as

$$y = a(1 + r)^x$$

EXPONENTIAL MODEL

If a set of data has initial value a and a constant percent change r (written as a decimal) for each equally spaced input x, the data can be modeled by the exponential function

$$y = a(1 + r)^x$$

To illustrate this, suppose we know that the present population of a city is 100,000 and that it will grow at a constant percent of 10% per year. Then we know that the future population can be modeled by an exponential function with initial value $a = 100{,}000$ and constant percent change $r = 10\% = 0.10$ per year. That is, the population can be modeled by

$$y = 100{,}000(1 + .10)^x = 100{,}000(1.10)^x$$

where x is the number of years from the present.

EXAMPLE 4 Inflation

Suppose inflation averages 4% per year for each year from 2000 to 2010. This means that an item that costs \$1 one year will cost \$1.04 one year later. In the second year, the \$1.04 cost will increase by a factor of 1.04, to $(1.04)(1.04) = (1.04)^2$.

a. Write an expression that gives the cost t years after 2000 of an item costing \$1 in 2000.

b. Write an exponential function that models the cost of an item t years from 2000 if its cost was \$100 in 2000.

c. Use the model to find the cost of the item from part (b) in 2005.

Solution

a. The cost of an item is \$1 in 2000, and the cost increases at a constant percent of $4\% = 0.04$ per year, so the cost after t years will be

$$1(1 + 0.04)^t = (1.04)^t \text{ dollars}$$

b. The inflation rate is $4\% = 0.04$. Thus, if an item costs \$100 in 2000, the function that gives the cost after t years is

$$f(t) = 100(1 + 0.04)^t = 100(1.04)^t \text{ dollars}$$

c. The year 2005 is 5 years after 2000, so the cost of an item costing \$100 in 2000 is

$$f(5) = 100(1.04)^5 = 121.67 \text{ dollars}$$

■

Comparison of Models

Sometimes it is hard to determine what type of model is the best fit for a set of data. When a scatter plot exhibits a single curvature without a visible high or low point, it is sometimes difficult to determine whether a power, quadratic, or exponential function should be used to model the data. Suppose after graphing the data points we are unsure which of these models is more appropriate. Then we can find the three models, graph them on the same axes as the data points, and inspect the graphs to see which is a better visual fit.*

Consider the exponential model for insurance premiums as a function of age that we found in Example 1. The curvature indicated by the scatter plot suggests that either a power, quadratic, or exponential model may fit the data. To compare the goodness of fit of the models, we find each model and compare its graph with the scatter plot (Figure 3.26).

*Statistical measures of goodness of fit exist, but caution must be used in applying these measures. For example, the correlation coefficient r found for linear fits to data is different from the coefficient of determination R^2 found for quadratic fits to data.

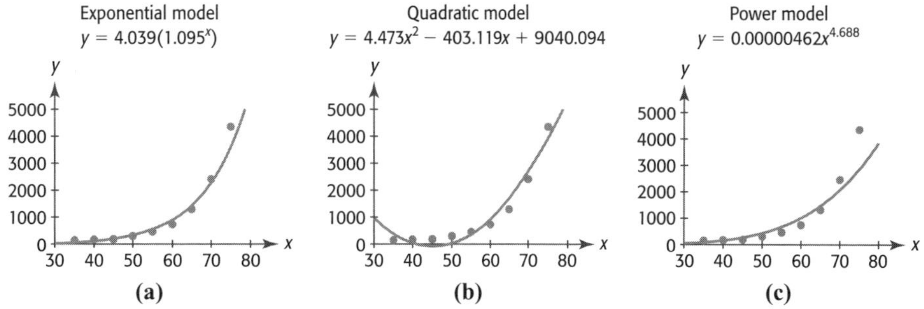

FIGURE 3.26

The exponential function appears to provide a better fit than the quadratic function because the data points seem to be approaching the x-axis as the input becomes closer to zero from the right. (Recall that the x-axis is a horizontal asymptote for the basic exponential function.) The exponential function also appears to be a better fit than the power function, especially for larger values of x.

When fitting exponential functions to data using technology, it is important to know that problems arise if the input values are large. For instance, when the input values are years (like 1995 and 1998), a technology-determined model may appear to be of the form $y = 0 \cdot b^x$, a function that certainly does not make sense as a model for exponential data. At other times when large inputs are used, your calculator or computer software may give an error message and not even produce an equation. To avoid these problems, it is helpful to align the inputs by converting years to the number of years from a specified year (for example, inputting years from 1990 rather than all possible years reduces the size of the inputs).

TECHNOLOGY NOTE

When using technology to fit an exponential model to data, you should align the inputs to reasonably small values.

Logarithmic Models

As with other functions we have studied, we can create models involving logarithms. Data that exhibit an initial rapid increase and then have a slow rate of growth can often be described by the function

$$f(x) = a + b \ln x \qquad (\text{for } b > 0)$$

Note that the parameter a is the vertical shift of the graph of $y = \ln x$ and the parameter b affects how much the graph of $y = \ln x$ is stretched. If $b < 0$, the graph of $f(x) = a + b \ln x$ is reflected about the line $y = a$ and decreases rather than increases. Most graphing utilities will create logarithmic models in the form $y = a + b \ln x$.

EXAMPLE 5 Mothers in the Workforce

Table 3.11 gives the percent of moms who returned to the workforce within 1 year after having a child (for the years 1976–1998). Find a logarithmic model for these data, with x equal to the number of years after 1970.

TABLE 3.11

Year	Returning Mothers (%)	Year	Returning Mothers (%)
1976	31	1988	51
1978	36	1990	52
1980	39	1992	53
1982	43	1994	52
1984	48	1995	55
1986	50	1998	59

(Source: Associated Press)

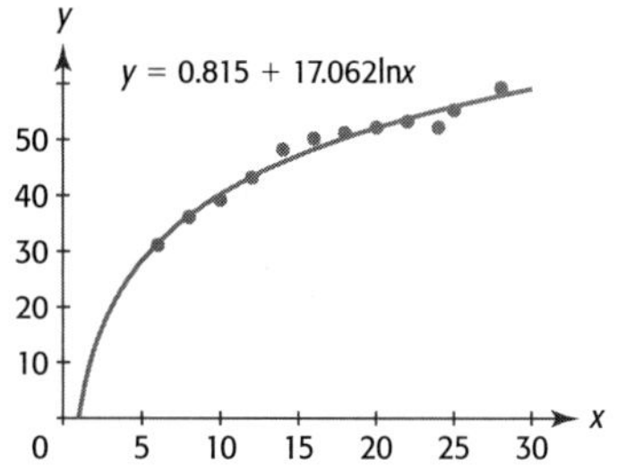

FIGURE 3.27

Solution

Entering the aligned input data (the number of years after 1970) as the values of x and the percents as the values of y, we use logarithmic regression on a graphing utility to find the function that models the data. This function, rounded to three decimal places, is

$$y = 0.815 + 17.062 \ln x$$

where x is the number of years after 1970. The graphs of this function and the data are shown in Figure 3.27. ■

Note that the input data in Example 5 was not aligned as the number of years after 1976 because the first aligned input value would be 0, and the logarithm of 0 does not exist.

> **TECHNOLOGY NOTE**
>
> When using technology to fit a logarithmic model to data, you must align the data so that all input values are positive.

EXAMPLE 6 International Visitors to the United States

The number of international visitors to the United States for each of the years 1986–1997 is given in Table 3.12.

a. Find a logarithmic function that models these data. Align the input to be the number of years after 1980.

b. Graph the equation and the aligned data points. Comment on how the model fits the data.

c. Assuming that the unrounded model $y = f(x)$ is valid in 2002, use it to estimate the number of international visitors to the United States in 2002.

d. Why might this model not apply for 2002?

TABLE 3.12

Years	International Visitors (millions)	Years	International Visitors (millions)
1986	26	1992	47.3
1987	29.5	1994	45.5
1988	34.1	1995	44
1989	36.6	1996	46.3
1990	39.5	1997	48.9
1991	43		

(Source: *1998 World Almanac*)

Solution

a. A logarithmic function that models these data is

$$y = -9.644 + 20.907 \ln x$$

million international visitors, where x is the number of years from 1980.

b. The scatter plot of the data and the graph of the logarithmic function that models the data are shown in Figure 3.28. The fit is fairly good except for 1991 and 1992. The number of international visitors during those 2 years was higher than what is given by the model.

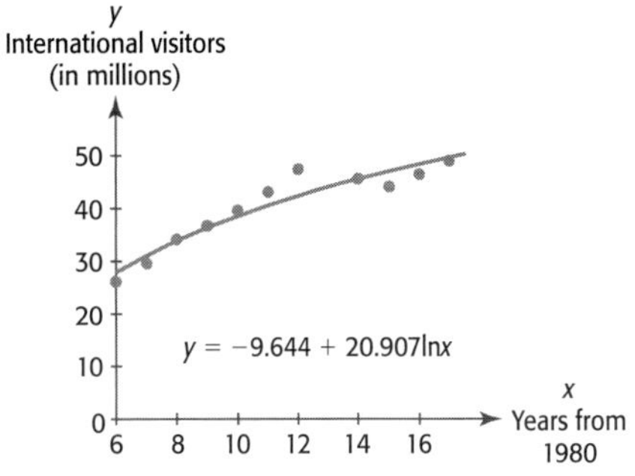

FIGURE 3.28

c. The number of international visitors in 2002 is estimated to be $f(22) = 54.98$, or approximately 55 million.

d. Immigration has been restricted since 9/11 of 2001. ■

Spreadsheet Solution

We can use software programs and spreadsheets to find the logarithmic function that is the best fit for data. Table 3.13 shows the Excel spreadsheet for the data of Example 6. Selecting the cells containing the data, using Chart Wizard to get the scatter plot of the data, selecting Add Trendline, and picking Logarithmic gives the equation of the logarithmic function that is the best fit for the data, along with the scatter plot and the graph of the best-fitting function. The equation and graph are shown in Figure 3.29.

TABLE 3.13

	A	B
1	Year x	International Visitors (millions) y
2	6	26
3	7	29.5
4	8	34.1
5	9	36.6
6	10	39.5
7	11	43
8	12	47.3
9	14	45.5
10	15	44
11	16	46.3
12	17	48.9

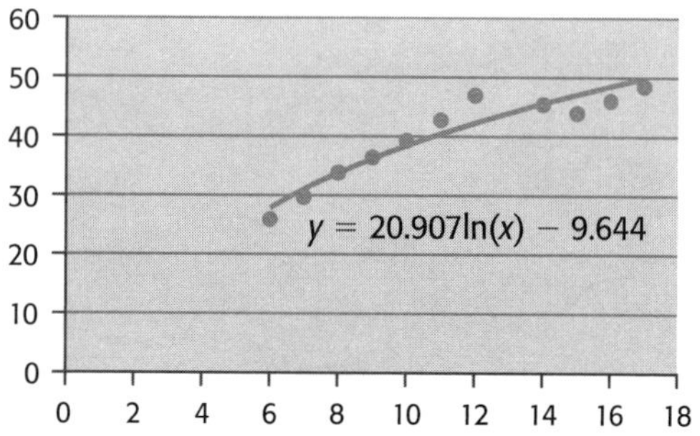

FIGURE 3.29

Exponents, Logarithms, and Linear Regression

Now that we have knowledge of the properties of logarithms and how exponential functions are related to logarithmic functions, we can show that linear regression can be used to create exponential models and logarithmic models that fit data. (The development of linear regression is discussed in Section 1.6.)

Logarithmic Property 3 states that $\log_b b^x = x$ and in particular that $\ln e^x = x$, so if we have data that can be approximated by an exponential function, we can convert the data to a linear form by taking the logarithm, base e, of the outputs. We can use linear regression to find the linear function that is the best fit for the converted data and then use the linear function as the exponent of e, which gives the exponential function that is the best fit for the original data. Consider Table 3.14, which gives the number y of paramecia in a population after x hours. We have shown that the population can be modeled by $y = 2^x$ in Section 3.1.

TABLE 3.14

x (hours)	0	1	2	3	4	5	6	7	8	9
y (paramecia)	1	2	4	8	16	32	64	128	256	512
$\ln y$	0	.6931	1.3863	2.0794	2.7726	3.4657	4.1589	4.8520	5.5452	6.2383

The third row of Table 3.14 has the logarithms, base e, of the numbers (y-values) in the second row, and the relationship between x and $\ln y$ is linear. The first differences of the $\ln y$ values are constant, with the difference approximately equal to 0.6931. Using linear regression on the x and $\ln y$ values (to 10 decimal places) gives the equation of the linear model:*

$$\ln y = 0.6931471806x$$

Because we seek the equation solved for y, we can write this equation in its exponential form, getting

$$y = e^{0.6931471806x}$$

To show that this model for the data is equal to $y = 2^x$, which we found in Section 3.1, we use properties of exponents:

$$y = e^{0.6931471806x} = \left(e^{0.6931471806}\right)^x \approx 2^x$$

Fortunately, most graphing utilities have combined these steps to give the exponential model directly from the data.

3.4 SKILLS CHECK

1. Find the exponential function that models the data in the table below.

x	−2	−1	0	1	2	3	4	5
y	2/9	2/3	2	6	18	54	162	486

2. The following table has the inputs, x, and the outputs for three functions, f, g, and h. Test the percent change of the outputs to determine which function is exactly exponential, which is approximately exponential, and which is not exponential.

x	$f(x)$	$g(x)$	$h(x)$
1	4	2.5	1.5
2	16	6	2.25
3	64	8.5	3.8
4	256	10	5
5	1024	8	11
6	4096	6	17

3. Find the exponential function that is the best fit for $f(x)$ defined by the table in Exercise 2.

4. Find the exponential function that is the best fit for $h(x)$ defined by the table in Exercise 2.

5. a. Make a scatter plot of the data in the table below.
b. Does it appear that a linear model or an exponential model is the better fit for the data?

x	1	2	3	4	5	6
y	2	3.1	4.3	5.4	6.5	7.6

6. Find the exponential function that models the data in the table below. Round the model with three-decimal place accuracy.

x	−2	−1	0	1	2	3
y	5	30	150	1000	4000	20,000

7. Compare the first differences and the percent change of the outputs to determine if the data in the table below should be modeled by a linear or an exponential function.

x	1	2	3	4	5
y	2	6	14	34	81

*This equation could be found directly, in the form $\ln y = ax + b$, where $b = 0$ and a equals the constant difference. This is because the exponential function is a perfect fit for the data in this application.

8. Use a scatter plot to determine if a linear or exponential function is the better fit for the data in Exercise 7.

9. Find the linear *or* exponential function that is the better fit for the data in Exercise 7. Round the model to three-decimal-place accuracy.

10. Find the logarithmic function that models the data in the table below. Round the model to two-decimal-place accuracy.

x	1	2	3	4	5	6	7
y	2	4.08	5.3	6.16	6.83	7.38	7.84

11. a. Make a scatter plot of the data in the table below.
b. Does it appear that a linear model or a logarithmic model is the better fit for the data?

x	1	3	5	7	9
y	−2	1	3	4	5

12. a. Find a logarithmic function that models the data in the table in Exercise 11. Round the model to three-decimal-place accuracy.
b. Find a linear function that models the data.
c. Visually determine which model is the better fit for the data.

3.4 EXERCISES

Report models accurate to three decimal places unless otherwise specified. Use the unrounded function to calculate and to graph the function.

Use the exponential form $y = a(1 + r)^x$ *to model the information in Exercises 13–16.*

13. *Inflation* Suppose the retail price of an automobile is \$30,000 in 1995 and that it increases at 4% per year.
a. Write the equation of the exponential function that models the retail price of the automobile t years after 1995.
b. Use the model to predict the retail price of the automobile in 2010.

14. *Population* Suppose the population of a city is 190,000 in 2000 and that it grows at 3% per year.
a. Write the equation of the exponential function that models the annual growth.
b. Use the model to find the population of this city in 2010.

15. *Sales Decay* At the end of an advertising campaign, weekly sales amounted to \$20,000 and then decreased by 2% each week after the end of the campaign.
a. Write the equation of the exponential function that models the weekly sales.
b. Find the sales 5 weeks after the end of the advertising campaign.

16. *Inflation* The average price of a house in a certain city is \$220,000 in 2000, and it increases at 3% per year.
a. Write the equation of the exponential function that models the average price of a house t years after 2000.
b. Use the model to predict the average price of a house in 2005.

17. *World Population* The following table gives the world population for selected years from 1650 to 2001.
a. Create an exponential function that models these data, with x representing the years after 1600 and y the population in millions. Round the model to four-decimal-place accuracy.
b. Graph the data and the exponential function that models the data on the same axes with window [0, 402] by [0, 6500].

Years	Population (millions)	Years	Population (millions)
1650	503	1950	2406
1750	711	1996	5771
1800	913	1999	6000
1850	1131	2001	6200
1900	1590		

(Source: UN Population Reference Bureau)

18. *Electric Cooperatives* The following figure gives the taxes paid by South Carolina's electric cooperatives for the years 1995–1999, in millions of dollars.

a. Find an exponential function that models the taxes paid in these years. Use the number of years after 1990 as your input.
b. Use the model to estimate the taxes paid in 2005. Discuss the reliability of this estimation.
c. Graph the data and the function on the same axes.

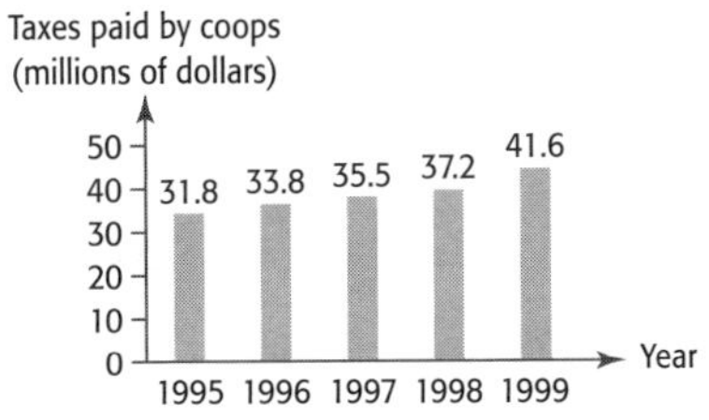

(Source: "Cooperative Pride," *Living in South Carolina*, October 2000)

19. *Insurance Premiums* The table below gives the annual premiums required for a \$250,000 20-year term-life insurance policy on female nonsmokers of different ages.
a. Find an exponential function that models the monthly premium as a function of the age of the female nonsmoking policyholder.
b. Find the quadratic function that is the best fit for this data.
c. Graph each function on the same axes with the data points to determine visually which model is the better fit for the data. Use the window [30, 80] by [−10, 6800].

Age	Monthly Premium for a 20-Year Policy (\$)	Age	Monthly Premium for a 20-Year Policy (\$)
35	145	60	845
40	185	65	1593
45	253	70	2970
50	363	75	5820
55	550		

(Source: Quotesmith.com, December 1999)

20. *World Population* The United Nations has projected the world population from 1995 to 2150 to be as shown in the following table.
a. Find the exponential function that models the data, using $x = 0$ in 1990.
b. Find the quadratic function that is the best fit for the data, using $x = 0$ in 1990.
c. Graph each function on the same axes with the data points to determine which model is the better fit for the data.

Year	Projected Population (millions)	Year	Projected Population (millions)
1995	5666	2075	26,048
2000	6113	2100	52,508
2025	9069	2125	113,302
2050	14,421	2150	255,846

(Source: UN Department of Economic and Social Affairs)

21. *Smokers* The percent of all U.S. persons 18 years of age and older who smoked in selected years from 1965 to 1995 is given in the table below.
a. Graph the data, using $x = 0$ in 1960.
b. Can an exponential function be used to describe the data?
c. Find an exponential function that models the data, using an input of the number of years after 1960.
d. Find a power function that models the data, using as input the number of years after 1960.
e. Which model is the better fit for the data?

Year	1965	1974	1979	1983	1985	1987
Percent	42.3	37.2	33.5	32.2	30.0	28.7

Year	1990	1991	1992	1993	1994	1995
Percent	25.4	25.4	26.4	25.0	25.5	24.7

(Source: Centers for Disease Control, www.cdc.gov/nchs/datawh/statabl/)

22. *Political Contributions* The figure below gives the total contributions, in millions of dollars, to federal political campaigns from the venture-capital industry for the years 1990–2000.

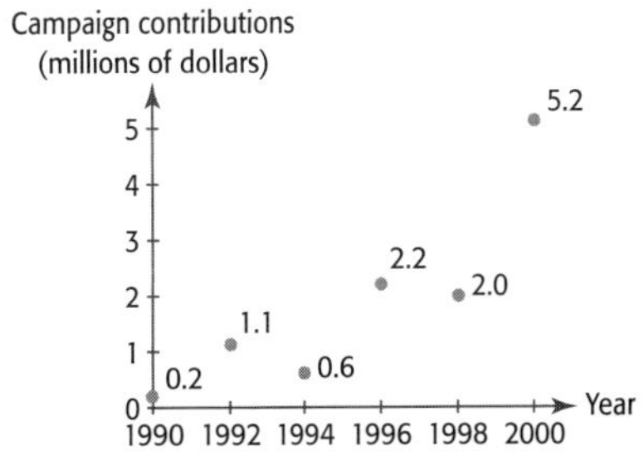

(Source: *The Wall Street Journal*, October 31, 2000)

a. Find an exponential function that models the taxes paid for these years. Use the number of years after 1990 as the input.
b. Find the quadratic function that is the best fit for these data.
c. Graph each function on the same axes with the data points to determine which model is the better fit for the data.

23. *Life Span* The table below gives the life expectancy for people in the United States for the birth years 1910–1998.
a. Find the logarithmic function that models these data, with x equal to 0 in 1900.
b. Find the quadratic function that is the best fit for the data. Round the quadratic coefficient to five decimal places.
c. Graph each of these functions on the same axes with the data points to determine visually which function is the better model for the data for the years 1910–1998.
d. Evaluate both models for the birth year 2010. Which model is better for prediction of life span after 2010?

Birth Year	Life Span (years)	Birth Year	Life Span (years)	Birth Year	Life Span (years)
1910	50.0	1981	74.2	1990	75.4
1920	54.1	1982	74.5	1991	75.5
1930	59.7	1983	74.6	1992	75.5
1940	62.9	1984	74.7	1993	75.5
1950	68.2	1985	74.7	1994	75.7
1960	69.7	1986	74.8	1995	75.8
1970	70.8	1987	75.0	1996	76.1
1975	72.6	1988	74.9	1997	76.5
1980	73.7	1989	75.2	1998	76.7

(Source: National Center for Health Statistics)

24. *Smokers* The percent of persons over age 18 who are smokers for selected years between 1965 and 1995 is given in the following table.
a. Find a logarithmic function that models the data, using an input equal to the number of years after 1960.
b. Use the model to estimate the percent of smokers in 1989.
c. A power model for this set of data was found to be $y = 74.575x^{-.298}$ in Exercise 21. Is this model as good a fit as the logarithmic function, or not?

Year	1965	1974	1979	1983	1985	1987
Smokers (%)	42.3	37.2	33.5	32.2	30.0	28.7

Year	1990	1991	1992	1993	1994	1995
Smokers (%)	25.4	25.4	26.4	25.0	25.5	24.7

(Source: Centers for Disease Control, www.cdc.gov/nchs/datawh/statabl/)

25. *Sexually Active Girls* The percent of girls of age x or younger who have been sexually active is given in the table below.
a. Create a logarithmic function that models the data, using an input equal to the age of the girls.
b. Use the model to estimate the percent for girls age 17 or younger who have been sexually active.
c. Find the quadratic function that is the best fit for the data.
d. Graph each of these functions on the same axes with the data points to determine which function is the better model for the data.

Age	Cumulative Percent Sexually Active Girls	Cumulative Percent Sexually Active Boys
15	5.4	16.6
16	12.6	28.7
17	27.1	47.9
18	44.0	64.0
19	62.9	77.6
20	73.6	83.0

(Source: "National Longitudinal Survey of Youth," *Risking the Future.* Washington D.C.: National Academy Press, 1987)

26. *Sexually Active Boys* The percent of boys age x or younger who have been sexually active is given in the table in Exercise 25.
a. Create a logarithmic function that models the data, using an input equal to the age of the boys.
b. Use the model to estimate the percent for boys age 17 or younger who have been sexually active.
c. Compare the percent that are sexually active for the two genders (see Exercise 25). What do you conclude?

3.5 Exponential Functions and Investing

Key Concepts
- Future value of investments
- Interest compounded annually
- Interest compounded k times per year
- Interest compounded continuously
  The number e
- Present value of an investment
- Investment models

Section Preview Investments

If \$1000 is invested in an account that earns 6% interest, with the interest added to the account at the end of each year, the interest is said to be **compounded annually**. The amount to which an investment grows over a period of time is called its **future value**. In this section, we use general formulas to find the future value of money invested when the interest is compounded at regular intervals of time and when it is compounded continuously. We also find the lump sum that must be invested to have it grow to a specified amount in the future. This is called the **present value** of the investment.

Compound Interest

In Section 3.4, we found that if a set of data has initial value a and a constant percent change r for equally spaced inputs x, the data can be modeled by the exponential function

$$y = a(1 + r)^x$$

Thus, if an investment of \$1000 earns 6% interest compounded annually, the future value at the end of t years will be

$$S = \$1000(1.06)^t \text{ dollars}$$

EXAMPLE 1 Future Value of an Account

To numerically and graphically view how \$1000 grows at 6% interest compounded annually, do the following:

a. Construct a table that gives the future values $S = 1000(1 + .06)^t$ for $t = 0, 1, 2, 3, 4$, and 5 years after the \$1000 was invested.

b. Graph $S = 1000(1 + .06)^t$ for the values of t given in part (a).

c. Graph the equation $S = 1000(1 + .06)^t$ as a continuous function of t for $0 \le t \le 5$.

d. Is there another graph that more accurately represents the amount that would be in the account at any time during the first 5 years that the money is invested?

TABLE 3.15

t (years)	Future Value S (\$)
0	1000.00
1	1060.00
2	1123.60
3	1191.02
4	1262.48
5	1338.23

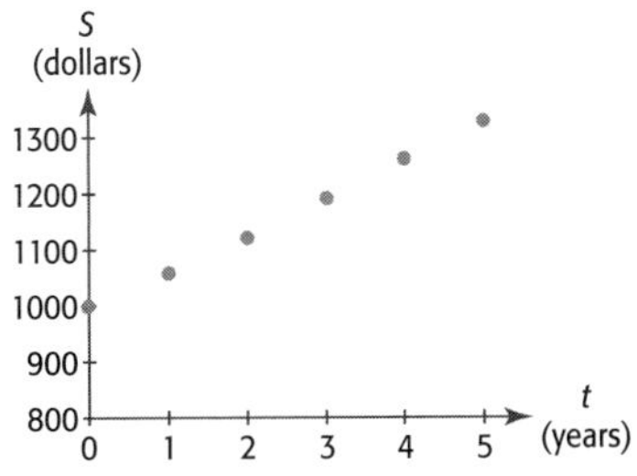

FIGURE 3.30

Solution

a. Substitute the values of t into the equation $S = 1000(1 + .06)^t$. Because the output units for S are dollars, we round to the nearest cent to obtain the values in Table 3.15.

b. Figure 3.30 shows the scatter plot of the data in Table 3.15.

c. Figure 3.31(a) shows a continuous graph of the function S for $0 \le t \le 5$.

d. The scatter plot in Figure 3.30 shows the amount in the account at the end of each year, but not at any other time during the 5-year period. The continuous graph in Figure 3.31(a) shows the amount continually increasing, but interest is paid only at the end of each year and the amount in the account is constant at the previous level until more interest is added. So the graph in Figure 3.31(b) is a more accurate graph

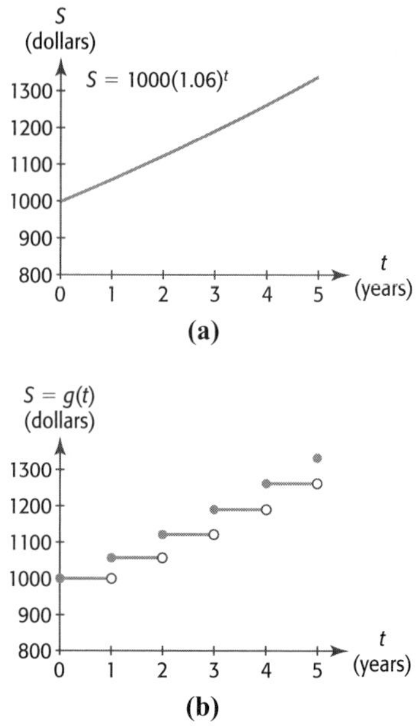

FIGURE 3.31

of the amount that would be in this account because it shows the amount remaining constant until the end of each year, when interest is added. ■

The function g graphed in Figure 3.31(b) is a step function:

$$g(t) = \begin{cases} 1000.00 & \text{if } 0 \le t < 1 \\ 1060.00 & \text{if } 1 \le t < 2 \\ 1123.60 & \text{if } 2 \le t < 3 \\ 1191.02 & \text{if } 3 \le t < 4 \\ 1262.48 & \text{if } 4 \le t < 5 \\ 1338.23 & \text{if } t = 5 \end{cases} \quad \text{dollars in } t \text{ years}$$

This step function is rather cumbersome to work with, so we will find the future value by using the graph in Figure 3.31(a) and the function $S = 1000(1 + .06)^t$ *interpreted discretely*. That is, we draw the graph of this function as it appears in Figure 3.31(a) and evaluate the formula for S with the understanding that the only inputs that make sense in this investment example are nonnegative integers. In general, we have the following.

FUTURE VALUE OF AN INVESTMENT WITH ANNUAL COMPOUNDING

If $\$P$ is invested at an interest rate r per year, compounded annually, the future value S at the end of t years is

$$S = P(1 + r)^t$$

The **annual interest rate** r is also called the **nominal interest rate**, or simply the **rate**. Remember that the interest rate is usually stated as a percent and that it is converted to a decimal when computing future values of investments.

If the interest is compounded more than once per year, then the additional compounding will result in a larger future value. For example, if the investment is compounded twice per year (semiannually), there would be twice as many compounding periods with the interest rate in each period equal to

$$r\left(\frac{1}{2}\right) = \frac{r}{2}$$

Thus, the future value is found by doubling the number of periods and halving the interest rate if the compounding is done twice per year. In general, we use the following model, interpreted discretely, to find the amount of money that results when P dollars are invested and earn compound interest.

FUTURE VALUE OF AN INVESTMENT WITH PERIODIC COMPOUNDING

If $\$P$ is invested for t years at the annual interest rate r, where the interest is compounded k times per year, then the interest rate per period is $\frac{r}{k}$, the number of compounding periods is kt, and the future value that results is given by

$$S = P\left(1 + \frac{r}{k}\right)^{kt} \text{ dollars}$$

For example, if \$1000 is placed in an account that earns 6% interest per year, with the interest added to the account at the end of every month (compounded monthly), the future value of this investment in 5 years is given by

$$S = 1000\left(1 + \frac{0.06}{12}\right)^{12(5)} = \$1348.85$$

Recall from Table 3.15 that compounding annually gives \$1338.23, so compounding monthly rather than yearly results in an additional \$1348.85 − \$1338.23 = \$10.62 from the 5-year investment.

EXAMPLE 2 Daily Versus Annual Compounding of Interest

a. Write the equation that gives the future value of \$1000 invested for t years at 8% compounded annually.

b. Write the equation that gives the future value of \$1000 invested for t years at 8% compounded daily.

c. Graph the equations from parts (a) and (b) on the same axes with t between 0 and 30.

d. What is the additional amount of interest earned in 30 years from compounding daily rather than annually?

Solution

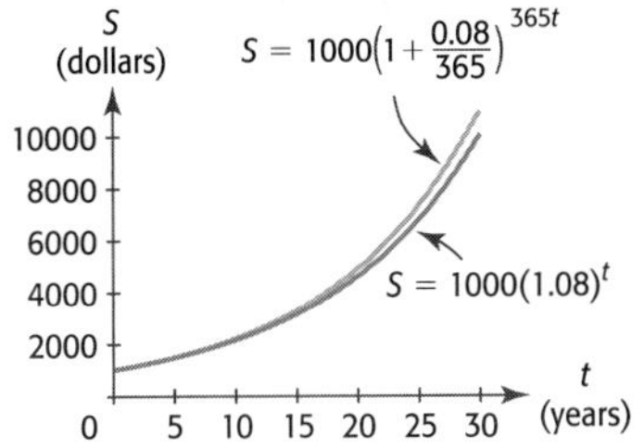

FIGURE 3.32

a. Substituting $P = 1000$ and $r = 0.08$ in $S = P(1 + r)^t$ gives

$$S = 1000(1 + 0.08)^t \quad \text{or} \quad S = 1000(1.08)^t$$

b. Substituting $P = 1000$, $r = 0.08$, and $k = 365$ in $S = P\left(1 + \frac{r}{k}\right)^{kt}$ gives

$$S = 1000\left(1 + \frac{0.08}{365}\right)^{365t}$$

c. The graphs of both functions are shown in Figure 3.32.

d. Evaluating the two functions in parts (a) and (b) at $t = 30$, we calculate the future value of \$1000 compounded annually at 8% to be \$10,062.66 and the future value of \$1000 at 8%, compounded daily, to be \$11,020.28. The additional amount earned in 30 years from compounding daily rather than annually is the difference between these two amounts:

$$\$11{,}020.28 - \$10{,}062.66 = \$957.62.$$

■

Continuous Compounding and the Number *e*

In Section 3.1, we stated that if \$$P$ is invested for t years at interest rate r compounded continuously, the future value of the investment is

$$S = Pe^{rt}$$

To see how the number e becomes part of this formula, we consider the future value of \$1 invested at an annual rate of 100%, compounded for 1 year with different compounding periods. If we denote the number of periods per year by k, the model that gives the future value is

$$S = \left(1 + \frac{1}{k}\right)^k \text{ dollars}$$

Table 3.16 shows the future value of this investment for different compounding periods.

TABLE 3.16

Type of Compounding	Number of Compounding Periods per Year	Future Value ($)
Annually	1	$\left(1 + \frac{1}{1}\right)^1 = 2$
Quarterly	4	$\left(1 + \frac{1}{4}\right)^4 = 2.44140625$
Monthly	12	$\left(1 + \frac{1}{12}\right)^{12} \approx 2.61303529022$
Daily	365	$\left(1 + \frac{1}{365}\right)^{365} \approx 2.71456748202$
Hourly	8760	$\left(1 + \frac{1}{8760}\right)^{8760} \approx 2.71812669063$
Each minute	525,600	$\left(1 + \frac{1}{525{,}600}\right)^{525{,}600} \approx 2.7182792154$
x times per year	x	$\left(1 + \frac{1}{x}\right)^x$

As Table 3.16 indicates, the future value increases (but not very rapidly) as the number of compounding periods during the year increases. As x gets very large, the future value approaches the number e, which is 2.718281828 to nine decimal places. In general, the outputs that result from larger and larger inputs in the function

$$f(x) = \left(1 + \frac{1}{x}\right)^x$$

approach the number e. (In calculus, e is defined as the *limit* of $\left(1 + \frac{1}{x}\right)^x$ as x approaches ∞.)

This definition of e permits us to create a new model for the future value of an investment when interest is compounded continuously. We can let $m = \frac{k}{r}$ to rewrite the formula

$$S = P\left(1 + \frac{r}{k}\right)^{kt} \quad \text{as} \quad S = P\left(1 + \frac{1}{m}\right)^{mrt} \quad \text{or} \quad P\left[\left(1 + \frac{1}{m}\right)^m\right]^{rt}$$

Now as the compounding periods k increase without bound, $m = \frac{k}{r}$ increases without bound and

$$\left(1 + \frac{1}{m}\right)^m$$

approaches e, so $S = P\left[\left(1 + \frac{1}{m}\right)^m\right]^{rt}$ approaches Pe^{rt}.

FUTURE VALUE OF AN INVESTMENT WITH CONTINUOUS COMPOUNDING

If $\$P$ is invested for t years at an annual interest rate r compounded continuously, then the future value S is given by

$$S = Pe^{rt} \text{ dollars}$$

For a given principal and interest rate, the function $S = Pe^{rt}$ is a continuous function whose domain consists of real numbers greater than or equal to 0. Unlike the other compound interest functions, this one is not discretely interpreted because of the continuous compounding.

EXAMPLE 3 Future Value and Continuous Compounding

a. What is the future value of \$2650 invested for 8 years at 12% compounded continuously?

b. How much will be earned on this investment?

Solution

a. The future value of this investment is $S = 2650e^{0.12(8)} = 6921.00$

b. The interest earned on this investment is the future value minus the original investment:

$$\$6921 - \$2650 = \$4271$$

■

EXAMPLE 4 Continuous Versus Annual Compounding of Interest

a. For each of 9 years, compare the future value of an investment of \$1000 at 8% compounded annually and of \$1000 at 8% compounded continuously.

b. Graph the functions for annual compounding and for continuous compounding on the same axes.

c. What conclusion can be made regarding compounding annually and compounding continuously?

Solution

a. The future value of \$1000 compounded annually is given by the function $S = 1000(1 + 0.08)^t$, and the future value of \$1000 compounded continuously is given by $S = 1000e^{0.08t}$. By entering these formulas in an Excel spreadsheet and evaluating them for each of 9 years (Table 3.17), we can compare the future value for each of these 9 years. At the end of the 9 years, we see that compounding continuously results in \$2054.43 − \$1999.00 = \$55.43 more than compounding annually yields.

TABLE 3.17

	A	B	C
1	YEAR	S = 1000*(1.08)^t	S = 1000*e^(.08t)
2	1	1080	1083.287068
3	2	1166.4	1173.510871
4	3	1259.712	1271.249150
5	4	1360.48896	1377.127764
6	5	1469.328077	1491.824698
7	6	1586.874323	1616.07440
8	7	1713.824269	1750.672500
9	8	1850.93021	1896.480879
10	9	1999.004627	2054.433211
11			

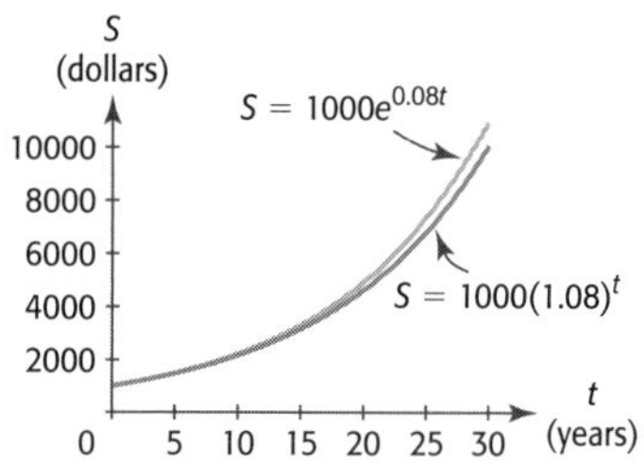

FIGURE 3.33

b. The graphs are shown in Figure 3.33.

c. The value of the investment increases more under continuous compounding than under annual compounding. ■

Present Value of an Investment

We can write the formula for the future value S when a lump sum P is invested for $n = kt$ periods at interest rate $i = \frac{r}{k}$ per compounding period as

$$S = P(1 + i)^n$$

We sometimes need to know the lump sum investment P that will give an amount S in the future. This is called the **present value** P, and it can be found by solving $S = P(1 + i)^n$ for P. Solving for P gives the present value as follows.

PRESENT VALUE

The lump sum that will give future value S in n compounding periods at rate i per period is the present value

$$P = \frac{S}{(1 + i)^n} \quad \text{or} \quad P = S(1 + i)^{-n}$$

EXAMPLE 5 Present Value

What lump sum must be invested at 10% compounded semiannually for the investment to grow to \$15,000 in 7 years?

Solution

We have the future value $S = 15{,}000$, $i = \frac{0.10}{2} = 0.05$, and $n = 2 \cdot 7 = 14$, so the present value of this investment is

$$P = \frac{15{,}000}{(1 + 0.05)^{14}} = 7576.02 \text{ dollars}$$

■

Investment Models

We have developed and used investment formulas. We can also use graphing utilities to model actual investment data.

EXAMPLE 6 Mutual Fund Growth

TABLE 3.18

Time (years)	Average Annual Total Return (\$)
1	1.23
3	2.24
5	3.11
10	6.82

(Source: *FundStation*.)

The data in Table 3.18 gives the annual return on an investment of \$1.00 made on March 31, 1990, in the AIM Value Fund, Class A Shares. The values reflect reinvestment of all distributions and changes in net asset value but exclude sales charges.

a. Use an exponential function to model these data.

b. Use the model to find what the fund will amount to on March 31, 2001 if \$10,000 is invested March 31, 1990 and the fund continues to follow this model after 2000.

c. Is it likely that this fund continued to grow at this rate in 2002?

Solution

a. Using technology to create an exponential model for the data gives

$$A(x) = 1.1609(1.2004)^x$$

Thus, \$1.00 invested in this fund grows to $A(x) = 1.1609(1.2004)^x$ dollars x years after March 31, 1990.

b. March 31, 2001 is 11 years after March 31, 1990, so using the equation from part (a) with $x = 11$ gives $A(11) = 1.1609(1.2004)^{11}$ as the future value of an investment of \$1.00. Thus, the future value of an investment of \$10,000 in 11 years is

$$\$10{,}000[1.1609(1.2004)^{11}] = \$86{,}572.64$$

c. No. Most stocks and mutual funds decreased in value in 2002 because of terrorist attacks in 2001.

■

3.5 SKILLS CHECK

Evaluate the expressions in Exercises 1–10. Write approximate answers rounded to two decimal places.

1. $15{,}000e^{0.06(20)}$

2. $8000e^{0.05(10)}$

3. $3000(1.06^x)$ for $x = 30$

4. $20{,}000(1.07^x)$ for $x = 20$

5. $12{,}000\left(1 + \frac{0.10}{k}\right)^{kn}$ for $k = 4$ and $n = 8$

6. $23{,}000\left(1 + \frac{0.08}{k}\right)^{kn}$ for $k = 12$ and $n = 20$

7. $P\left(1 + \frac{r}{k}\right)^{kn}$ for $P = 3000, r = 8\%, k = 2, n = 18$

8. $P\left(1 + \frac{r}{k}\right)^{kn}$ for $P = 8000, r = 12\%, k = 12, n = 8$

9. $300\left[\frac{1.02^n - 1}{0.02}\right]$ for $n = 240$

10. $2000\left[\frac{1.10^n - 1}{0.10}\right]$ for $n = 12$

11. Find $g(2.5)$, $g(3)$, and $g(3.5)$ if

$$g(x) = \begin{cases} 1000.00 & \text{if } 0 \le x < 1 \\ 1060.00 & \text{if } 1 \le x < 2 \\ 1123.60 & \text{if } 2 \le x < 3 \\ 1191.00 & \text{if } 3 \le x < 4 \\ 1262.50 & \text{if } 4 \le x < 5 \\ 1338.20 & \text{if } x = 5 \end{cases}$$

12. Find $f(2)$, $f(1.99)$, and $f(2.1)$ if

$$f(x) = \begin{cases} 100 & \text{if } 0 \le x \le 1 \\ 200 & \text{if } 1 < x < 2 \\ 300 & \text{if } 2 \le x \le 3 \end{cases}$$

13. Solve $S = P\left(1 + \frac{r}{k}\right)^{kn}$ for P.

14. Solve $S = P(1 + i)^n$ for P.

3.5 EXERCISES

15. *Future Value* If \$8800 is invested for x years at 8% interest compounded annually, find the future value that results in

a. 8 years.

b. 30 years.

16. *Investments* Suppose \$6400 is invested for x years at 7% interest compounded annually. Find the future value of this investment at the end of

a. 10 years.

b. 30 years.

17. *Future Value* If \$3300 is invested for x years at 10% interest compounded annually, the future value that results is $S = 3300(1.10)^x$ dollars.

a. Graph the function for $x = 0$ to $x = 8$.

b. Use the graph to estimate when the money in the account will double.

18. *Investments* If \$5500 is invested for x years at 12% interest compounded quarterly, the future value that results is $S = 5500(1.03)^{4x}$ dollars.

a. Graph this function for $0 \le x \le 8$.

b. Use the graph to estimate when the money in the account will double.

19. *Future Value* If \$10,000 is invested at 12% interest compounded quarterly, find the future value in 10 years.

20. *Future Value* If \$8800 is invested at 6% interest compounded semiannually, find the future value in 10 years.

21. *Future Value* An amount of \$10,000 is invested at 12% interest compounded daily.

a. Find the future value in 10 years.

b. How does this future value compare with the future value in Exercise 19? Why are they different?

22. *Future Value* A total of \$8800 is invested at 6% interest compounded daily.

a. Find the future value in 10 years.

b. How does this future value compare with the future value in Exercise 20? Why are they different?

23. *Compound Interest* If \$10,000 is invested at 12% interest compounded monthly, find the interest earned in 15 years.

24. *Compound Interest* If \$20,000 is invested at 8% interest compounded quarterly, find the interest earned in 25 years.

25. *Continuous Compounding* Suppose \$10,000 is invested for t years at 6% interest compounded continuously. Give the future value at the end of
a. 12 years.
b. 18 years.

26. *Continuous Compounding* If \$42,000 is invested for t years at 7% interest compounded continuously, find the future value in
a. 10 years.
b. 20 years.

27. *Continuous versus Annual Compounding*
a. If \$10,000 is invested at 6% compounded annually, find the future value in 18 years.
b. How does this compare with the result from Exercise 25(b)?

28. *Continuous versus Annual Compounding*
a. If \$42,000 is invested at 7% compounded annually, find the future value in 20 years.
b. How does this compare with the result from Exercise 26(b)?

29. *Doubling Time* Use a spreadsheet, a table, or a graph to estimate how long it takes for an investment to double if it is invested at 10% interest
a. Compounded annually.
b. Compounded continuously.

30. *Doubling Time* Use a spreadsheet, a table, or a graph to estimate how long it takes for an investment to double if it is invested at 6% interest
a. Compounded annually.
b. Compounded continuously.

31. *Future Value* Suppose \$2000 is invested in an account paying 5% interest compounded annually. What is the future value of this investment
a. After 8 years?
b. After 18 years?

32. *Future Value* If \$12,000 is invested in an account that pays 8% interest compounded quarterly, find the future value of this investment
a. After 2 quarters.
b. After 10 years.

33. *Investments* Suppose \$3000 is invested in an account that pays 6% interest compounded monthly. What is the future value of this investment after 12 years?

34. *Investments* If \$9000 is invested in an account that pays 8% interest compounded quarterly, find the future value of this investment
a. After 0.5 year.
b. After 15 years.

35. *Doubling Time* If money is invested at 10% interest compounded quarterly, the future value of the investment doubles approximately every 7 years.
a. Use this information to complete the table below for an investment of \$1000 at 10% interest compounded quarterly.
b. Create an exponential function, rounded to three decimal places, that models the discrete function defined by the table.
c. Because the interest is compounded quarterly, this model must be interpreted discretely. Use the rounded function to find the value of the investment in 5 years and in $10\frac{1}{2}$ years after the money was invested.

Years	0	7	14	21	28
Future Value (\$)	1000				

36. *Doubling Time* If money is invested at 11.6% interest compounded monthly, the future value of the investment doubles approximately every 6 years.
a. Use this information to complete the table below for an investment of \$1000 at 11.6% interest compounded monthly.
b. Create an exponential function, rounded to three decimal places, that models the discrete function defined by the table.
c. Because the interest is compounded monthly, this model must be interpreted discretely. Use the rounded model to find the value of the investment in 2 months, in 4 years, and in $12\frac{1}{2}$ years.

Years	0	6	12	18	24
Future Value (\$)	1000				

3.6 Annuities; Loan Repayment

Key Concepts

- Annuity
  Ordinary annuity
- Present value of an investment
  Present value of a lump sum investment
  Present value of an ordinary annuity
- Loan repayment
  Amortization

Section Preview Annuities

To prepare for future retirement, people frequently invest a fixed amount of money at the end of each month into an account that pays interest that is compounded monthly. Such an investment plan, or any other characterized by regular payments, is called an annuity. In this section, we develop a model to find future values of annuities.

A person planning retirement may also want to know how much money is needed to provide regular payments to him or her during retirement, and a couple may want to know what lump sum to put in an investment to pay future college expenses for their child. The amount of money they need to invest to receive a series of payments in the future is the present value of an annuity.

We can also use the present value concept to determine the payments that are necessary to repay a loan with equal payments made on a regular schedule, which is called amortization. The formulas used to determine the future value and the present value of annuities and the payments necessary to repay loans are applications of exponential functions.

Future Value of an Annuity

An **annuity** is a financial plan characterized by regular payments. We can view an annuity as a savings plan where regular payments are made to an account, and we can use an exponential function to determine what the future value of the account will be. One type of annuity, called an **ordinary annuity**, is a financial plan where equal payments are contributed at the end of each period to an account that pays a fixed rate of interest compounded at the same time as payments are made. For example, suppose \$1000 is invested at the end of each year for 5 years in an account that pays interest at 10% compounded annually. To find the future value of this annuity, we can think of it as the sum of 5 investments of \$1000, one that draws interest for 4 years (from the end of the first year to the end of the fifth year), one that draws interest for 3 years, and so on (Table 3.19).

TABLE 3.19

Investment	Future Value
Invested at end of 1st year	$1000(1.10)^4$
Invested at end of 2nd year	$1000(1.10)^3$
Invested at end of 3rd year	$1000(1.10)^2$
Invested at end of 4th year	$1000(1.10)^1$
Invested at end of 5th year	1000

Note that the last investment is at the end of the fifth year, so it earned no interest. The future value of the annuity is the sum of the separate future values. It is

$$\begin{aligned} S &= 1000 + 1000(1.10)^1 + 1000(1.10)^2 + 1000(1.10)^3 + 1000(1.10)^4 \\ &= 6105.10 \text{ dollars} \end{aligned}$$

In Section 6.4, we will develop the following formula to find the future value of an annuity.

FUTURE VALUE OF AN ORDINARY ANNUITY

If R dollars are contributed at the end of each period for n periods into an annuity that pays interest at rate i at the end of the period, the future value of the annuity is

$$S = R\left[\frac{(1+i)^n - 1}{i}\right] \text{ dollars}$$

Note that the interest rate used in the above formula is the rate per period, not the rate for a year. We can use this formula to find the future value of an annuity in which \$1000 is invested at the end of each year for 5 years into an account that pays interest at 10% compounded annually. This future value, which we found earlier without the formula, is

$$S = R\left[\frac{(1+i)^n - 1}{i}\right] = 1000\left[\frac{(1+0.10)^5 - 1}{0.10}\right] = 6105.10 \text{ dollars}$$

EXAMPLE 1 Future Value of an Ordinary Annuity

Find the 5-year future value of an ordinary annuity with a contribution of \$500 per quarter into an account that pays 12% per year compounded quarterly.

Solution

The payments and interest compounding occur quarterly, so the interest rate per period is $\frac{0.12}{4} = 0.03$ and the number of compounding periods is $4(5) = 20$. Substituting the information into the formula for the future value of an annuity gives

$$S = 500\left[\frac{(1+0.03)^{20} - 1}{0.03}\right] = 13{,}435.19$$

Thus, the future value of this investment is \$13,435.19. ■

EXAMPLE 2 Future Value of an Annuity

Suppose \$200 is deposited at the end of each month into an account that pays interest 12% per year, compounded monthly. Find the future value for every 4-month period, for up to 36 months.

Solution

To find the future value of this annuity, we note that the interest rate per period (month) is $\frac{12\%}{12} = .01$, so the future value at the end of n months is given by the function

$$S = 200\left(\frac{1.01^n - 1}{0.01}\right) \text{ dollars}$$

This model for the future value of an ordinary annuity is a continuous function with discrete interpretation because the future value changes only at the end of each period. A graph of the function and the table of future values for every 4 months up to 36 months are shown in Figure 3.34 and Table 3.20, respectively. Keep in mind that the graph should be interpreted discretely; that is, only positive integer inputs and corresponding outputs make sense.

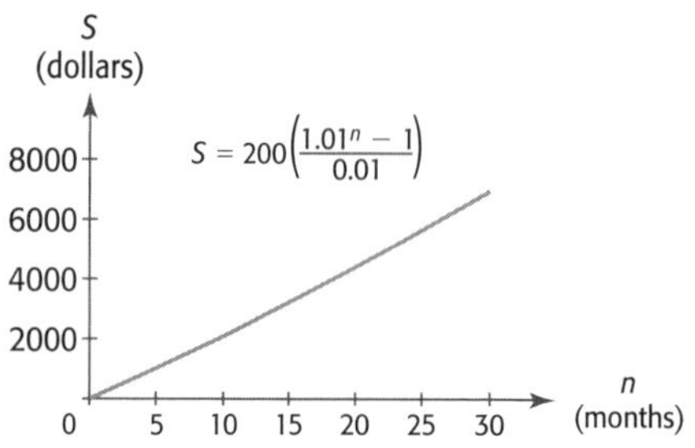

FIGURE 3.34

TABLE 3.20

Number of Months	Value of Annuity ($)
4	812.08
8	1657.13
12	2536.50
16	3451.57
20	4403.80
24	5394.69
28	6425.82
32	7498.81
36	8615.38

■

Present Value of an Annuity

Many retirees purchase annuities that give them regular payments through a period of years. Suppose a lump sum of money is invested to return payments of $\$R$ at the end of each of n periods, with the investment earning interest at rate i per period, after which \$0 will remain in the account. This lump sum is the **present value** of this ordinary annuity, and it is equal to the sum of the present values of each of the future payments. Table 3.21 shows the present value of each payment of this annuity.

TABLE 3.21

Payment	Present Value of This Payment
Paid at end of 1st period	$\$R\,(1+i)^{-1}$
Paid at end of 2nd period	$\$R\,(1+i)^{-2}$
Paid at end of 3rd period	$\$R\,(1+i)^{-3}$
.	.
.	.
Paid at end of $(n-1)$st period	$\$R\,(1+i)^{-(n-1)}$
Paid at end of nth period	$\$R\,(1+i)^{-n}$

We must have enough money in the account to provide the present value of each of these payments, so the present value of the annuity is

$$A = R(1+i)^{-1} + R(1+i)^{-2} + R(1+i)^{-3} + \cdots + R(1+i)^{-(n-1)} + R(1+i)^{-n}$$

We will show in Section 6.4 that this sum can be written in the form

$$A = R\left[\frac{1-(1+i)^{-n}}{i}\right]$$

PRESENT VALUE OF AN ORDINARY ANNUITY

If a payment of $\$R$ is to be made at the end of each period for n periods from an account that earns interest at a rate of i per period, then the account is an **ordinary annuity**, and the **present value** is

$$A = R\left[\frac{1 - (1 + i)^{-n}}{i}\right]$$

EXAMPLE 3 Present Value of an Annuity

Suppose a retiring couple wants to establish an annuity that will provide \$2000 at the end of each month for 20 years. If the annuity earns 6%, compounded monthly, how much must the couple put in the account to establish the annuity?

Solution

We seek the present value of an annuity that pays \$2000 at the end of each month for $12(20) = 240$ months, with interest at $\frac{6\%}{12} = \frac{0.06}{12} = 0.005$ per month. The present value is

$$A = 2000\left[\frac{1 - (1 + 0.005)^{-240}}{0.005}\right] = 279{,}161.54 \text{ dollars}$$

Thus, the couple can receive a payment of \$2000 at the end of each month for 20 years if they put a lump sum of \$279,161.54 in an annuity. (Note that the sum of the money they will receive over the 20 years is $\$2000 \cdot 20 \cdot 12 = \$480{,}000$.) ■

EXAMPLE 4 Home Mortgage

A couple who wants to purchase a home has \$30,000 for a down payment and wants to make monthly payments of \$2200. If the interest rate for a 25-year mortgage is 6% per year on the unpaid balance, how much can they spend on a house?

Solution

The amount of money that they can pay for the house is the sum of the down payment and the present value of the payments that they can afford to pay. The present value of the $12(25) = 300$ monthly payments of \$2200 with interest at $\frac{6\%}{12} = \frac{0.06}{12} = 0.005$ per month is

$$A = 2200\left[\frac{1 - (1 + .005)^{-300}}{.005}\right] = 341{,}455.10$$

The total amount that they can spend on a house is $\$341{,}455 + 30{,}000 = \$371{,}455.10$.

■

Loan Repayment

When money is borrowed, the borrower must repay the total amount that was borrowed (the debt) plus interest on that debt. Prior to the passage of a truth-in-lending law, several types of repayment plans were available, some of which were misleading to consumers. Most loans (including those for houses, for cars, and for other consumer goods) now require regular payments on the debt plus payment of interest on the unpaid balance of the loan. That is, loans are now paid off by a series of partial payments with interest charged on the unpaid balance at the end of each period. The stated interest rate (the **nominal rate**) is the *annual rate*. The rate per period is the nominal rate divided by the number of payment periods per year.

There are two popular repayment plans for these loans. One plan applies an equal amount to the debt each payment period plus the interest for the period, which is the interest on the unpaid balance. For example, a loan of \$240,000 for 10 years at 12% could be repaid with 120 monthly payments of \$2000 plus 1% $\left(\frac{1}{12} \text{ of } 12\%\right)$ of the unpaid balance each month. When this payment method is used, the payments will decrease as the unpaid balance decreases. A few sample payments are shown in Table 3.22.

TABLE 3.22

Payment Number (month)	Unpaid Balance (\$)	Interest on the Unpaid Balance (\$)	Balance Reduction (\$)	Monthly Payment (\$)	New Balance (\$)
1	240,000	2400	2000	4400	238,000
2	238,000	2380	2000	4380	236,000
.					
51	140,000	1400	2000	3400	138,000
.					
101	40,000	400	2000	2400	38,000
.					
118	6000	60	2000	2060	4000

A loan of this type can also be repaid by making all payments (including the payment on the balance and the interest) of equal size. The process of repaying the loan in this way is called **amortization**. If a bank makes a loan of this type, it uses an amortization formula that gives the size of the equal payments. This formula is developed by considering the loan as an annuity purchased from the borrower by the bank, with the annuity paying a fixed return to the bank each payment period. The lump sum that the bank gives to the borrower (the principal of the loan) is the present value of this annuity, and each payment that the bank receives from the borrower is a payment from this annuity. To find the size of these equal payments, we solve the formula for the present value of an annuity, $A = R\left[\frac{1 - (1 + i)^{-n}}{i}\right]$, for R. This gives the following formula.

AMORTIZATION FORMULA

If a debt of \$$A$, with interest at a rate of i per period, is amortized by n equal periodic payments made at the end of each period, then the size of each payment is

$$R = A\left[\frac{i}{1 - (1 + i)^{-n}}\right]$$

We can use this formula to find the *equal* monthly payments that will amortize the loan of \$240,000 for 10 years at 12%. The interest rate per month is .01, and there are 120 periods.

$$R = A\left[\frac{i}{1-(1+i)^{-n}}\right] = 240{,}000\left[\frac{0.01}{1-(1+0.01)^{-120}}\right] \approx 3443.303$$

Thus, repaying this loan requires 120 equal payments of \$3443.31. If we compare this payment with the selected payments in Table 3.22, we see that the first payment plan in Table 3.22 is nearly \$1000 more than the equal amortization payment and that every payment after the 101st is \$1000 less than the amortization payment. The advantage of the amortization method of repaying the loan is that the borrower can budget better by knowing what the payment will be each month. Of course, the balance of the loan will not be reduced as fast with this payment method, because most of the monthly payment will be needed to pay interest in the first few months. The partial spreadsheet in Table 3.23 shows how the outstanding balance is reduced.

TABLE 3.23

	A	B	C	D	E	F
1	Payment	Unpaid		Monthly	Balance	New
2	Number	Balance	Interest	Payment	Reduction	Balance
3	1	240,000	2400	3443.31	1043.31	238,956.69
4	2	238,956.69	2389.57	3443.31	1053.74	237,902.95
	.					
53	51	172,745.38	1727.45	3443.31	1715.86	171,029.52
	.					
103	101	63,136.30	631.36	3443.31	2811.95	60,324.35

EXAMPLE 5 Home Mortgage

A couple that wants to purchase a home with a price of \$230,000 has \$50,000 for a down payment. If they can get a 25-year mortgage at 9% per year on the unpaid balance,

a. What will be their equal monthly payments?

b. What is the total amount they will pay before they own the house outright?

c. How much interest will they pay?

Solution

a. The amount of money that they must borrow is \$230,000 − \$50,000 = \$180,000. The number of monthly payments is 12(25) = 300, and the interest rate is $\frac{9\%}{12} = \frac{0.09}{12} = 0.0075$ per month. The monthly payment is

$$R = 180{,}000\left[\frac{.0075}{1-(1.0075)^{-300}}\right] \approx 1510.553$$

so the required payment would be \$1510.56.

b. The amount that they must pay before owning the house is the down payment plus the total of the 300 payments:

$$\$50{,}000 + 300(\$1510.56) = \$503{,}168$$

c. The total interest is the total amount paid minus the price paid for the house. The interest is

$$\$503{,}168 - \$230{,}000 = \$273{,}168$$

When the interest paid is based on the periodic interest rate on the unpaid balance of the loan, the nominal rate is also called the **annual percentage rate** (APR). You will occasionally see an advertisement for a loan with a stated interest rate and a different, larger APR. This can happen if a lending institution charges fees (sometimes called points) in addition to the stated interest rate. Federal law states that all these charges must be included in computing the APR, which is the true interest rate that is being charged on the loan.

3.6 SKILLS CHECK

Give answers to two decimal places.

1. Solve $S = P(1 + i)^n$ for P with positive exponent n.
2. Solve $S = P(1 + i)^n$ for P, with no denominator.
3. Solve $Ai = R[1 - (1 + i)^{-n}]$ for A.
4. Evaluate $2000\left[\frac{1 - (1 + .01)^{-240}}{.01}\right]$.
5. Solve $A = R\left[\frac{1 - (1 + i)^{-n}}{i}\right]$ for R.
6. Evaluate $240{,}000\left[\frac{.01}{1 - (1 + .10)^{-120}}\right]$.

3.6 EXERCISES

7. *Present Value* What lump sum investment will grow to \$10,000 in 10 years if it is invested at 6% compounded annually?
8. *Present Value* What lump sum investment will grow to \$30,000 if it is invested for 15 years at 7% compounded annually?
9. *College Tuition* New parents want to put a lump sum into a money market fund to provide \$30,000 in 18 years, to help pay for college tuition for their child. If the fund averages 10% per year compounded monthly, how much should they invest?
10. *Trust Fund* Grandparents decide to put a lump sum of money into a trust fund on their granddaughter's 10th birthday so that she will have \$1,000,000 on her 60th birthday. If the fund pays 11% compounded monthly, how much money must they put in the account?
11. *Annuities* Find the present value of an annuity that will pay \$1000 at the end of each year for 10 years if the interest rate is 7% compounded annually.
12. *Annuities* Find the present value of an annuity that will pay \$500 at the end of each year for 20 years if the interest rate is 9% compounded annually.
13. *Lottery Winnings* The winner of a "million dollar" lottery is to receive \$50,000 plus \$50,000 at the end of each year for 19 years, or the present value of this annuity in cash. How much cash would she receive if money is worth 8% compounded annually?
14. *College Tuition* A couple wants to establish a fund that will provide \$3000 for tuition at the end of each 6-month period for 4 years. If a lump sum can be placed in an account that pays 8% compounded semiannually, what lump sum is required?

15. *Insurance Payment* A man is disabled in an accident and wants to receive an insurance payment that will provide him with $3000 at the end of each month for 30 years. If the payment can be placed in an account that pays 9% compounded monthly, what size payment should he seek?

16. *Auto Leasing* A woman wants to lease rather than buy a car but does not want to make monthly payments. A dealer has the car she wants, which leases for $400 at the end of each month for 48 months. If money is worth 8% compounded monthly, what lump sum should she offer the dealer to keep the car for 48 months?

17. *Business Sale* A man can sell his Thrifty Electronics business for $800,000 cash or for $100,000 plus $122,000 at the end of each year for 9 years.

a. Find the present value of the annuity that is offered if money is worth 10% compounded annually.

b. If he takes the $800,000, spends $100,000 of it, and invests the rest in a 9-year annuity at 10% compounded annually, what size annuity payment will he receive at the end of each year?

c. Which is better, taking the $100,000 and the annuity or taking the cash settlement? Discuss the advantages of your choice.

18. *Sale of a Practice* A physician can sell her practice for $1,200,000 cash or for $200,000 plus $250,000 at the end of each year for 5 years.

a. Find the present value of the annuity that is offered if money is worth 7%, compounded annually.

b. If she takes the $1,200,000, spends $200,000 of it, and invests the rest in a 5-year annuity at 7% compounded annually, what size annuity payment will she receive at the end of each year?

c. Which is better, taking the $200,000 and the annuity or taking the cash settlement? Discuss the advantages of your choice.

19. *Home Mortgage* A couple wants to buy a house and can afford to pay $1600 per month.

a. If they can get a loan for 30 years with interest at 9% per year on the unpaid balance, how much can they pay for a house?

b. What is the total amount paid over the life of the loan?

c. What is the total interest paid on the loan?

20. *Auto Loan* A man wants to buy a car and can afford to pay $400 per month.

a. If he can get a loan for 48 months with interest at 12% per year on the unpaid balance, how much can he pay for a car?

b. What is the total amount paid over the life of the loan?

c. What is the total interest paid on the loan?

21. *Loan Repayment* A loan of $10,000 is to be amortized with quarterly payments over 4 years. If the interest on the loan is 8% per year, paid on the unpaid balance,

a. What is the interest rate charged each quarter on the unpaid balance?

b. How many payments are made to repay the loan?

c. What payment is required quarterly to amortize the loan?

22. *Loan Repayment* A loan of $36,000 is to be amortized with monthly payments over 6 years. If the interest on the loan is 6% per year, paid on the unpaid balance,

a. What is the interest rate charged each month on the unpaid balance?

b. How many payments are made to repay the loan?

c. What payment is required each month to amortize the loan?

23. *Home Mortgage* A couple who wants to purchase a home with a price of $350,000 has $100,000 for a down payment. If they can get a 30-year mortgage at 6% per year on the unpaid balance,

a. What will be their monthly payments?

b. What is the total amount they will pay before they own the house outright?

c. How much interest will they pay over the life of the loan?

24. *Business Loan* Business partners want to purchase a restaurant that costs $750,000. They have $300,000 for a down payment, and they can get a 25-year business loan for the remaining funds at 8% per year on the unpaid balance, with quarterly payments.

a. What will be the payments?

b. What is the total amount that they will pay over the 25-year period?

c. How much interest will they pay over the life of the loan?

CHAPTER 3 Summary

In this chapter, we discussed exponential and logarithmic functions and their applications. We explored the fact that logarithmic and exponential functions are inverses of each other. Many applications in today's world involve exponential and logarithmic functions, including the pH of substances, stellar magnitude, intensity of earthquakes, the growth of bacteria, the decay of radioactive isotopes, and compound interest. Logarithmic and exponential equations are solved by using both analytical and graphical solution methods.

Key Concepts and Formulas

Section 3.1 Exponential Functions

Exponential function

If a and b are real numbers, $b \neq 1$, then the function $f(x) = a(b^x)$ is an exponential function.

Exponential growth

Whenever the base b of $f(x) = a(b^x)$ is greater than 1 and $a > 0$, the exponential function is increasing and can be used to model exponential growth. The graph of an exponential growth model resembles the graph at the right. The x-axis (on the left) is a horizontal asymptote for the graph.

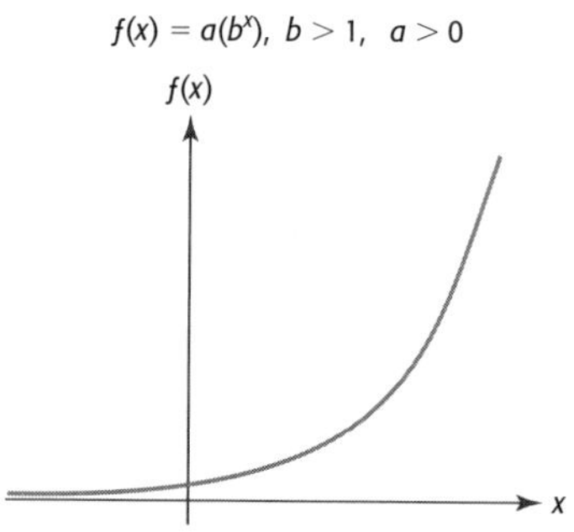

Exponential decay

Whenever the base b of $f(x) = a(b^{-x})$ is greater than 1 and $a > 0$, the exponential function is decreasing and can be used to model exponential decay. The graph of an exponential decay model resembles the graph at the right. The x-axis (on the right) is a horizontal asymptote for the graph. The function $f(x) = a(c^x)$, $a > 0$, $0 < c < 1$, is also an exponential decay function.

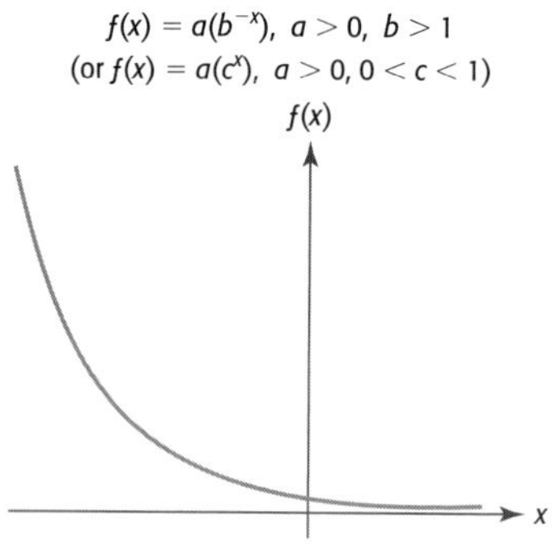

Number e

Many real applications involve exponential functions with the base e. The number e is an irrational number with decimal approximation 2.718281828 (to nine decimal places).

Section 3.2 Logarithmic Functions

Logarithmic function

For $x > 0$, $a > 0$, and $a \neq 1$, the logarithmic function to the base a is $y = \log_a x$, which is defined by $x = a^y$.

The graph of $y = \log_a x$ for the base $a > 0$ is shown at the right. The y-axis is a vertical asymptote for the graph.

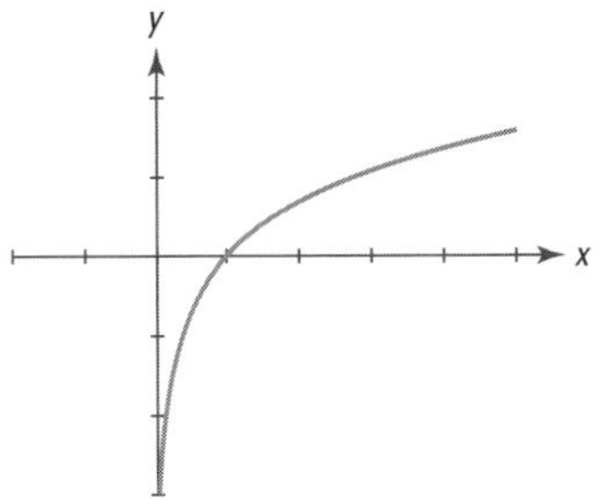

Logarithmic and exponential forms

These forms are equivalent:

$$y = \log_a x \text{ and } x = a^y$$

Common logarithms

Logarithms with a base of 10; $\log_{10} x$ is denoted $\log x$.

Natural logarithms

Logarithms with a base of e; $\log_e x$ is denoted $\ln x$.

Solving logarithmic equations

Many logarithmic equations can be solved by converting to exponential form; that is, $y = \log_a x$ is equivalent to $x = a^y$.

Section 3.3 Exponential and Logarithmic Equations; Properties of Logarithms

Solving exponential equations

Many exponential equations can be solved by converting to logarithmic form; that is, $x = a^y$ is equivalent to $y = \log_a x$.

Properties of logarithms

For $a > 0$, $a \neq 1$, k a real number, and M and N positive real numbers,

Property 1. $\log_a a = 1$

Property 2. $\log_a 1 = 0$

Property 3. $\log_a a^x = x$

Property 4. $a^{\log_a x} = x$

Change of base

If $a > 0$, $a \neq 1$, $b > 0$, $b \neq 1$, and $x > 0$, then $\log_b x = \dfrac{\log_a x}{\log_a b}$.

Solving exponential and logarithmic equations

Some exponential and logarithmic equations can be solved by using properties of logarithms to rewrite the equation in a form that can be more easily solved.

Exponential inequalities

Solving inequalities involving exponential functions involves rewriting the related equation in logarithmic form and solving the related logarithmic equation. Then graphical methods are used to find the values of the variable that satisfy the inequality.

Section 3.4 Exponential and Logarithmic Models

Exponential models

Many sources provide data that can be modeled by exponential growth and decay functions according to the model $y = ab^x$ for real numbers a and b, $b \neq 1$.

Constant percent changes

If the percent change of the outputs of a set of data is constant for equally spaced inputs, an exponential function will be a perfect fit for the data. If the percent change of the outputs is approximately constant for equally spaced inputs, an exponential function will be an approximate fit for the data.

Exponential model

If a set of data has initial value a and has a constant percent change r, the data can be modeled by the exponential function

$$y = a(1 + r)^x$$

Comparison of models

To determine which type of model is the best fit for a set of data, we can look at the first differences and second differences of the outputs and the percent change of the outputs.

Logarithmic models

Models can be created involving $\ln x$. Data that exhibit an initial rapid increase and then have a slow rate of growth can often be described by the function $f(x) = a + b \ln x$ for $b > 0$.

Section 3.5 Exponential Functions and Investing

Future value with annual compounding

If P dollars are invested for n years at an interest rate r compounded annually, then the future value of P is given by $S = P(1 + r)^n$ dollars.

Future value of an investment

If P dollars are invested for t years at the annual interest rate r, where the interest is compounded k times per year, then the interest rate per period is $\frac{r}{k}$, the number of compounding periods is kt, and the future value that results is given by $S = P\left(1 + \frac{r}{k}\right)^{kt}$ dollars.

Future value with continuous compounding

If P dollars are invested for t years at an annual interest rate r compounded continuously, then the future value S is given by $S = Pe^{rt}$ dollars.

Present value of an investment

The lump sum P that is necessary to have it grow to the future value S if invested for n periods at interest rate i per compounding period is

$$P = \frac{S}{(1 + i)^n} \quad \text{or} \quad P = S(1 + i)^{-n}$$

Section 3.6 Annuities; Loan Repayment

Future value of an ordinary annuity

If R dollars are contributed at the end of each period for n periods to an annuity that pays at rate i at the end of the period, the future value of the annuity is

$$S = R\left[\frac{(1 + i)^n - 1}{i}\right]$$

Present value of an ordinary annuity

If a payment of $\$R$ is to be made at the end of each period for n periods from an account that earns interest at a rate of i per period, then the account is an ordinary annuity, and the present value is

$$A = R\left[\frac{1 - (1 + i)^{-n}}{i}\right]$$

Amortization formula

If a debt of $\$A$, with interest at a rate of i per period, is amortized by n equal periodic payments made at the end of each period, then the size of each payment is

$$R = A\left[\frac{i}{1 - (1 + i)^{-n}}\right]$$

Chapter 3 Skills Check

1. **a.** Graph the function $f(x) = 4e^{-0.3x}$.
   b. Find $f(-10)$ and $f(10)$.

2. Is $f(x) = 4e^{-0.3x}$ an increasing or a decreasing function?

3. Graph $f(x) = 3^x$ on $[-10, 10]$ by $[-10, 20]$.

4. Graph $y = 3^{(x-1)} + 4$ on $[-10, 10]$ by $[-10, 20]$.

5. How does the graph of Exercise 4 compare with the graph of Exercise 3?

6. Is $y = 3^{(x-1)} + 4$ an increasing or a decreasing function?

7. **a.** If $y = 1000(2^{-0.1x})$, find the value of y when $x = 10$.
   b. Use graphical or numerical methods to determine the value of x that gives $y = 250$ if $y = 1000(2^{-0.1x})$.

In Exercises 8 and 9, write the following exponential equations in logarithmic form.

8. $x = 6^y$
9. $y = 7^{3x}$

In Exercises 10–12, write the following logarithmic equations in exponential form.

10. $y = \log_4 x$
11. $y = \log(x)$
12. $y = \ln x$
13. Write the inverse of $y = 4^x$ in logarithmic form.

In Exercises 14–16, evaluate the following logarithms, if possible. Give approximate solutions to four decimal places.

14. $\log 22$
15. $\ln 56$
16. $\log 10$

In Exercises 17–19, find the value of the following logarithms without *using a calculator.*

17. $\log_2 16$
18. $\ln(e^4)$
19. $\log 0.001$

In Exercises 20 and 21, use a change of base formula to evaluate each of the following logarithms. Give approximate solutions to four decimal places.

20. $\log_3(54)$
21. $\log_8(56)$

In Exercises 22 and 23, graph the following functions with technology.

22. $y = \ln(x - 3)$
23. $y = \log_3 x$

In Exercises 24–27, solve the following exponential equations. Give approximate solutions to four decimal places.

24. $340 = e^x$
25. $1500 = 300e^{8x}$
26. $9200 = 23(2^{3x})$
27. $4(3^x) = 36$

28. Rewrite $\ln\dfrac{(2x-5)^3}{x-3}$ as the sum, difference, or product of logarithms, and simplify if possible.

29. Rewrite $6\log_4 x - 2\log_4 y$ as a single logarithm.

30. Determine whether a linear or an exponential function is the better fit for the data in the table. Then find the equation of the function with three-decimal-place accuracy.

x	1	2	3	4	5
y	2	5	12	30	75

31. Evaluate $P\left(1 + \dfrac{r}{k}\right)^{kn}$ for $P = 1000$, $r = 8\%$, $k = 12$, $n = 20$, to two decimal places.

32. Evaluate $1000\left[\dfrac{1 - 1.03^{-240+n}}{0.03}\right]$ for $n = 120$, to two decimal places.

33. **a.** Graph $f(t) = \dfrac{2000}{1 + 8e^{-0.8t}}$ for $t = 0$ to $t = 10$.
    b. Find $f(0)$ and $f(8)$, to two decimal places.
    c. What is the limiting value of this function?

34. **a.** Graph $y = 500(0.1^{0.2^x})$ for $x = 0$ to $x = 10$.
    b. What is the initial value (the value of y when $x = 0$) of this function?
    c. What is the limiting value of this function?

Chapter 3 Review

35. *Prisoners* The total number of prisoners (in thousands) in state and federal prisons from 1970 to 1997 can be modeled by the function $y = 84.7518(1.0746^x)$, where x is the number of years after 1960. According to this model, how many prisoners were incarcerated in 1990? (Source: Index of Leading Cultural Indicators)

36. *Sales Decay* At the end of an advertising campaign, the daily sales (in dollars) declined, with daily sales given by the equation $y = 2000(2^{-0.1x})$, where x is the number of weeks after the end of the campaign. According to this model, what will sales be 4 weeks after the end of the ad campaign?

37. *Corporate Revenue* Prior to the November 1994 \$1.7 billion takeover proposal by Quaker Oats, Snapple Beverage Corporation's annual revenues were given by the function $B(t) = 1.337e^{0.718t}$ million dollars, where t is the number of years after 1985. When does this model indicate that the annual revenue is more than \$10 million? (Source: *U.S. News and World Report*, November 14, 1994)

38. *Investments* The number of years needed for an investment of \$10,000 to grow to \$30,000, when money is invested at 12% compounded annually, is given by

$$t = \log_{1.12}\frac{30{,}000}{10{,}000} = \log_{1.12}3$$

In how many years will this occur?

39. *Investments* If \$1000 is invested at 10% compounded quarterly, the future value of the investment is given by $S = 1000(2^{x/7})$, where x is the number of years after the investment is made.
    a. Use the logarithmic form of this function to solve the equation for x.
    b. Find when the future value of the investment is \$19,504.

40. *Sales Decay* At the end of an advertising campaign, the weekly sales (in dollars) declined, with weekly sales given by the equation $y = 2000(2^{-0.1x})$, where x is the number of weeks after the end of the campaign. In how many weeks will sales be half of the sales at the end of the ad campaign?

41. *Prisoners* The rate (number per 100,000 people in the U.S. population) of prisoners in state and federal prisons from 1970 to 1997 can be modeled by the function $y = 45.6786(1.06386^x)$, where x is the number of years from 1960. In what year does this model indicate a rate of 275 per 100,000?

42. *Mobile Home Sales* A company that buys and sells used mobile homes estimates its cost to be given by $C(x) = 2x + 50$ thousand dollars when x mobile homes are purchased. The same company estimates that its revenue from the sale of x mobile homes is given by $R(x) = 10(1.26)^x$ thousand dollars.
    a. Combine C and R into a single function that gives the profit for the company when x used mobile homes are bought and sold.
    b. How many mobile homes must the company sell so that revenue is at least \$30,000 more than cost?

43. *Sales Decay* At the end of an advertising campaign, the sales (in dollars) declined, with weekly sales given by the equation $y = 40{,}000(3^{-0.1x})$, where x is the number of weeks after the end of the campaign. In how many weeks will sales be less than half of the sales at the end of the ad campaign?

44. *Purchasing Power* If a retired couple has a fixed income of \$60,000 per year, the purchasing power (adjusted value of the money) after t years of 5% inflation is given by the equation $P = 60{,}000e^{-0.05t}$. In how many years will the purchasing power of their income fall below half of their current income?

45. *Investments* If \$2000 is invested at 8% compounded continuously, the future value of the investment after t years is given by $S = 2000e^{0.08t}$ dollars. What is the future value of this investment in 10 years?

46. *Investments* If \$3300 is invested for x years at 10% compounded annually, the future value that will result is $S = 3300(1.10)^x$ dollars. In how many years will the investment result in \$13,784.92?

47. *Smokers* The table below gives the percent of black females over 18 years of age who are smokers in each of the years 1965–1995.
   a. Find an exponential equation that models this percent, with x equal to 0 in 1960.
   b. Use the model to estimate the percent of black females who are smokers in 1972 and in 1996.

Years	Percent
1965	38.1
1974	34.1
1979	33.8
1983	35.5
1985	30.4
1987	26.1
1990	22.5
1991	22.4
1992	24.9
1993	22.9
1994	25.2
1995	21.8

(Source: Centers for Disease Control, www.cdc.gov/nchs/datawh/statabl/)

48. *Smokers* The percents of the U.S. population who are smokers for the years 1965–1995 are given in the table.
   a. Find an exponential model for the data, where x is the number of years from 1960.
   b. Use the unrounded model to estimate the percent in 2003.

Year	Percent	Year	Percent
1965	51.6	1990	28
1974	42.9	1991	27.5
1979	37.2	1992	28.2
1983	34.7	1993	27.5
1985	32.1	1994	27.8
1987	31	1995	26.7

(Source: Centers for Disease Control, www.cdc.gov/nchs/datawh/statabl/)

49. *Out-of-Wedlock Births* The percent of all teen mothers who were unmarried during the years 1960–1996 is shown in the table below.
   a. Find an exponential function to fit the data, using as input the number of years after 1950.
   b. Find a logistic function to fit the data, using as input the number of years after 1950.
   c. Graph both models with the data to determine which model is the better fit for the data.

Year	Teen Mothers Who Are Unwed (%)	Year	Teen Mothers Who Are Unwed (%)
1960	15	1991	69
1970	30	1992	71
1980	48	1993	72
1986	61	1994	76
1990	68	1996	76

50. *Per Pupil Expenditures* The per pupil expenditures by public elementary and secondary schools, in constant 1997–1998 dollars, are given in the table.
   a. Find an exponential function to fit the data, using as input the number of years after 1950.
   b. Find a logistic function to fit the data, using as input the number of years after 1950.
   c. Graph both models with the data to determine which model is the better fit for the data.

Years	Spending per Pupil	Years	Spending per Pupil
1960	2422	1992	6587
1965	3077	1993	6587
1970	3764	1994	6633
1975	4444	1995	6676
1980	4770	1996	6745
1985	5285	1997	6814
1990	6591	1998	6943
1991	6626		

(Source: U.S. Department of Education)

51. *Mothers in the Workforce* The data shown in the table below give the percent of moms who returned to the workforce within 1 year after they had a child (for selected years 1976–1998). Find a logarithmic model for the data, with x equal to the number of years after 1975.

Year	Returning Mothers (%)	Year	Returning Mothers (%)
1976	31	1988	51
1978	36	1990	52
1980	39	1992	53
1982	43	1994	52
1984	48	1995	55
1986	50	1998	59

(Source: Associated Press)

52. *Japan's Population* The following table gives the population of Japan, in millions, for the years 1984–1999.

a. Find a logarithmic function that models the data, with input equal to the number of years after 1980.

b. What does the model predict as the population in 2004?

Year	Population (millions)	Year	Population (millions)
1984	120.235	1992	124.452
1985	121.049	1993	124.764
1986	121.672	1994	125.034
1987	122.264	1995	125.570
1988	122.783	1996	125.864
1989	123.255	1997	126.166
1990	123.611	1998	126.486
1991	124.043	1999	126.686

(Source: www.jinjapan.org/stat/)

53. *Hospital Discharges* The numbers of discharges per 1000 people in nonfederal short-stay hospitals are given in the table below.

a. Align the data as the number of years after 1979 and find a logarithmic model to fit the data.

b. Graph the model and the data on the same graph.

c. Use the model in part (a) to estimate the number of hospital discharges per 1000 people in 2005.

Year	Discharges per 1000 People
1980	158.5
1985	137.7
1988	117.6
1990	113.0
1992	110.5
1994	106.5
1996	102.3

(Source: National Hospital Discharge Survey)

54. *Investment* Suppose $12,500 is invested in an account earning 5% annual interest compounded continuously. What is the future value in 10 years?

55. *Investments* Find the 7-year future value of an investment of $20,000 placed into an account that pays 6% compounded annually.

56. *Annuities* At the end of each quarter, $1000 is placed into an account that pays 12%, compounded quarterly. What is the future value of this annuity in 6 years?

57. *Annuities* Find the 10-year future value of an ordinary annuity with a contribution of $1500 at the end of each month, placed into an account that pays 8% compounded monthly.

58. *Present Value* Find the present value of an annuity that will pay $2000 at the end of each month for 15 years if the interest rate is 8% compounded monthly.

59. *Present Value* Find the present value of an annuity that will pay $500 at the end of each 6-month period for 12 years if the interest rate is 10% compounded semiannually.

60. *Loan Amortization* A debt of $2000 with interest at 12% compounded monthly is amortized by equal monthly payments for 36 months. What is the size of each payment?

61. *Loan Amortization* A debt of \$120,000 with interest at 6% compounded monthly is amortized by equal monthly payments for 25 years. What is the size of each monthly payment?

62. *Out-of-Wedlock Births* The percent of all teen mothers who were unmarried during the years 1960–1996 can be modeled by the logistic function

$$y = \frac{96.3641}{1 + 12.3313e^{-0.0844x}}$$

where x is the number of years after 1950.

a. Use this model to estimate the percent in 1990 and in 1996.

b. What is the upper limit of the percent of teen mothers who were unmarried according to this model?

63. *Spread of Disease* A student brings a contagious disease to an elementary school of 2000 students. The number of students infected by the disease is given by

$$n = \frac{1400}{1 + 200e^{-0.5x}}$$

where x is the number of days after the student brings the disease.

a. How many students will be infected in 14 days?

b. How many days will it take for 1312 students to be infected?

64. *Organizational Growth* The president of a new campus of a university predicts that the student body will grow rapidly after the campus is open, with the number of students at the beginning of year t given by

$$N = 4000\left(.06^{0.4^{t-1}}\right)$$

a. How many students does this model predict for the beginning of the second year ($t = 2$)?

b. How many students are predicted for the beginning of the tenth year?

c. What is the limit on the number of students that can attend this campus, according to the model?

65. *Sales Growth* The number of units of a new product that were sold each month after the product was introduced is given by

$$N = 18{,}000\left(.03^{0.4^{t}}\right)$$

where t is the number of months.

a. How many units were sold 10 months after the product was introduced?

b. What is the limit on sales if this model is accurate?

66. *Cell Phones* The table shows the numbers of cellular telephone sites at the ends of the indicated years and the numbers of people working in the cellular telephone industry.

Year	Cell Sites (thousands)	Employees
1990	5.62	21,382
1992	10.31	34,348
1994	17.92	53,902
1996	30.05	84,191
1998	65.89	134,754

(Source: Cellular Telecommunications Industry Association)

a. Fit an exponential function $C(t)$ with input t equal to the number of years after 1990 and output the cell sites (in thousands). Round to three decimal places.

b. Find a quadratic function $E(C)$ with input C equal to the number of cell sites (in thousands) and output the number of employees. Round to three decimal places.

c. Find and interpret the function $E(C(t))$, using the functions reported in parts (a) and (b).

d. Use the result of part (c) and the table to find $E(C(7))$.

e. Estimate the number of cellular telephone employees in 1997.

GROUP ACTIVITIES/EXTENDED APPLICATIONS

1. Chain Letters

Suppose you receive a chain letter asking you to send \$1 to the person at the top of the list of six names, then to add your name to the bottom of the list, and finally to send the revised letter to six people. The promise is that when the letters have been sent to each of the people above you on the list, you will be at the top of the list and you will receive a large amount of money. To investigate whether this is worth the dollar that you are asked to spend, create the requested models and answer the following questions.

1. Suppose each of the six persons on the original list sent six letters. How much money would the person on the top of the list receive?
2. Consider the person who is second on the original list. How much money would this person receive?
3. Complete the partial table of amounts that will be sent to the person on the top of the list during each "cycle" of the chain letter.

Cycle Number (after original six names)	Money Sent to Person on Top of List (\$)
1	$6 \times 6 = 36$
2	$6 \times 36 = 216$
3	
4	
5	

4. If you were going to start such a chain letter (don't, it's illegal), would you put your name at the top of the first 6 names or as number 5?
5. Find the best quadratic, power, and exponential models for the data in the table. Which of these function types gives the best model for the data?
6. Use the exponential function found in part (5) to determine the amount of money the person who was at the bottom of the original list would receive if all people contacted send the \$1 and mail the letter to 6 additional people.
7. How many people will have to respond to the chain letter for the sixth person on the original list to receive all the money that was promised?
8. If you receive the letter in its tenth cycle, how many other people have been contacted, along with you, assuming that everyone who receives the letter cooperates with its suggestions?
9. Who remains in the United States for you to send your letter to, if no one sends a letter to someone who has already received it?
10. Why do you think the federal government has made it illegal to send chain letters in the U.S. mail?

2. Modeling

Your mission is to find real-world data for company sales, a stock price, biological growth, or some sociological trend over a period of years that fits an exponential, logistic, or logarithmic function. To do this, you can look at a statistical graph called a histogram which describes the situation or at a graph or table describing it. Once you have found data consisting of numerical values, you are to find the equation that is the best exponential, logistic, or logarithmic fit for the data and then write the model that describes the relationship. Some helpful steps for this process are given in the instructions below.

1. Look in newspapers, in periodicals, on the Internet, or in statistical abstracts for a graph or table of data containing at least six data points. The data must have numerical values for the independent and dependent variables.
2. If you decide to use a relation determined by a graph that you have found, read the graph very carefully to determine the data points or (for full credit) contact the source of the data to obtain the data from which the graph was drawn.

3. If you have a table of values for different years, create a graph of the data to determine the type of function that will be the best fit for the data. To do this, first align the independent variable by letting the input represent the number of years from some convenient year and then enter the data into your graphing technology and draw a scatter plot. Check to see if the points on the scatter plot lie near some curve that could be described by an exponential, logistic, or logarithmic function. If not, save the data for possible later use and renew your search. You can also test to see if the percent change of the outputs is nearly constant for equally spaced inputs. If so, the data can be modeled by an exponential function.
4. Use your calculator to create the equation of the function that is the best fit for the data. Graph this equation and the data points on the same axes to see if the equation is reasonable.
5. Write some statements about the data that you have been working with. For example, describe how the quantity is increasing or decreasing over periods of years, why it is changing as it is, or when this model is no longer appropriate, and why.
6. Your completed project should include the following:
   a. A proper bibliographical reference for the source of your data.
   b. An original copy or photocopy of the data being used.
   c. A scatter plot of the data and reasons why you chose the model that you did to fit it.
   d. The equation that you have created and labeled with appropriate units of measure and variable descriptions.
   e. A graph of the scatter plot and the model on the same axes.
   f. One or more statements about how you think the model you have created could actually be used. Include any restrictions that should be placed on the model.

APPENDIX A

CHAPTER 1: ALGEBRA *Toolbox*

Key Concepts

- Sets
- Real numbers
- Inequalities and intervals
- Absolute value
- Algebraic expressions
- Removing parentheses
- Solving equations
- Subscripts
- The coordinate system

The algebra toolbox is designed to review prerequisite skills needed for success in each chapter. This toolbox discusses the real numbers, the coordinate system, algebraic expressions, equations, inequalities, absolute values, and subscripts.

Sets

A **set** is a well-defined collection of objects, including but not limited to numbers. In this section, we will discuss sets of real numbers, including natural numbers, integers, and rational numbers, and we will discuss the set of complex numbers later in the text. There are two ways to define a set. One way is by listing the **elements** (or **members**) of the set (usually between braces). For example, we may say that a set A contains 2, 3, 5, and 7 by writing $A = \{2, 3, 5, 7\}$. To say that 5 is an element of the set A, we write $5 \in A$. To indicate that 6 is not an element of the set, we write $6 \notin A$.

If all the elements of the set can be listed, the set is said to be a **finite set**. If all elements of a set cannot be listed, the set is called an **infinite set**. To indicate that a set continues with the established pattern, we use three dots. For example, $B = \{1, 2, 3, 4, 5, \ldots, 100\}$ describes the finite set of whole numbers from 1 through 100, and the set $N = \{1, 2, 3, 4, 5, \ldots\}$ describes the infinite set of all whole numbers beginning with 1. This set is called the **natural numbers**.

Another way to define a set is to give its description. For example, we may write $\{x | x \text{ is a math book}\}$ to define the set of math books. This is read as "the set of all x such that x is a math book." $N = \{x | x \text{ is a natural number}\}$ defines the natural numbers, which was also defined by $N = \{1, 2, 3, 4, 5, \ldots\}$ above. The set that contains no elements is called the **empty set**, and is denoted by $\varnothing$.

EXAMPLE 1

Write the following sets in two ways.

a. The set A containing the natural numbers less than 7.

b. The set B of natural numbers that are at least 7.

Solution

a. $A = \{1, 2, 3, 4, 5, 6\}, A = \{x | x \in N, x < 7\}$

b. $B = \{7, 8, 9, 10, \ldots\}, B = \{x | x \in N, x \geq 7\}$ ■

The relations that can exist between two sets follow.

RELATIONS BETWEEN SETS

1. Sets X and Y are **equal** if they contain exactly the same elements.
2. Set A is called a **subset** of set B if each element of A is an element of B. This is denoted $A \subseteq B$.
3. If set C and D have no elements in common, they are called **disjoint**.

EXAMPLE 2

For the sets $A = \{x | x \leq 9, x \text{ is a natural number}\}$, $B = \{2, 4, 6\}$, $C = \{3, 5, 8, 10\}$:

a. Which of the sets A, B, and C are subsets of A?

b. Which pairs of sets are disjoint?

c. Are any of these three sets equal?

Solution

a. Every element of B is contained in A. Thus, set B is a subset of A. Because every element of A is contained in A, A is a subset of A.

b. Sets B and C have no elements in common, so they are disjoint.

c. None of these sets have exactly the same elements, so none are equal. ■

The Real Numbers

Because most of the mathematical applications you will encounter in an applied nontechnical setting use real numbers, the emphasis in this text is the **real number system**.* Real numbers can be rational or irrational. **Rational numbers** include integers, fractions containing only integers (with no 0 in a denominator), and decimals that either terminate or repeat. Some examples of rational numbers are

$$-9, \quad \frac{1}{2}, \quad 0, \quad 12, \quad -\frac{4}{7}, \quad 6.58, \quad -7.\overline{3}$$

Irrational numbers are real numbers that are not rational. Some examples of irrational numbers are π (a number familiar to us from the study of circles), $\sqrt{2}$, $\sqrt[3]{5}$, and $\sqrt[3]{-10}$.

The types of real numbers are described in Table A.1.

TABLE A.1

Types of Real Numbers	Descriptions
Natural numbers	1, 2, 3, 4, . . .
Integers	Natural numbers, zero, and the negatives of the natural numbers: . . . , −3, −2, −1, 0, 1, 2, 3, . . .
Rational numbers	All numbers that can be written in the form $\frac{p}{q}$, where p and q are both integers with $q \neq 0$. Rational numbers can be written as terminating or repeating decimals.
Irrational numbers	All real numbers that are not rational numbers. Irrational numbers cannot be written as terminating or repeating decimals.

* The complex number system will be discussed in the Chapter 2 toolbox.

We can represent real numbers on a **real number line**. Exactly one real number is associated with each point on the line, and we say there is a one-to-one correspondence between the real numbers and the points on the line. That is, the real number line is a graph of the real numbers (see Figure A.1).

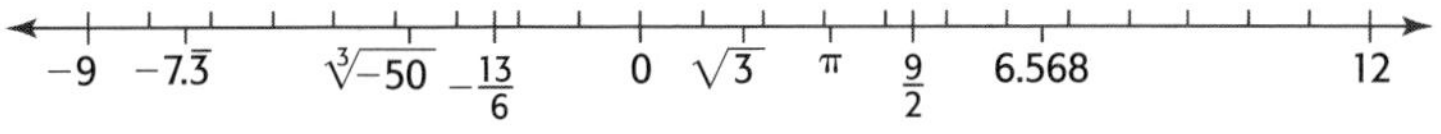

FIGURE A.1

Notice the number π on the real number line in Figure A.1. This special number, which can be approximated by 3.14, results when the circumference of (distance around) any circle is divided by the diameter of the circle. Another special real number is e; it is denoted by

$$e \approx 2.71828$$

We will discuss this number, which is important in financial and biological applications, later in the text.

Inequalities and Intervals on the Number Line

An **inequality** is a statement that one quantity is greater (or less) than another quantity. We say that a is less than b (written $a < b$) if the point representing a is to the left of the point representing b on the real number line. We may indicate that the number a is greater than or equal to b by writing $a \geq b$. The subset of real numbers x that lie between a and b (excluding a and b) can be denoted by the **double inequality** $a < x < b$ or by the **open interval** (a, b). This is called an open interval because neither of the endpoints is included in the interval. The **closed interval** $[a, b]$ represents the set of all real numbers satisfying $a \leq x \leq b$. Intervals containing one endpoint, such as $[a, b)$ or $(a, b]$, are called **half-open intervals**. We can represent the inequality $x \geq a$ by the interval $[a, \infty)$, and we can represent the inequality $x < a$ by the interval $(-\infty, a)$. Note that ∞ and $-\infty$ are not numbers, but ∞ is used in $[a, \infty)$ to represent the fact that x increases without bound and $-\infty$ is used in $(-\infty, a)$ to indicate that x decreases without bound. Table A.2 shows the graphs of different types of intervals.

TABLE A.2

Interval Notation	Inequality Notation	Verbal Description	Number Line Graph
(a, ∞)	$x > a$	x is greater than a	
$[a, \infty)$	$x \geq a$	x is greater than or equal to a	
$(-\infty, b)$	$x < b$	x is less than b	
$(-\infty, b]$	$x \leq b$	x is less than or equal to b	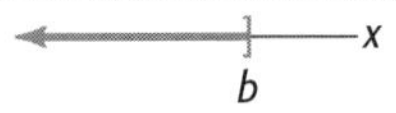
(a, b)	$a < x < b$	x is between a and b, not including either a or b	

TABLE A.2 *continued*

Interval Notation	Inequality Notation	Verbal Description	Number Line Graph
$[a, b)$	$a \le x < b$	x is between a and b, including a but not including b	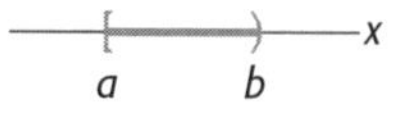
$(a, b]$	$a < x \le b$	x is between a and b, not including a but including b	
$[a, b]$	$a \le x \le b$	x is between a and b, including both a and b	

Note that open circles may be used instead of parentheses and closed circles may be used instead of brackets in the number line graphs.

EXAMPLE 3 Intervals

Write the interval corresponding to each of the inequalities in parts (a)–(e), then graph the inequality.

a. $-1 \le x \le 2$ **b.** $2 < x < 4$ **c.** $-2 < x \le 3$ **d.** $x \ge 3$ **e.** $x < 5$

Solution

a. $[-1, 2]$

b. $(2, 4)$

c. $(-2, 3]$

d. $[3, \infty)$

e. $(-\infty, 5)$

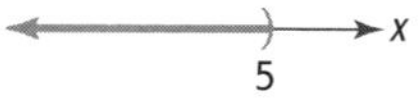

■

Absolute Value

The distance the number a is from 0 on a number line is the **absolute value** of a, denoted by $|a|$. The absolute value of any nonzero number is positive, and the absolute value of 0 is 0. For example, $|5| = 5$ and $|-8| = 8$. Note that if a is a nonnegative number, then $|a| = a$, but if a is negative, then $|a|$ is the positive number $-a$. Formally, we say

$$|a| = \begin{cases} a & \text{if } a \ge 0 \\ -a & \text{if } a < 0 \end{cases}$$

For example, $|5| = 5$ and $|-5| = -(-5) = 5$.

Algebraic Expressions

In algebra we deal with a combination of real numbers and letters. Generally, the letters are symbols used to represent unknown quantities or fixed but unspecified constants. Letters representing unknown quantities are usually called **variables**, and letters representing fixed but unspecified numbers are called **literal constants**. An expression created by performing additions, subtractions, or other arithmetic operations with one or more real numbers and variables is called an **algebraic expression**. Unless otherwise specified, the variables represent real numbers for which the algebraic expression is a real number. Examples of algebraic expressions include

$$5x - 2y, \frac{3x - 5}{12 + 5y}, \text{ and } 7z + 2$$

A term of an algebraic expression is the product of one or more variables and a real number; the real number is called a **numerical coefficient**, or simply a **coefficient**. A constant is also considered a term of an algebraic expression and is called a **constant term**. For instance, the term $5yz$ is the product of the factors 5, y, and z; this term has coefficient 5. An expression containing a finite number of additions, subtractions, and multiplications of constants and nonnegative integer powers of variables is called a **polynomial**. A polynomial can be written so that it does not contain negative powers of variables, variables in a denominator, or variables inside a radical. The expressions $5x - 2y$ and $7z^3 + 2y$ are polynomials, but $\frac{3x - 5}{12 + 5y}$ and $3x^2 - 6\sqrt{x}$ are not polynomials.

Terms that contain exactly the same variables with exactly the same exponents are called **like terms**. For example, $3x^2y$ and $7x^2y$ are like terms, but $3x^2y$ and $3xy$ are not. We can *simplify* an expression by adding or subtracting the coefficients of the like terms. For example, the simplified form of

$$3x + 4y - 8x + 2y \quad \text{is} \quad -5x + 6y$$

and the simplified form of

$$3x^2y + 7xy^2 + 6x^2y - 4xy^2 - 5xy \quad \text{is} \quad 9x^2y + 3xy^2 - 5xy$$

Removing Parentheses

We often need to remove parentheses when simplifying algebraic expressions and when solving equations. Removing parentheses frequently requires use of the **distributive property**, which says that for real numbers a, b, and c, $a(b + c) = ab + ac$. Care must be taken to avoid mistakes with signs when using the distributive property. Multiplying a sum in parentheses by a negative number changes the sign of each term in the parentheses. For example,

$$-3(x - 2y) = -3(x) + (-3)(-2y) = -3x + 6y \quad \text{and} \quad -(3xy - 5x^3) = -3xy + 5x^3$$

We add or subtract (**combine**) algebraic expressions by combining the like terms. For example, the sum of the expressions $5sx - 2y + 7z^3$ and $2y + 5sx - 4z^3$ is

$$\begin{aligned}(5sx - 2y + 7z^3) + (2y + 5sx - 4z^3) &= 5sx - 2y + 7z^3 + 2y + 5sx - 4z^3 \\ &= 10sx + 3z^3\end{aligned}$$

and the difference of these two expressions is

$$\begin{aligned}(5sx - 2y + 7z^3) - (2y + 5sx - 4z^3) &= 5sx - 2y + 7z^3 - 2y - 5sx + 4z^3 \\ &= -4y + 11z^3\end{aligned}$$

Solving Equations

An **equation** is a statement that two quantities or expressions are equal. To **solve** an equation means to find the value(s) of the variable(s) that makes the equation a true statement. For example, the equation $3x = 6$ is true when $x = 2$, so 2 is a solution of this equation. Equations of this type are sometimes called **conditional equations** because they are true only for certain values of the variable. Equations that are true for all values of the variables for which both sides are defined are called **identities**. For example, the equation $7x - 4x = 5x - 2x$ is an identity.

We frequently can solve an equation for a given variable by rewriting the equation in an equivalent form whose solution is easy to find. Two equations are **equivalent** if and only if they have the same solutions. The following operations give equivalent equations:

PROPERTIES OF EQUATIONS

1. **Addition Property** Adding the same number to both sides of an equation gives an equivalent equation. For example, $x - 5 = 2$ is equivalent to $x - 5 + 5 = 2 + 5$ or to $x = 7$.
2. **Subtraction Property** Subtracting the same number from both sides of an equation gives an equivalent equation. For example, $z + 12 = -9$ is equivalent to $z + 12 - 12 = -9 - 12$ or to $z = -21$.
3. **Multiplication Property** Multiplying both sides of an equation by the same nonzero number gives an equivalent equation. For example, $\frac{y}{6} = 5$ is equivalent to $6\left(\frac{y}{6}\right) = 6(5)$ or to $y = 30$.
4. **Division Property** Dividing both sides of an equation by the same nonzero number gives an equivalent equation. For example, $17x = -34$ is equivalent to $\frac{17x}{17} = \frac{-34}{17}$ or to $x = -2$.

EXAMPLE 4 Solving Equations

Solve the following equations.

a. $3x = 6$ **b.** $4x - 5 = 7 + 2x$ **c.** $8x + 12 = 25 - 5x$

Solution

a. Dividing both sides of the equation by 3 gives the solution.

$$\frac{3x}{3} = \frac{6}{3}$$

$$x = 2$$

Thus, $x = 2$ is the solution to the original equation.

b. Adding 5 to both sides of the equation and subtracting $2x$ from both sides of the equation gives an equivalent equation.

$$4x - 5 = 7 + 2x$$

$$4x - 5 + 5 = 7 + 2x + 5$$

$$4x = 12 + 2x$$

$$4x - 2x = 12 + 2x - 2x$$

$$2x = 12$$

Dividing both sides by 2 gives the solution to the original equation.

$$\frac{2x}{2} = \frac{12}{2}$$

$$x = 6$$

Thus, $x = 6$ is the solution to the original equation.

c. Adding $5x$ to both sides of the equation and subtracting 12 from both sides gives an equivalent equation.

$$8x + 12 = 25 - 5x$$
$$8x + 12 + 5x = 25 - 5x + 5x$$
$$13x + 12 = 25$$
$$13x + 12 - 12 = 25 - 12$$
$$13x = 13$$

Dividing both sides by 13 gives the solution to the original equation.

$$\frac{13x}{13} = \frac{13}{13}$$
$$x = 1$$

The Coordinate System

Much of our work in algebra involves graphing. To graph in two dimensions, we use a rectangular coordinate system, or **Cartesian coordinate system**. Such a system allows us to assign a unique point in a plane to each ordered pair of real numbers. We construct the coordinate system by drawing a horizontal number line and a vertical number line so that they intersect at their origins (Figure A.2). The point of intersection is called the **origin** of the system, the number lines are called the coordinate **axes**, and the plane is divided into four parts called **quadrants**. In Figure A.3, we call the horizontal axis the ***x*-axis** and the vertical axis the ***y*-axis**, and we denote any point in the plane as the ordered pair (x, y).

The ordered pair (a, b) represents the point P that is $|a|$ units from the y-axis (right if a is positive, left if a is negative) and $|b|$ units from the x-axis (up if b is positive, down if b is negative). The values of a and b are called the **rectangular coordinates** of the point. Figure A.3 shows point P with coordinates (a, b). The point is in the second quadrant, where $a < 0$ and $b > 0$.

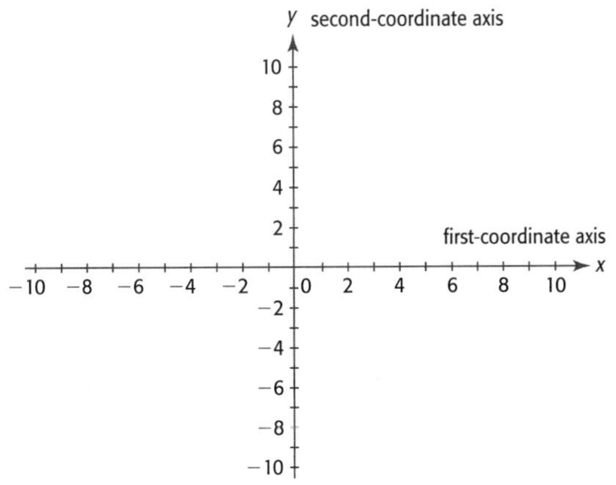

FIGURE A.2

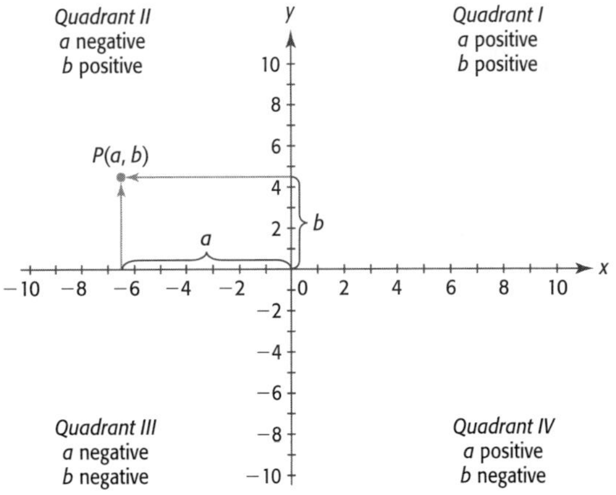

FIGURE A.3

Subscripts

We sometimes need to distinguish between two y-values and/or x-values in the same problem or on the same graph, or to designate literal constants. It is often convenient to do this by using **subscripts**. For example, if we have two fixed but unidentified points on a graph, we can represent one point as (x_1, y_1) and the other as (x_2, y_2). Subscripts also can be used to designate different equations entered in graphing utilities; for example, $y = 2x - 5$ may appear as $y_1 = 2x - 5$ when entered in the equation editor of a graphing calculator.

Toolbox EXERCISES

1. Write "the set of all natural numbers less than 9" in two different ways.

2. Is it true that $3 \in \{1, 3, 4, 6, 8, 9, 10\}$?

3. Is A a subset of B if $A = \{2, 3, 5, 7, 8, 9, 10\}$ and $B = \{3, 5, 8, 9\}$?

4. Is it true that $\frac{1}{2} \in N$, if N is the set of natural numbers?

5. Is the set of integers a subset of the set of rational numbers?

6. Are sets of rational numbers and irrational numbers disjoint sets?

Identify the sets of numbers in Exercises 7–9 as containing one or more of the following: integers, rational numbers, and/or irrational numbers.

7. $\{5, 2, 5, 8, -6\}$

8. $\left\{\frac{1}{2}, -4.1, \frac{5}{3}, 1\frac{2}{3}\right\}$

9. $\left\{\sqrt{3}, \pi, \frac{\sqrt[3]{2}}{4}, \sqrt{5}\right\}$

In Exercises 10–12, express each interval or graph as an inequality.

10. 

11. $[-3, 3]$

12. $(-\infty, 3]$

In Exercises 13–15, express each inequality or graph in interval notation.

13. $x \le 7$

14. $3 < x \le 7$

15. 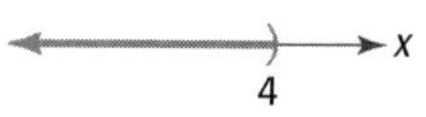

In Exercises 16–18, graph the inequality or interval on a real number line.

16. $(-2, \infty)$

17. $5 > x \ge 2$

18. $x < 3$

In Exercises 19–21, plot the points on a coordinate system.

19. $(-1, 3)$

20. $(4, -2)$

21. $(-4, 3)$

Find the absolute values in Exercises 22 and 23.

22. $|-6|$

23. $|7 - 11|$

For each algebraic expression in Exercises 24 and 25, give the coefficient of each term and give the constant term.

24. $-3x^2 - 4x + 8$

25. $5x^4 + 7x^3 - 3$

26. Find the sum of $z^4 - 15z^2 + 20z - 6$ and $2z^4 + 4z^3 - 12z^2 - 5$.

27. Simplify the expression

$$3x + 2y^4 - 2x^3y^4 - 119 - 5x - 3y^2 + 5y^4 + 110.$$

Remove the parentheses and simplify in Exercises 28–33.

28. $4(p + d)$

29. $-2(3x - 7y)$

30. $-a(b + 8c)$

31. $4(x - y) - (3x + 2y)$

32. $4(2x - y) + 4xy - 5(y - xy) - (2x - 4y)$

33. $2x(4yz - 4) - (5xyz - 3x)$

Solve the equations in Exercises 34–40.

34. $3x = 6$

35. $\frac{x}{3} = 6$

36. $x + 3 = 6$

37. $4x - 3 = 6 + x$

38. $3x - 2 = 4 - 7x$

39. $\frac{3x}{4} = 12$

40. $2x - 8 = 12 + 4x$

APPENDIX B

CHAPTER 2: ALGEBRA *Toolbox*

Key Concepts

- Properties of exponents
- Roots and radicals
- Radicals and rational exponents
- Polynomials
- Multiplying polynomials
- Factoring
- Solving literal equations
- Imaginary and complex numbers
- Operations with complex numbers

In this toolbox, we discuss the properties of exponents that are used in computations with exponents, and we discuss roots and radicals. In addition, we discuss polynomials, including multiplication, factoring, solving literal equations, and complex numbers.

Properties of Exponents

For the real numbers a and b and positive integers m and n,

1. $a^m \cdot a^n = a^{m+n}$

2. For $a \neq 0$, $\dfrac{a^m}{a^n} = \begin{cases} a^{m-n} & \text{if } m > n \\ 1 & \text{if } m = n \\ \dfrac{1}{a^{n-m}} & \text{if } m < n \end{cases}$

3. $(ab)^m = a^m b^m$

4. $(a^m)^n = a^{mn}$

5. $\left(\dfrac{a}{b}\right)^m = \dfrac{a^m}{b^m} \quad (b \neq 0)$

EXAMPLE 1 Using the Properties of Exponents

Use properties of exponents to simplify each of the following. Assume that denominators are nonzero.

a. $\dfrac{5^6}{5^4}$ **b.** $\dfrac{y^2}{y^5}$ **c.** $(3xy)^3$ **d.** $\left(\dfrac{y}{z}\right)^4$ **e.** $3^{15-2m} \cdot 3^{2m}$

Solution

a. $\dfrac{5^6}{5^4} = 5^{6-4} = 5^2 = 25$ **b.** $\dfrac{y^2}{y^5} = \dfrac{1}{y^{5-2}} = \dfrac{1}{y^3}$ **c.** $(3xy)^3 = 3^3x^3y^3 = 27x^3y^3$

d. $\left(\dfrac{y}{z}\right)^4 = \dfrac{y^4}{z^4}$ **e.** $3^{15-2m} \cdot 3^{2m} = 3^{(15-2m)+2m} = 3^{15-2m+2m} = 3^{15}$ ■

We can also simplify and evaluate expressions involving powers of powers.

EXAMPLE 2 Powers of Powers

a. Simplify $(x^4)^5$. **b.** Evaluate $(2^3)^5$. **c.** Simplify $(x^2y^3)^5$. **d.** Evaluate 2^{3^2}.

Solution

a. $(x^4)^5 = x^{4 \cdot 5} = x^{20}$

b. We can evaluate $(2^3)^5$ in two ways:

$$(2^3)^5 = 2^{15} = 32{,}768 \quad \text{or} \quad (2^3)^5 = (8)^5 = 32{,}768$$

c. $(x^2y^3)^5 = (x^2)^5(y^3)^5 = x^{10}y^{15}$ **d.** $2^{3^2} = 2^9 = 512$ ■

Roots and Radicals

We know that if each side of a square is a units, then the area of the square can be written as a^2 square units, and if we know that the area of a square is S square units, then we can write the length of each edge of the square as the **square root** of S, denoted $\sqrt{S}$.

Any positive real number has two square roots. The positive, or **principal, square root** of a positive number a is written $\sqrt{a}$, and the **negative square root** is written $-\sqrt{a}$. For example,

$$\sqrt{9} = 3$$
$$-\sqrt{9} = -3$$

The nth root of a number a is written $\sqrt[n]{a}$, and $\sqrt[n]{a} = b$ only if $b^n = a$. For example, $\sqrt[3]{64} = 4$ because $4^3 = 64$. In $\sqrt[n]{a}$, the symbol $\sqrt[n]{}$ is called a **radical**, n is the **index of the radical**, and a is the **radicand**. If the index of a radical is even, such as $\sqrt{}$ or $\sqrt[4]{}$, the radicand must be nonnegative for the root to be a real number. For example, $\sqrt{-64}$ is not a real number because no real number squared gives -64. We summarize some important facts regarding roots below.

> **RADICAL FACTS**
>
> 1. If a is a positive real number and n is even, there are two real n^{th} roots of a, $\sqrt[n]{a}$ and $-\sqrt[n]{a}$, where $\sqrt[n]{a}$ is the principal root.
> 2. If a is a negative real number and n is even, there is no real n^{th} root of a.
> 3. If a is a real number and n is odd, $\sqrt[n]{a^n} = a$.
> 4. If a is a real number and n is even, $\sqrt[n]{a^n} = |a|$; in particular, $\sqrt{a^2} = |a|$.
> 5. $\sqrt[n]{0} = 0$, because 0 raised to any nonzero power is 0.

EXAMPLE 3

Find the following roots, if they are real numbers.

a. $\sqrt[4]{16}$ **b.** $\sqrt[3]{-8}$ **c.** $-\sqrt{64}$ **d.** $\sqrt[4]{-16}$ **e.** $\sqrt[4]{(-2)^4}$

Solution

a. $\sqrt[4]{16} = 2$ because $2^4 = 16$.

b. $\sqrt[3]{-8} = -2$ because $(-2)^3 = -8$.

c. $-\sqrt{64} = -(8) = -8$.

d. $\sqrt[4]{-16}$ is not a real number because no real number raised to the fourth power gives -16.

e. $\sqrt[4]{(-2)^4} = |-2| = 2$. Note also that $\sqrt[4]{(-2)^4} = \sqrt[4]{16} = 2$. ■

> **PRODUCT RULE AND QUOTIENT RULE FOR RADICALS**
>
> If $\sqrt[n]{a}$ and $\sqrt[n]{b}$ are real numbers, $\sqrt[n]{a} \cdot \sqrt[n]{b} = \sqrt[n]{ab}$.
>
> If $\sqrt[n]{a}$ and $\sqrt[n]{b}$ are real numbers, $\dfrac{\sqrt[n]{a}}{\sqrt[n]{b}} = \sqrt[n]{\dfrac{a}{b}}$ $(b \neq 0)$.

We can frequently simplify radicals by removing from the radicand all factors whose indicated roots are present. For example, we can remove square factors from a square root by using this procedure. Writing the product rule ($\sqrt[n]{a} \cdot \sqrt[n]{b} = \sqrt[n]{ab}$) in the form $\sqrt[n]{ab} = \sqrt[n]{a} \cdot \sqrt[n]{b}$ is a useful step in this process.

EXAMPLE 4

Simplify the following radicals, assuming that the expressions are real.

a. $\sqrt{72}$ **b.** $\sqrt{b^2 - 4ac}$ if $a = 7, b = -10, c = -1$

Solution

a. Because 72 is divisible by the square 36, we can simplify the radical as follows:

$$\sqrt{72} = \sqrt{36 \cdot 2} = \sqrt{36}\sqrt{2} = 6\sqrt{2}$$

b. Substituting the values of a, b, and c gives $\sqrt{(-10)^2 - 4(7)(-1)} = \sqrt{128}$. We seek squares that are factors of 128 to simplify the radical.

$$\sqrt{128} = \sqrt{4 \cdot 32} = \sqrt{4} \cdot \sqrt{16 \cdot 2} = \sqrt{4}\sqrt{16}\sqrt{2} = 2 \cdot 4\sqrt{2} = 8\sqrt{2}$$

This simplification could be done more quickly by noting that the square 64 divides 128.

$$\sqrt{128} = \sqrt{64 \cdot 2} = \sqrt{64} \cdot \sqrt{2} = 8\sqrt{2}$$ ■

Radicals containing variables can be simplified in the same manner as those containing numbers.

EXAMPLE 5

Simplify the following radicals, assuming that the expressions are real.

a. $\sqrt[3]{x^6y^5}$ **b.** $\sqrt[4]{32x^5y^{10}}$

Solution

a. To simplify $\sqrt[3]{x^6y^5}$, we rewrite the radicand with powers of 3.

$$\sqrt[3]{x^6y^5} = \sqrt[3]{(x^2)^3y^3 \cdot y^2} = \sqrt[3]{(x^2)^3} \cdot \sqrt[3]{y^3} \cdot \sqrt[3]{y^2} = x^2y\sqrt[3]{y^2}$$

b. To simplify $\sqrt[4]{32x^5y^{10}}$, we seek powers of 4 in the radicand.

$$\begin{aligned}\sqrt[4]{32x^5y^{10}} &= \sqrt[4]{16 \cdot 2 \cdot x^4 \cdot x \cdot y^8 \cdot y^2} = \sqrt[4]{16x^4 \cdot (y^2)^4 \cdot 2xy^2} \\ &= \sqrt[4]{16}\sqrt[4]{x^4}\sqrt[4]{(y^2)^4}\sqrt[4]{2xy^2} = 2|x|y^2\sqrt[4]{2xy^2}\end{aligned}$$ ■

Radicals and Rational Exponents

To use technology to graph equations involving radicals, it is sometimes easier to convert expressions involving radicals to equivalent expressions involving fractional exponents. Note that for $a \geq 0$ and $b \geq 0$

$$\sqrt{a} = b \text{ only if } a = b^2$$

Thus,

$$\left(\sqrt{a}\right)^2 = b^2 = a, \text{ so}$$
$$\left(\sqrt{a}\right)^2 = a$$
$$\text{We define } a^{1/2} = \sqrt{a}, \text{ so } (a^{1/2})^2 = a \text{ for } a \geq 0$$

The following definitions show the connection between rational exponents and radicals.

RATIONAL EXPONENTS

If a is a real number, variable, or algebraic expression and n is a positive integer for which the principal root exists, then

$$a^{1/n} = \sqrt[n]{a}$$

EXAMPLE 6

Write the following expressions with exponents rather than radicals.

a. $\sqrt[3]{x^2}$ **b.** $\sqrt[4]{x^3}$ **c.** $\sqrt{(3xy)^5}$ **d.** $3\sqrt{(xy)^5}$

Solution

a. $\sqrt[3]{x^2} = x^{2/3}$ **b.** $\sqrt[4]{x^3} = x^{3/4}$ **c.** $\sqrt{(3xy)^5} = (3xy)^{5/2}$ **d.** $3(xy)^{5/2}$ ■

EXAMPLE 7

Write the following in radical form.

a. $y^{1/2}$ **b.** $(3x)^{3/7}$ **c.** $12x^{3/5}$

Solution

a. $y^{1/2} = \sqrt{y}$ **b.** $(3x)^{3/7} = \sqrt[7]{(3x)^3} = \sqrt[7]{27x^3}$ **c.** $12x^{3/5} = 12\sqrt[5]{x^3}$ ■

Polynomials

An algebraic expression containing a finite number of additions, subtractions, and multiplications of constants and nonnegative integer powers of variables is called a **polynomial**. When simplified, a polynomial cannot contain negative powers of variables, fractional powers of variables, variables in a denominator, or variables inside a radical. The expressions $5x - 2y$ and $7z^3 + 2y$ are polynomials, but $\frac{3x - 5}{12 + 5y}$ and $3x^2 - 6\sqrt{x}$ are not polynomials. If the only variable in the polynomial is x, then the polynomial is called a **polynomial in *x***. The general form of a polynomial in x is

$$a_n x^n + a_{n-1}x^{n-1} + \cdots + a_1 x + a_0$$

where a_0 and each coefficient $a_n, a_{n-1}, \ldots$ is a real number and each exponent $n, n - 1, \ldots$ is a positive integer.

For a polynomial in the single variable x, the power of x in each term is the **degree** of that term, with the degree of a constant term equal to 0. The term that has the highest power of x is called the **leading term** of the polynomial, the coefficient of this term is the **leading coefficient**, and the degree of this term is the **degree of the polynomial**. Thus, $5x^4 + 3x^2 - 6$ is a fourth-degree polynomial with leading coefficient 5. If a term contains two or more variables, the **degree of the term** is the sum of the powers of the variables in that term. Thus, the degree of $8x^4y^3$ is $4 + 3 = 7$. The degree of a polynomial containing one or more variables is the degree of the term that has the highest degree in the polynomial. Polynomials with one term are called **monomials**, those with two terms are called **binomials**, and those with three terms are called **trinomials**.

EXAMPLE 8

For each polynomial, state the constant term, the degree of the highest-degree term, the leading coefficient, and the degree of the polynomial.

a. $5x^2 - 8x + 2x^4 - 3$ **b.** $5xy^2 - 6x^3y + 3x^4y^2 + 7$

Solution

a. The constant term is -3; the degree of the highest degree term is 4, so the leading coefficient is 2 and the degree of the polynomial is 4.

b. The constant term is 7; the degree of the highest-degree term is $4 + 2 = 6$, so the leading coefficient is 3 and the degree of the polynomial is 6. ■

Multiplication of Polynomials

We multiply two monomials by multiplying the coefficients and adding the exponents of the respective variables that are in both monomials. For example,

$$(3x^3y^2)(4x^2y) = 3 \cdot 4 \cdot x^3 \cdot x^2 \cdot y^2 \cdot y = 3 \cdot 4x^{3+2}y^{2+1} = 12x^5y^3$$

We can multiply more than two monomials in the same manner.

EXAMPLE 9

Find the product:

$$(-2x^4z)(4x^2y^3)(yz^5)$$

Solution

$$(-2x^4z)(4x^2y^3)(yz^5) = -2 \cdot 4x^{4+2}y^{3+1}z^{1+5} = -8x^6y^4z^6$$ ■

We can use the **distributive property**,

$$a(b + c) = ab + ac$$

to multiply a monomial times a polynomial. For example,

$$x(3x + y) = x \cdot 3x + x \cdot y = 3x^2 + xy$$

We can extend the property $a(b + c) = ab + ac$ to multiply a monomial times any polynomial. For example,

$$3x(2x + xy + 6) = 3x \cdot 2x + 3x \cdot xy + 3x \cdot 6 = 6x^2 + 3x^2y + 18x$$

EXAMPLE 10

Find the following products:

a. $-2ab(3ax + 12bc - 2abx)$ **b.** $(3st^2 - 2tx + 5s^3t)6st$

Solution

a. $$\begin{aligned}-2ab(3ax + 12bc - 2abx) &= -2ab(3ax) + (-2ab)(12bc) - (-2ab)(2abx)\\ &= -6a^2bx - 24ab^2c + 4a^2b^2x\end{aligned}$$

b. Multiplying by $6st$ on the right gives

$$\begin{aligned}(3st^2 - 2tx + 5s^3t)6st &= 3st^2(6st) - 2tx(6st) + 5s^3t(6st)\\ &= 18s^2t^3 - 12st^2x + 30s^4t^2\end{aligned}$$

■

The product of two binomials can be found by using the distributive property as follows:

$$(a + b)(c + d) = a(c + d) + b(c + d) = ac + ad + bc + bd$$

Note that this product can be remembered as the sum of the products of the **F**irst, **O**uter, **I**nner, and **L**ast terms of the binomials, and we use the word **FOIL** to denote this method.

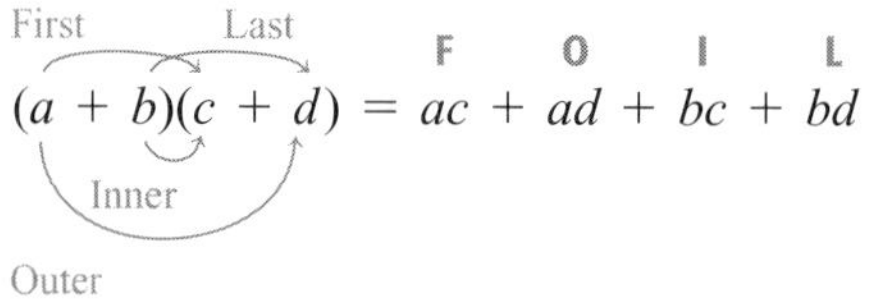

EXAMPLE 11

Find the following products:

a. $(x - 4)(x - 5)$ **b.** $(2x - 3)(3x + 2)$ **c.** $(3x - 5y)(3x + 5y)$

Solution

a. $$\begin{aligned}(x - 4)(x - 5) &= x \cdot x + x(-5) + (-4)x + (-4)(-5)\\ &= x^2 - 5x - 4x + 20 = x^2 - 9x + 20\end{aligned}$$

b. $$\begin{aligned}(2x - 3)(3x + 2) &= (2x)(3x) + (2x)2 + (-3)(3x) + (-3)2\\ &= 6x^2 + 4x - 9x - 6 = 6x^2 - 5x - 6\end{aligned}$$

c. $$\begin{aligned}(3x - 5y)(3x + 5y) &= (3x)(3x) + (3x)(5y) + (-5y)(3x) + (-5y)(5y)\\ &= 9x^2 + 15xy - 15xy - 25y^2 = 9x^2 - 25y^2\end{aligned}$$

■

To multiply two polynomials, multiply each term of the first polynomial by each term of the second polynomial and then combine like terms.

EXAMPLE 12

Multiply $(2x + 3)(4x^2 - x + 7)$

Solution

$$\begin{aligned}(2x + 3)(4x^2 - x + 7) &= 2x(4x^2) + 2x(-x) + 2x(7) + 3(4x^2) + 3(-x) + 3(7)\\ &= 8x^3 - 2x^2 + 14x + 12x^2 - 3x + 21\\ &= 8x^3 + 10x^2 + 11x + 21\end{aligned}$$ ■

Certain products and powers involving binomials occur frequently, so the following special products should be remembered.

SPECIAL BINOMIAL PRODUCTS	
I. $(x + a)(x - a) = x^2 - a^2$	(difference of two squares)
II. $(x + a)^2 = x^2 + 2ax + a^2$	(perfect square trinomial)
III. $(x - a)^2 = x^2 - 2ax + a^2$	(perfect square trinomial)

EXAMPLE 13

Find the following products by using the special binomial products formulas.

a. $(5x + 1)^2$ **b.** $(2x - 5)(2x + 5)$ **c.** $(3x - 4)^2$

Solution

a. $(5x + 1)^2 = (5x)^2 + 2(5x)(1) + 1^2 = 25x^2 + 10x + 1$

b. $(2x - 5)(2x + 5) = (2x)^2 - 5^2 = 4x^2 - 25$

c. $(3x - 4)^2 = (3x)^2 - 2(3x)(4) + 4^2 = 9x^2 - 24x + 16$ ■

Factoring

Factoring is the process of writing a number or an algebraic expression as the product of two or more numbers or expressions. For example, the distributive property justifies factoring of monomials from polynomials. For example,

$$5x^2 - 10x = 5x(x - 2)$$

Factoring out the **greatest common factor** (gcf) from a polynomial is the first step in factoring.

EXAMPLE 14

Factor out the greatest common factor.

a. $4x^2y^3 - 18xy^4$ **b.** $3x(a - b) - 2y(a - b)$

Solution

a. The gcf of 4 and 18 is 2. The gcf of x^2 and x is the lowest power of x, which is x. The gcf of y^3 and y^4 is the lowest power of y, which is y^3. Thus, the gcf of $4x^2y^3$ and $18xy^4$ is $2xy^3$. Factoring out the gcf gives

$$4x^2y^3 - 18xy^4 = 2xy^3(2x - 9y)$$

b. The gcf of $3x(a - b)$ and $2y(a - b)$ is the binomial $a - b$. Factoring out $a - b$ from each term gives

$$3x(a - b) - 2y(a - b) = (a - b)(3x - 2y)$$ ■

By recognizing that a polynomial has the form of one of the special products given above, we can factor that polynomial.

EXAMPLE 15

Use knowledge of binomial products to factor the following algebraic expressions.

a. $9x^2 - 25$ **b.** $4x^2 - 12x + 9$

Solution

a. Both terms are squares, so the polynomial can be recognized as the **difference of two squares**. It will then factor as the product of the sum and the difference of the square roots of the terms (see Special Binomial Products, Formula I).

$$9x^2 - 25 = (3x + 5)(3x - 5)$$

b. Recognizing that the second-degree term and the constant term are squares leads us to investigate whether $12x$ is twice the product of the square roots of these two terms (see Special Binomial Products, Formula III). The answer is yes, so the polynomial is a **perfect square**, and it can be factored as follows:

$$4x^2 - 12x + 9 = (2x - 3)^2$$

This can be verified by expanding $(2x - 3)^2$. ■

The first step in factoring is to look for common factors. All applicable factoring techniques should be applied to factor a polynomial completely.

EXAMPLE 16

Factor the following polynomials completely.

a. $3x^2 - 33x + 72$ **b.** $6x^2 - x - 1$

Solution

a. The number 3 can be factored from all three terms, giving

$$3x^2 - 33x + 72 = 3(x^2 - 11x + 24)$$

If the trinomial can be factored into the product of two binomials, the first term of each binomial must be x, and we seek two numbers whose product is 24 and whose sum is -11. Because -3 and -8 satisfy these requirements, we get

$$3(x^2 - 11x + 24) = 3(x - 3)(x - 8)$$

b. The four possible factorizations of $6x^2 - x - 1$ that give $6x^2$ as the product of the first terms and -1 as the product of the last terms follow:

$$(6x - 1)(x + 1) \qquad (6x + 1)(x - 1)$$
$$(2x - 1)(3x + 1) \qquad (2x + 1)(3x - 1)$$

The factorization that gives a product with middle term $-x$ is the correct factorization.

$$(2x - 1)(3x + 1) = 6x^2 - x - 1$$ ■

Some polynomials, such as $6x^2 + 9x - 8x - 12$, can be factored by **grouping**. To do this, we factor out common factors from pairs of terms and then factor out a common binomial expression if it exists. For example,

$$6x^2 + 9x - 8x - 12 = 3x(2x + 3) - 4(2x + 3) = (2x + 3)(3x - 4)$$

When a second-degree trinomial can be factored but there are many possible factors to test to find the correct one, an alternate method of factoring can be used. The steps used to factor the trinomial using factoring by grouping techniques follow.

Factoring a Trinomial into the Product of Two Binomials Using Grouping

Steps	Example
To factor a quadratic trinomial in the variable x:	Factor $5x - 6 + 6x^2$
1. Arrange the trinomial with the powers of x in descending order.	1. $6x^2 + 5x - 6$
2. Form the product of the second-degree term and the constant term (first and third terms).	2. $6x^2(-6) = -36x^2$
3. Determine if there are two factors of the product in step 2 that will sum to the middle (first-degree) term. (If there are no such factors, the trinomial will not factor into two binomials.)	3. $-36x^2 = (-4x)(9x)$ and $-4x + 9x = 5x$
4. Replace the middle term from step 1 with the *sum* of the two factors from step 3.	4. $6x^2 + 5x - 6 = 6x^2 - 4x + 9x - 6$
5. Factor the four-term polynomial from step 4 by grouping.	5. $6x^2 + 5x - 6 = 2x(3x - 2) + 3(3x - 2) = (3x - 2)(2x + 3)$

EXAMPLE 17

Factor $10x^2 + 23x - 5$, using grouping.

Solution

To use this method to factor $10x^2 + 23x - 5$, we

1. **Note that the trinomial has powers of x in descending order.** $10x^2 + 23x - 5$
2. **Multiply the second-degree and constant terms:** $10x^2(-5) = -50x^2$
3. **Factor $-50x^2$ so the sum of factors is $23x$:** $-50x^2 = 25x(-2x)$, $25x + (-2x) = 23x$

4. Rewrite the middle term of the expression as the sum of the factors from step 3. $10x^2 + 23x - 5 = 10x^2 + 25x - 2x - 5$

5. Factor by grouping:

$$\begin{aligned} &10x^2 + 25x - 2x - 5 \\ &= 5x(2x + 5) - (2x + 5) \\ &= (2x + 5)(5x - 1) \end{aligned}$$

■

Solving Literal Equations

To solve a literal equation for one of two or more variables, it is sometimes necessary to use factoring. Consider the following example.

EXAMPLE 18 Solving an Equation for a Specified Variable

Solve the equation $2(2x - b) = \frac{5cx}{3}$ for x.

Solution

We solve the equation for x by treating the other variables as constants:

$$2(2x - b) = \frac{5cx}{3}$$

$$6(2x - b) = 5cx$$ Clear the equation of fractions by multiplying by the LCD, 3.

$$12x - 6b = 5cx$$ Multiply to remove parentheses.

$$12x - 5cx = 6b$$ Get all terms containing x on one side and all other terms on the other side.

$$x(12 - 5c) = 6b$$ Factor x from the expression. The remaining factor is the coefficient of x.

$$\frac{x(12 - 5c)}{(12 - 5c)} = \frac{6b}{12 - 5c}$$ Divide both sides by the coefficient of x.

$$x = \frac{6b}{12 - 5c}$$

■

Imaginary and Complex Numbers

The numbers discussed up to this point are real numbers (either rational numbers such as 2, -3, $\frac{5}{8}$ and $-\frac{2}{3}$ or irrational numbers such as $\sqrt{3}$, $\sqrt[3]{6}$, and π). But some equations do not have real solutions. For example, if

$$x^2 + 1 = 0, \text{ then } x^2 = -1$$

but there is no real number that, when squared, will equal -1. However, we can denote one solution to this equation as the **imaginary unit *i***, defined by

$$i = \sqrt{-1}$$

The set of **complex numbers** is formed by adding real numbers and multiples of i.

COMPLEX NUMBER

The number $a + bi$, in which a and b are real numbers, is said to be a **complex number** in **standard form**. The a is the real part of the number, and b is the imaginary part. If $b = 0$, the number $a + bi = a$ is a real number, and if $b \neq 0$, the number $a + bi$ is an **imaginary number**. If $a = 0$, bi is a **pure imaginary number**.

The complex number system includes the real numbers as well as the imaginary numbers (Figure B.1). Examples of imaginary numbers are $3 + 2i$, $5 - 4i$, and $\sqrt{3} - \frac{1}{2}i$; examples of pure imaginary numbers are $-i$, $2i$, $12i$, $-4i$, $i\sqrt{3}$, and πi.

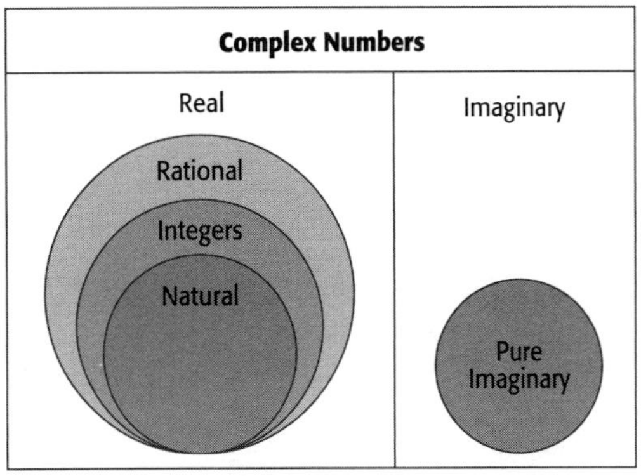

FIGURE B.1

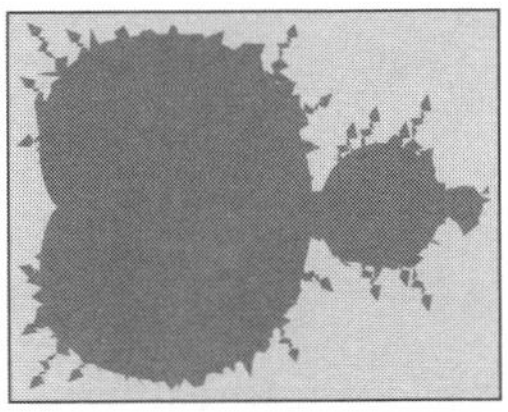

FIGURE B.2

From "The Fractal Geometry of Nature" by B.B. Mandelbrot. Copyright 1982. Used with permission of the author.

The complex number system is an extension of the real numbers that includes imaginary numbers. The term *imaginary number* seems to imply that the numbers do not exist, but in fact they do have important theoretical and technical applications. Complex numbers are used in the design of electrical circuits and airplanes, and they were used in the development of quantum physics. For example, one way of accounting for the amount as well as the phase of the current or voltage in an alternating current electrical circuit involves complex numbers. Special sets of complex numbers can be graphed on the **complex coordinate system** to create pictures called **fractal images**. (Figure B.2 shows a fractal image called the Mandelbrot set, which can be generated with the complex number i.)

Operations with Complex Numbers

Complex numbers can be added and subtracted by combining their real and imaginary parts.

ADDITION AND SUBTRACTION OF COMPLEX NUMBERS

The sum and difference of the complex numbers $a + bi$ and $c + di$ are

$$(a + bi) + (c + di) = (a + c) + (b + d)i$$

$$(a + bi) - (c + di) = (a - c) + (b - d)i$$

EXAMPLE 19 Sums and Differences of Complex Numbers

Find the sum and difference.

a. $(3 - 5i) + (-12 - 3i)$ **b.** $(2 + 6i) - (7 - 9i)$

Solution

a. $(3 - 5i) + (-12 - 3i) = (3 + (-12)) + (-5i) + (-3i)$
$= -9 + (-5 - 3)i = -9 - 8i$

b. $(2 + 6i) - (7 - 9i) = (2 - 7) + (6i + 9i) = -5 + (6 + 9)i = -5 + 15i$ ■

Because of the definition of i, $i \cdot i = i^2 = (\sqrt{-1})^2 = -1$. The product of two complex numbers of the form $a + bi$ can be found by multiplying them as if they were binomials (using the FOIL method) and substituting -1 for i^2. The general procedure follows:

$$\begin{aligned}(a + bi)(c + di) &= ac + adi + bci + bdi^2 \\ &= ac + (ad + bc)i + bd(-1) \\ &= (ac - bd) + (ad + bc)i\end{aligned}$$

EXAMPLE 20 Products of Complex Numbers

Find the following products.

a. $(4 + 3i)(2 - 5i)$ **b.** $(3 + 2i)(3 - 2i)$

Solution

a. $(4 + 3i)(2 - 5i) = 8 - 20i + 6i - 15i^2 = 8 - 14i - 15(-1) = 23 - 14i$

b. $(3 + 2i)(3 - 2i) = 9 - 6i + 6i - 4i^2 = 9 - 4(-1) = 13$ ■

Notice that the two factors multiplied in Example 20(b) have identical real parts and that their imaginary parts are negatives; such a pair of complex numbers are called **conjugates**. Example 20(b) illustrates that the product of two conjugates is a real number. In fact, this is always the case.

COMPLEX CONJUGATES

Two complex numbers with the same real parts and imaginary parts with opposite signs are called **complex conjugates**.

The product of $a + bi$ and $a - bi$ is the real number $a^2 + b^2$; that is, $(a + bi)(a - bi) = a^2 + b^2$.

The conjugate of the divisor can be used to find the quotient of two complex numbers.

EXAMPLE 21 Quotient of Two Complex Numbers

Find the quotient $\frac{5 - 3i}{1 - 2i}$ by writing the fraction as a complex number in standard form.

Solution

The quotient is found by converting the denominator to a real number. This is done by multiplying the numerator and denominator by the conjugate of the denominator, $1 + 2i$, and then simplifying.

$$\frac{5 - 3i}{1 - 2i} = \frac{5 - 3i}{1 - 2i} \cdot \frac{1 + 2i}{1 + 2i} = \frac{5 + 10i - 3i - 6i^2}{1 + 2i - 2i - 4i^2} = \frac{5 + 7i + 6}{1 + 4}$$

$$= \frac{11 + 7i}{5} = \frac{11}{5} + \frac{7}{5}i$$

■

Some graphing calculators have a complex number mode that can be used to perform operations with complex numbers. Figure B.3 shows some of the computations from Examples 19 to 21 done with technology.

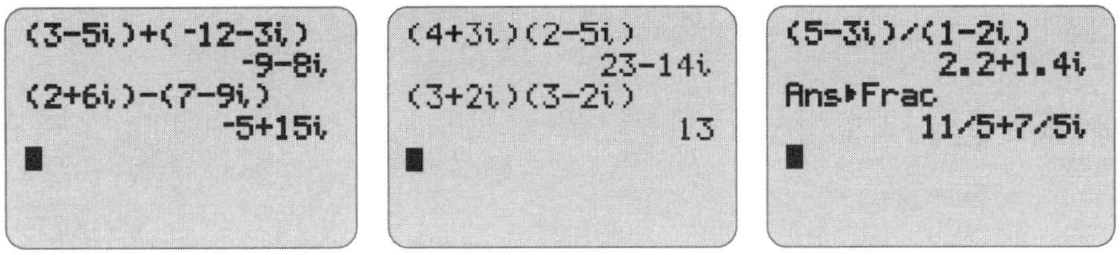

(a) (b) (c)

FIGURE B.3

Toolbox EXERCISES

In Exercises 1 and 2, use the rules of exponents to simplify.

1. **a.** $x^4 \cdot x^3$ **b.** $\frac{x^{12}}{x^7}$ **c.** $(4ay)^4$ **d.** $\left(\frac{3}{z}\right)^4$
e. $2^3 \cdot 2^2$ **f.** $(x^4)^2$

2. **a.** $y^5 \cdot y$ **b.** $\frac{w^{10}}{w^4}$ **c.** $(6bx)^3$
d. $\left(\frac{5z}{2}\right)^3$ **e.** $3^2 \cdot 3^3$ **f.** $(2y^3)^4$

3. Evaluate each of the following roots, if they are real numbers.
a. $\sqrt{16}$ **b.** $-\sqrt{16}$ **c.** $\sqrt{-16}$
d. $\sqrt[3]{27}$ **e.** $-\sqrt[5]{-32}$

4. Simplify the following radicals, assuming that the expressions are real.
a. $\sqrt{18}$ **b.** $\sqrt{216}$
c. $\sqrt{b^2 - 4ac}$ if $a = 2, b = -12, c = -3$
d. $\sqrt[3]{243a^7b^4}$

5. Write each of the following expressions in simplified exponential form.
a. $\sqrt{x^3}$ **b.** $\sqrt[4]{x^3}$ **c.** $\sqrt[5]{x^3}$
d. $\sqrt[6]{27y^9}$ **e.** $27\sqrt[6]{y^9}$

6. Write each of the following in radical form.
a. $a^{3/4}$ **b.** $-15x^{5/8}$
c. $(-15x)^{5/8}$

For each polynomial in Exercises 7–9, give the degree of the polynomial, give the leading coefficient, and decide if it is a polynomial in one or several variables.

7. $4x - 3x^2 + 8$

8. $7x^3 - 3 + 5x^4$

9. $3x + 2y^4 - 2x^3y^4 - 119$

In Exercises 10–14, find the products.

10. $(4x^2y^3)(-3a^2x^3)$

11. $2xy^3(2x^2y + 4xz - 3z^2)$

12. $(x - 7)(2x + 3)$

13. $(k - 3)^2$

14. $(4x - 7y)(4x + 7y)$

In Exercises 15–24, factor each of the polynomials completely.

15. $3x^2 - 12x$

16. $12x^5 - 24x^3$

17. $9x^2 - 25m^2$

18. $x^2 - 8x + 15$

19. $x^2 - 2x - 35$

20. $3x^2 - 5x - 2$

21. $8x^2 - 22x + 5$

22. $6n^2 + 18 + 39n$

23. $3y^4 + 9y^2 - 12y^2 - 36$

24. $18p^2 + 12p - 3p - 2$

25. Solve $5x - 4xy = 2y$ for y.

26. Solve $2xy = \dfrac{6}{y} + \dfrac{x}{3}$ for x.

In Exercises 27 and 28, identify each number as one or more of real, imaginary, or pure imaginary.

27. **a.** $2 - i\sqrt{2}$ **b.** $5i$ **c.** $4 + 0i$
d. $2 - 5i^2$

28. **a.** $3 + i\sqrt{5}$ **b.** $3 + 0i$ **c.** $8i$
d. $2i^2 - i$

In Exercises 29–31, find values for a and b that make the statement true.

29. $a + bi = 4$

30. $a + 3i = 15 - bi$

31. $a + bi = 2 + 4i$

In Exercises 32–40, perform the indicated operations and write the answer in the form $a + bi$.

32. $(3 + 7i) + (2i - 4)$

33. $(4 - 2i) - (2 - 3i)$

34. $2(3 - i) - (7 - 2i)$

35. $2(2i + 3) - (4i + 5)$

36. $(2 - i)(2 + i)$

37. $(6 - 3i)(4 + 5i)$

38. $(2i - 1)^2$

39. $\dfrac{4 - 3i}{3 - 3i}$

40. $\dfrac{\sqrt{3} - 2i}{\sqrt{5} + i}$

In Exercises 41–49, simplify the expressions by writing them in the form a, bi, or $a + bi$.

41. $\sqrt{-36}$

42. $\sqrt{-98}$

43. $(\sqrt{-4})^2$

44. $\sqrt{-6}\sqrt{-15}$

45. $(3 + \sqrt{-4}) + (2 - \sqrt{-9})$

46. $(2 + \sqrt{-2}) - (3 - \sqrt{-8})$

47. $(1 + \sqrt{-4})(2 - \sqrt{-9})$

48. $(2 - 3\sqrt{-1})^2$

49. $\dfrac{1 + \sqrt{-1}}{\sqrt{-1}}$

APPENDIX C

CHAPTER 3: ALGEBRA *Toolbox*

Key Concepts

- Zero and negative exponents
- Scientific notation
- Simple interest

In this toolbox, we discuss zero and negative exponents and properties of integer exponents. We also discuss scientific notation, which uses positive and negative powers of 10. To prepare us for the discussion of other financial functions, we also discuss the computation of simple interest.

Zero and Negative Exponents

Several applications in this chapter require the use of zero and negative exponents. To extend the rules for positive exponents to include the **zero exponent**, consider the following. For $a \neq 0$, we know that

$$\frac{a^n}{a^n} = 1 \quad \text{and} \quad \frac{a^n}{a^n} = a^{n-n}$$

so we define $a^{n-n} = a^0$ and $a^0 = 1$ for $a \neq 0$ (0^0 is undefined).

To extend the rules to **negative exponents**, we know that for $a \neq 0$ and n a positive integer

$$\frac{a^m}{a^{m+n}} = \frac{1}{a^{m+n-m}} = \frac{1}{a^n} \quad \text{and} \quad \frac{a^m}{a^{m+n}} = a^{m-(m+n)}$$

So

$$a^{m-(m+n)} = a^{-n} \quad \text{and} \quad a^{-n} = \frac{1}{a^n} \quad \text{for} \quad a \neq 0$$

Special cases of this definition are

$$a^{-1} = \frac{1}{a} \quad \text{for} \quad a \neq 0 \quad \text{and} \quad \left(\frac{a}{b}\right)^{-n} = \frac{1}{\left(\frac{a}{b}\right)^n} = \left(\frac{b}{a}\right)^n$$

These definitions are summarized below.

ZERO AND NEGATIVE EXPONENTS

For $a \neq 0, b \neq 0$

1. $a^0 = 1$
2. $a^{-1} = \dfrac{1}{a}$
3. $a^{-n} = \dfrac{1}{a^n}$
4. $\left(\dfrac{a}{b}\right)^{-n} = \left(\dfrac{b}{a}\right)^n$

The rules of exponents that we discussed for positive integer exponents are also true for zero and negative integer exponents. In addition, these rules apply if the exponents are rational numbers.

EXAMPLE 1 Zero and Negative Exponents

Simplify the following expressions by removing all zero and negative exponents, for nonzero a, b, and c.

a. $(4c)^0$ **b.** $4c^0$ **c.** $(5b)^{-1}$ **d.** $5b^{-1}$ **e.** $(6a)^{-3}$ **f.** $6a^{-3}$ **g.** -4^0

Solution

a. $(4c)^0 = 1$ **b.** $4c^0 = 4(1) = 4$ **c.** $(5b)^{-1} = \dfrac{1}{(5b)} = \dfrac{1}{5b}$

d. $5b^{-1} = 5 \cdot \dfrac{1}{b} = \dfrac{5}{b}$ **e.** $(6a)^{-3} = \dfrac{1}{(6a)^3} = \dfrac{1}{6^3a^3} = \dfrac{1}{216a^3}$

f. $6a^{-3} = 6 \cdot \dfrac{1}{a^3} = \dfrac{6}{a^3}$ **g.** $-4^0 = (-1)(4^0) = -1$

EXAMPLE 2 Applying the Rules for Exponents

If a, b, c and d are nonzero real numbers, write $\dfrac{5a^3b^{-2}}{(4c)^0d^{-1}}$ without zero or negative exponents, and simplify the expression.

Solution

We write $\dfrac{5a^3b^{-2}}{(4c)^0d^{-1}}$ without zero or negative exponents and then simplify the expression, as follows:

$$\frac{5a^3b^{-2}}{(4c)^0d^{-1}} = \frac{5a^3 \cdot \dfrac{1}{b^2}}{1 \cdot \dfrac{1}{d}} = \frac{\dfrac{5a^3}{b^2}}{\dfrac{1}{d}} = \frac{5a^3}{b^2} \cdot \frac{d}{1} = \frac{5a^3d}{b^2}$$

Multiplications and divisions with algebraic expressions involving zero and negative exponents are performed by using the properties of exponents.

EXAMPLE 3 Applying Exponent Rules to Products and Quotients

Compute the following products and quotients and write the answer with positive exponents.

a. $(-2x^{-2}y)(5x^{-2}y^{-3})$ **b.** $\dfrac{8xy^{-2}}{2x^4y^{-6}}$ **c.** $\dfrac{\dfrac{2x^{-1}y}{3a}}{\dfrac{6xy^{-2}}{5a}}$ **d.** $(-3x^{1/2})^3(2x^{-1/3})$

Solution

a. $(-2x^{-2}y)(5x^{-2}y^{-3}) = -10x^{-2+(-2)}y^{1+(-3)} = -10x^{-4}y^{-2}$

$$= -10 \cdot \frac{1}{x^4} \cdot \frac{1}{y^2} = \frac{-10}{x^4y^2}$$

b. $\dfrac{8xy^{-2}}{2x^4y^{-6}} = 4x^{1-4}y^{-2-(-6)} = 4x^{-3}y^4 = 4 \cdot \dfrac{1}{x^3} \cdot y^4 = \dfrac{4y^4}{x^3}$

c. To perform this division, we invert and multiply.

$$\dfrac{\dfrac{2x^{-1}y}{3a}}{\dfrac{6xy^{-2}}{5a}} = \dfrac{2x^{-1}y}{3a} \cdot \dfrac{5a}{6xy^{-2}} = \dfrac{10ax^{-1}y}{18axy^{-2}} = \dfrac{5x^{-1-1}y^{1-(-2)}}{9} = \dfrac{5x^{-2}y^3}{9} = \dfrac{5y^3}{9x^2}$$

d. $(-3x^{1/2})^3(2x^{-1/3}) = \left((-3)^3x^{(1/2)(3)}\right)(2x^{-1/3}) = (-27x^{3/2})(2x^{-1/3}) = -54x^{7/6}$

Scientific Notation

A convenient way to write very large (positive or negative) numbers or numbers close to zero is to express them with exponents. This can be done by converting the numbers to **scientific notation**, which has the form

$$N \times 10^p \qquad \text{(where } 1 \le N < 10 \text{ and } p \text{ is an integer)}$$

For instance, the scientific notation form of 2,654,000 is 2.654×10^6.

Calculators often automatically express numbers by using scientific notation when the numbers are very large (positive or negative) numbers or when the numbers are close to zero. Most calculators display a number in scientific notation as a number N, the symbol E, and the power of 10. Figure C.1 shows two numbers expressed in standard notation and in scientific notation, and the calculator display of their scientific notation.

Standard Notation	Scientific Notation	Calculator Display of Scientific Notation
62256 0.000235	6.2256×10^4 2.35×10^{-4}	62256 6.2256E4 0.000235 2.35E-4

FIGURE C.1

A number written in scientific notation can be converted to standard notation by multiplying (when the exponent on 10 is positive) or by dividing (when the exponent on 10 is negative.) For example, we convert 7.983×10^5 to standard notation by multiplying 7.983 by $10^5 = 100{,}000$:

$$7.983 \times 10^5 = 7.983 \cdot 100{,}000 = 798{,}300$$

and we convert 4.563×10^{-7} to standard notation by dividing 4.563 by $10^7 = 10{,}000{,}000$:

$$4.563 \times 10^{-7} = \frac{4.563}{10{,}000{,}000} = 0.0000004563$$

Multiplying two numbers in scientific notation involves adding the powers of 10, and dividing them involves subtracting the powers.

EXAMPLE 4 Scientific Notation

Compute the following and write the answers in scientific notation.

a. $(7.983 \times 10^5)(4.563 \times 10^{-7})$ **b.** $(7.983 \cdot 10^5)/(4.563 \times 10^{-7})$

Solution

a. $(7.983 \times 10^5)(4.563 \times 10^{-7}) = 36.426429 \times 10^{5+(-7)}$

$= 36.426429 \times 10^{-2} = (3.6426429 \times 10^1) \times 10^{-2} = 3.6426429 \times 10^{-1}$

The calculation using technology is shown in Figure C.2(a).

b. $(7.983 \times 10^5)/(4.563 \times 10^{-7}) = 1.749506903 \times 10^{5-(-7)}$

$= 1.749506903 \times 10^{12}$. Figure C.2(b) shows the calculation using technology.

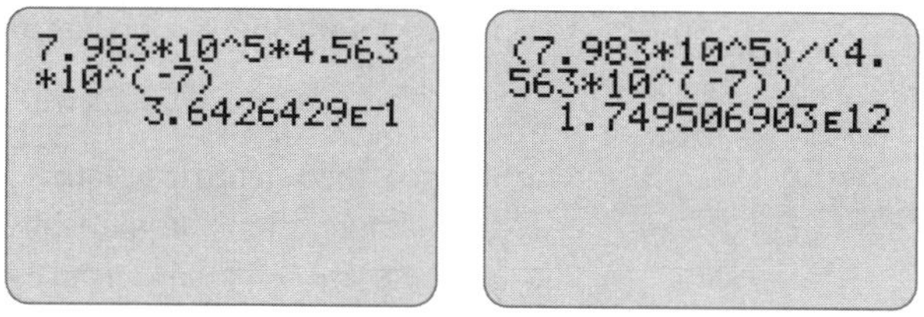

(a) (b)

FIGURE C.2

Simple Interest

The formula for the simple interest I earned on an investment of P dollars at rate r per year for t years is

$$I = Prt$$

The sum of the original investment and the earned interest is called the **future value** of the investment. The formula for calculating the future value S of this investment is

$$S = P + I = P + Prt = P(1 + rt)$$

EXAMPLE 5 Simple Interest

a. Find the simple interest on an investment of \$2000 earning interest at a rate of 10% per year for 3 years. Also find the future value of this investment.

b. Find the simple interest on an investment of \$2000 at 12% per year for 3 months, and find the future value of this investment.

Solution

a. The interest on the investment of \$2000 at 10% per year for 3 years is

$$I = Prt = (\$2000)(0.10)(3) = \$600$$

The future value of this investment is

$$S = P + I = \$2000 + \$600 = \$2600$$

b. The interest on an investment of \$2000 at 12% per year for 3 months (which is $\frac{3}{12}$ of a year) is

$$I = Prt = (\$2000)(0.12)\left(\frac{3}{12}\right) = \$60$$

The future value of this investment is

$$S = P + I = \$2000 + \$60 = \$2060$$

Toolbox EXERCISES

In Exercises 1–28, use the rules of exponents to simplify the following expressions and remove all zero and negative exponents. Assume that all variables are nonzero.

1. 3^{-2}

2. 2^{-3}

3. -4^{-3}

4. -4^{-2}

5. $\frac{1}{3^{-3}}$

6. $\frac{1}{5^{-2}}$

7. $\left(\frac{2}{3}\right)^{-2}$

8. $\left(\frac{3}{2}\right)^{-3}$

9. $10^{-2} \cdot 10^{0}$

10. $8^{-2} \cdot 8^{0}$

11. $(2^{-1})^{3}$

12. $(4^{-2})^{2}$

13. $10^{5^{0}}$

14. $4^{2^{2}}$

15. $x^{-4} \cdot x^{-3}$

16. $y^{-5} \cdot y^{-3}$

17. $(c^{-6})^{3}$

18. $(x^{-2})^{4}$

19. $\frac{a^{-4}}{a^{-5}}$

20. $\frac{b^{-6}}{b^{-8}}$

21. $(x^{-1/2})(x^{2/3})$

22. $(y^{-1/3})(y^{2/5})$

23. $(3a^{-3}b^{2})(2a^{2}b^{-4})$

24. $(4a^{-2}b^{3})(-2a^{4}b^{-5})$

25. $\left(\frac{2x^{-3}}{x^{2}}\right)^{-2}$

26. $\left(\frac{3y^{-4}}{2y^{2}}\right)^{-3}$

27. $\frac{28a^{4}b^{-3}}{-4a^{6}b^{-2}}$

28. $\frac{36x^{5}y^{-2}}{-6x^{6}y^{-4}}$

In Exercises 29–32, write the following numbers in scientific notation.

29. 46,000,000

30. 862,000,000,000

31. 0.000094

32. 0.00000278

In Exercises 33–36, write each of the following numbers in standard form.

33. 4.372×10^{5}

34. 7.91×10^{6}

35. 5.6294×10^{-4}

36. 6.3478×10^{-3}

In Exercises 37–38, multiply or divide, as indicated, and write the result in scientific notation.

37. $(6.250 \times 10^{7})(5.933 \times 10^{-2})$

38. $\frac{2.961 \times 10^{-2}}{4.583 \times 10^{-4}}$

In Exercises 39–42, compute the simple interest and future value of each of the following investments.

39. \$2000 invested for 5 years at 6% per year

40. \$8500 invested for 10 years at $7\frac{1}{2}$% per year

41. \$3500 invested for 6 months at 1% per month

42. \$5600 invested for 6 months at 6% per year

APPENDIX D

ANSWERS TO SELECTED EXERCISES

Chapter 1 Functions, Graphs, and Models; Linear Functions

1.1 Skills Check

1. a. Input **b.** Output **c.** D: $\{-9, -5, -7, 6, 12, 17, 20\}$; $R = \{5, 6, 7, 4, 9, 10\}$ **d.** Each input gives one output. **3.** 5; 9 **5.** No; input 9 for y gives two outputs for x. **7. a.** $f(2) = -1$ **b.** $f(2) = 10 - 3(2^2) = -2$ **c.** $f(2) = -3$ **9. a.** -7 **b.** 3 **c.** 18 **11.** Yes; $D = \{-1, 0, 1, 2, 3\}$; $R = \{5, 7, 2, -1, -8\}$ **13.** No **15.** No; one value of x (3) gives two values of y. **17.** Set b. **19.** b **21.** No **23.** $C = 2\pi r$ **25.** A rule that assigns to each input exactly one output **27.** The set of outputs of a function

1.1 Exercises

29. a. No. **b.** Yes **31.** Yes **33.** Yes **35.** Yes **37. a.** Yes **b.** The days 1–14 of May **c.** $\{171, 172, 173, 174, 175, 176, 177, 178\}$ **d.** May 1, May 3 **e.** May 14 **f.** 3 days **39. a.** \$16,115 **b.** 23,047; at the end of 2 years, the couple owes \$23,047. **c.** Year 2 **d.** 4 **41. a.** 4 to 1 **b.** 4; 4 is the projected ratio of the working-age population (25- to 64-year-olds) to the elderly in 2005. **c.** $\{1995, 2000, 2005, 2010, 2015, 2020, 2025, 2030\}$ **d.** Decreasing **43. a.** 492,671 **b.** The years 1985 through 1998 **c.** 581,697; 1993 **45. a.** Yes **b.** 68.6 **c.** 1920 **d.** The percent of U.S. workers in farm occupations was 6.1 in 1960. **e.** It is decreasing over the years shown. **47. a.** 3.4; in 1990, there were 3.4 workers for each retiree. **b.** 2030 **c.** Funding is in jeopardy. **49. a.** 6400, the revenue from the sale of 200 hats is \$6400. **b.** \$80,000; $R(2500) = 80{,}000$ **51. a.** 198,000; there is a profit of \$198,000 when 500 iPods are produced and sold **b.** \$198,000; $P(4000) = 198{,}000$ **53. a.** 110.8; the bill for 1000 KWh is \$110.80. **b.** \$163.30 **55. a.** 86; 1 second after thrown, the ball is 86 feet high **b.** 150 feet **c.** after 3 seconds **57. a.** yes **b.** $s \ge -1/4$ **c.** $s \ge 0$ **59. a.** $0 \le p < 100$ **b.** 355,500; 2,133,000 **61. a.** 8640; 11,664 **b.** $0 < x < 27$, so volume is positive and box exists **c.** Testing values in the table shows a maximum volume of 11,664 cubic inches when $x = 18$ inches. The dimensions of the box are 18 in. by 18 in. by 36 in.

x	y
10	6,800
15	10,800
20	11,200
21	10,584
19	11,552
18	11,664
17	11,560

1.2 Skills Check

1. a.

x	-3	-2	-1	0	1	2	3
$y = x^3$	-27	-8	-1	0	1	8	27

b.

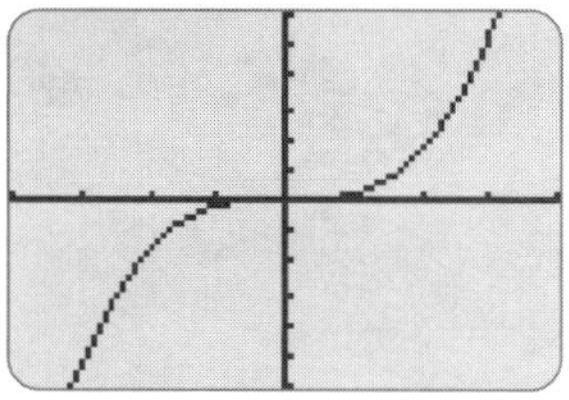

[−4, 4] by [−30, 30]

c. Graphs are the same.

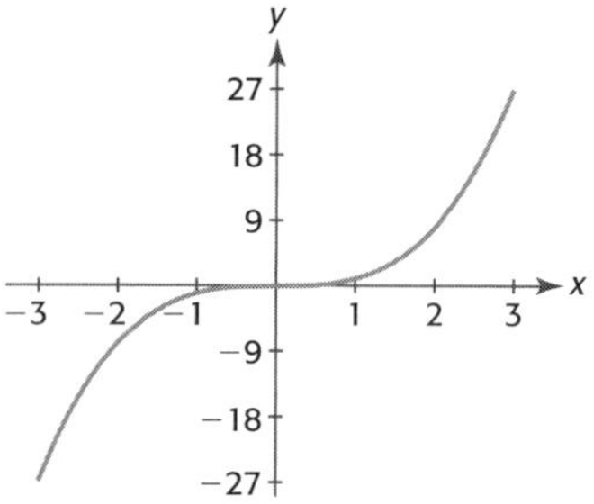

3.

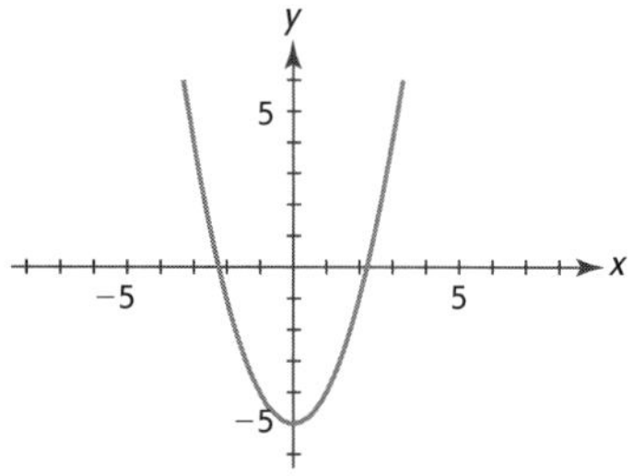

5.

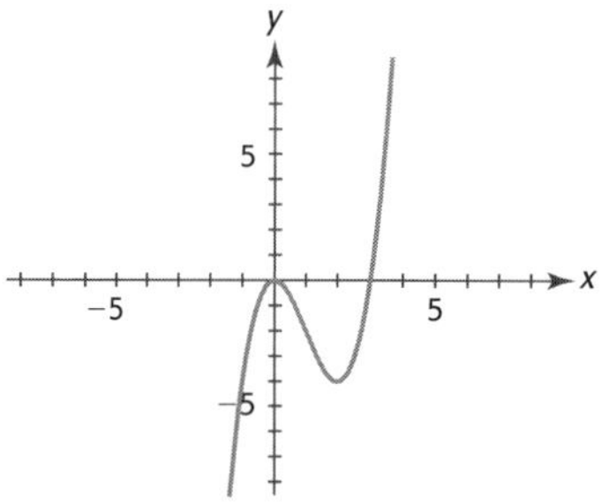

7.

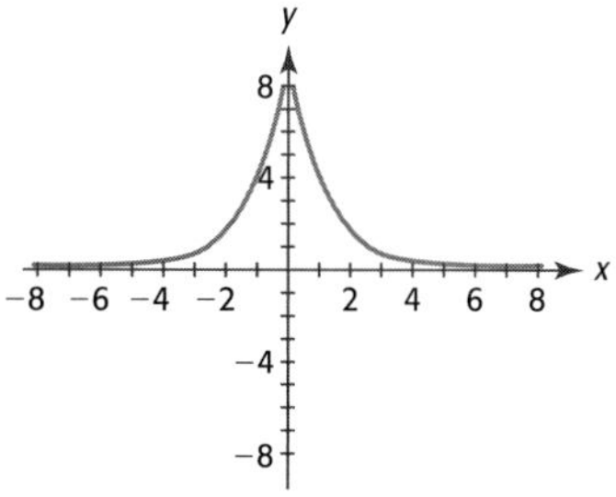

9.

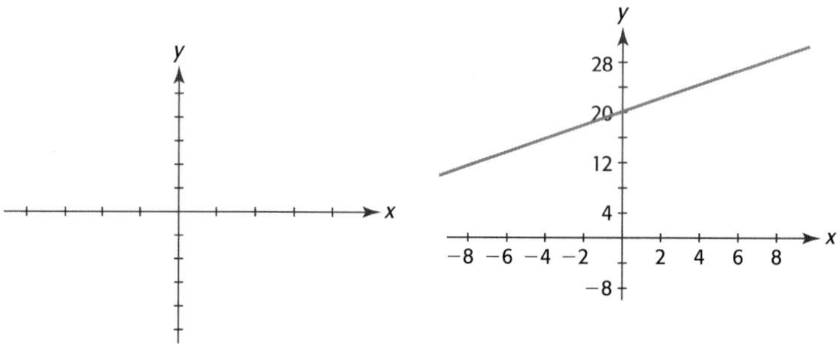

(a) **(b)**

(b) is the better view.

11.

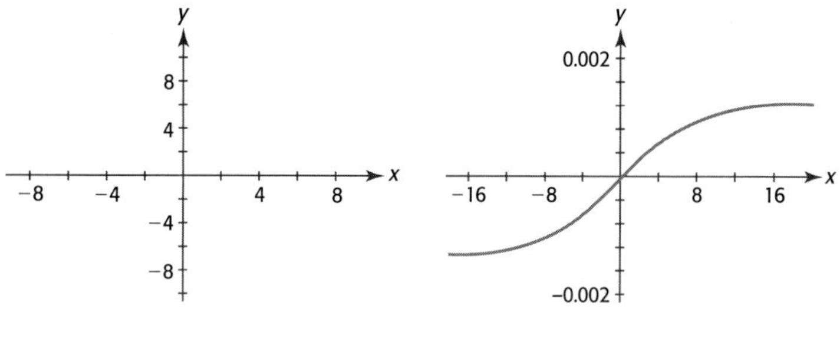

(a) **(b)**

(b) is the better view.

13. Letting y vary from 0 to 100 gives the view.

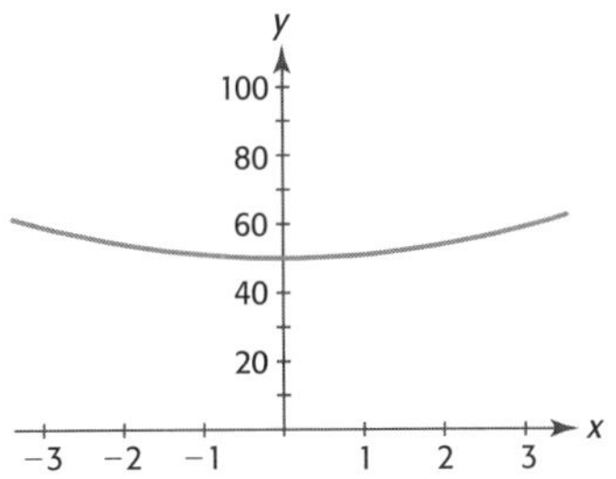

15. $y = x^3 + 3x^2 - 45x$ for values of x between -10 and 10.

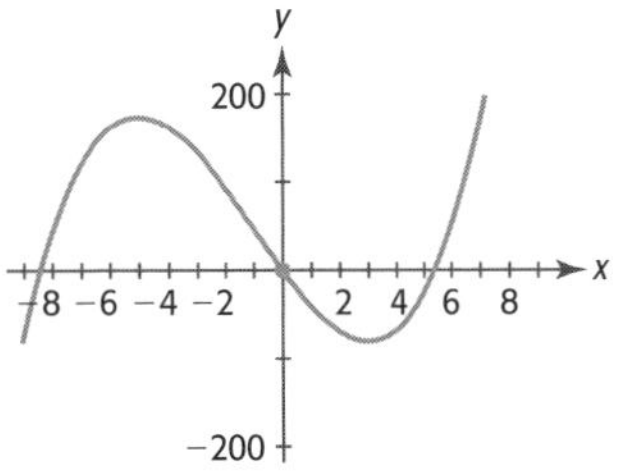

17.

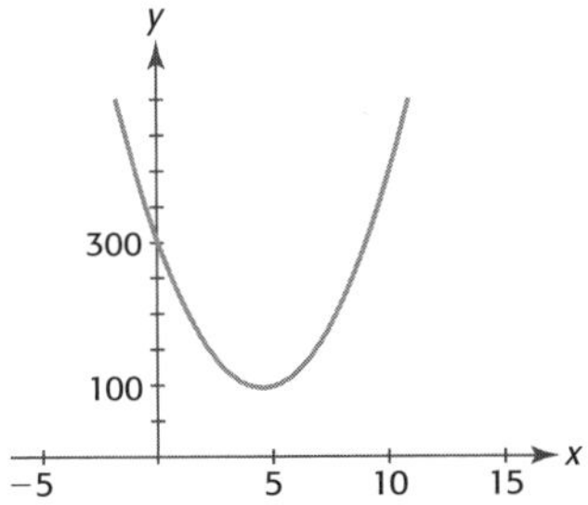

19.

t	s
12	51.9
16	72.7
28	135.1
43	213.1

21.

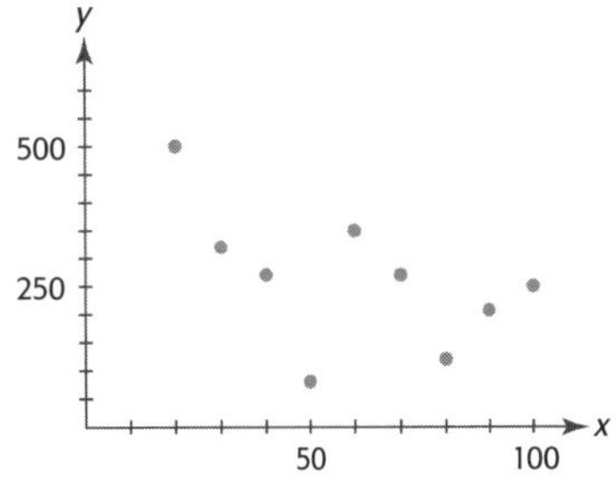

23. a.

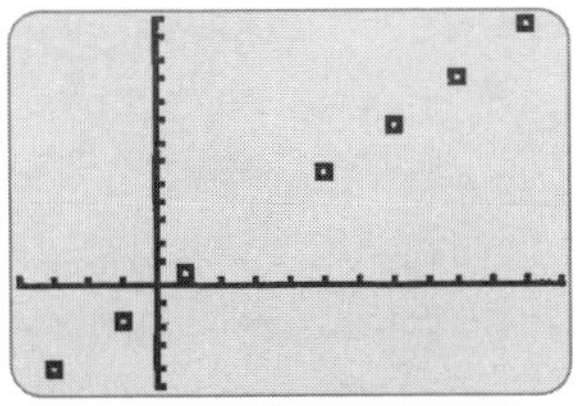

[−4, 12] by [−50, 130]

b.

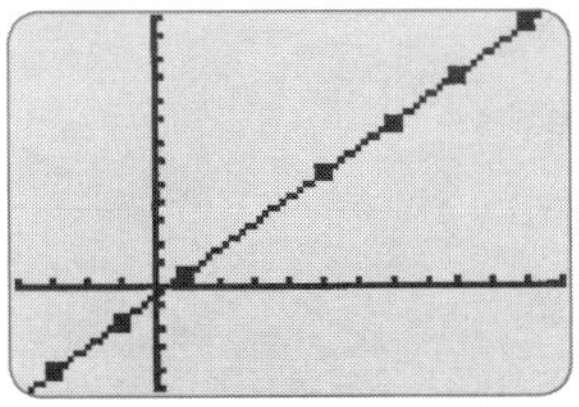

[−4, 12] by [−50, 130]

c. Yes; yes

25. a. 300 **b.** 2020

1.2 Exercises

27. a. 4 and 8 **b.** 7032; it means that when x is 8, y is 7032, or that in 1998, the number of welfare cases is 7032. **c.** 11,721 **29. a.** 1; 9 **b.** 2003; 51.69 **c.** 0; 10

31. a.

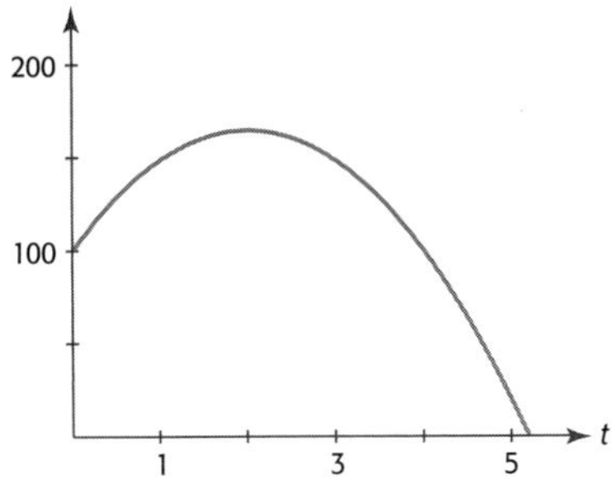

b. 148 ft and 148 ft; ball rising and falling **c.** 164 ft, in 2 seconds

33. a.

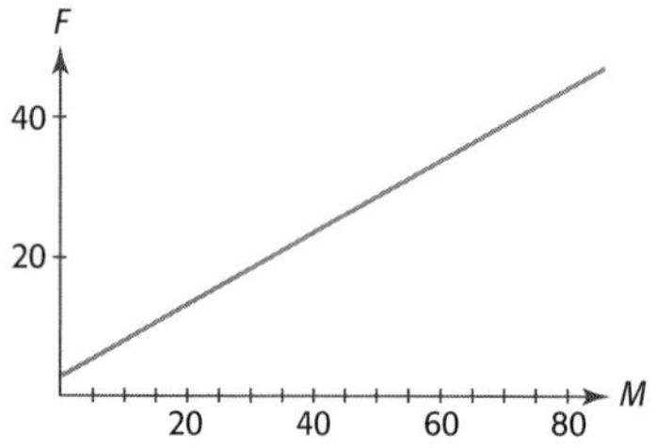

b. \$35,540

35. a.

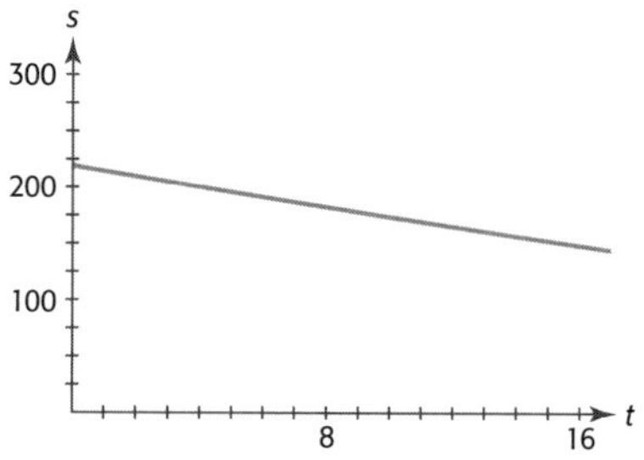

b. 152,255

c. 152,255

37. a.

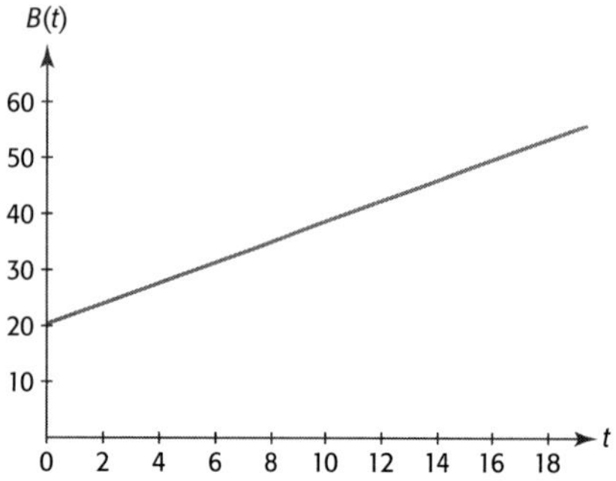

b. increase

39.

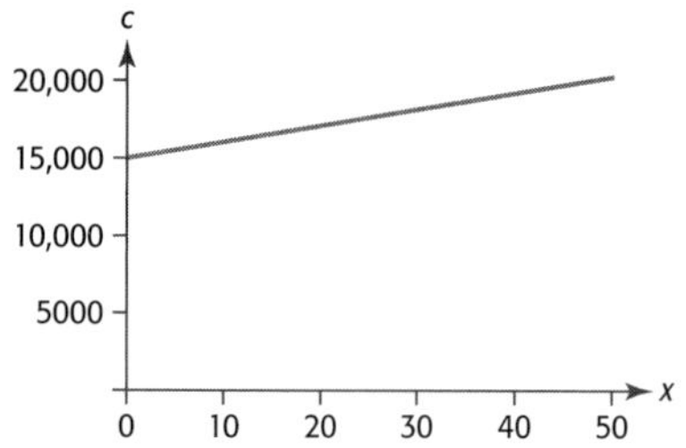

41.

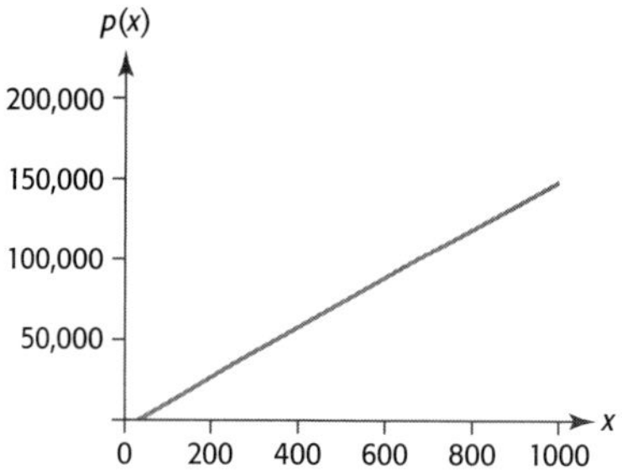

43. a. 0 through 15
b. 32,903.77; 47,634.67 **c.**

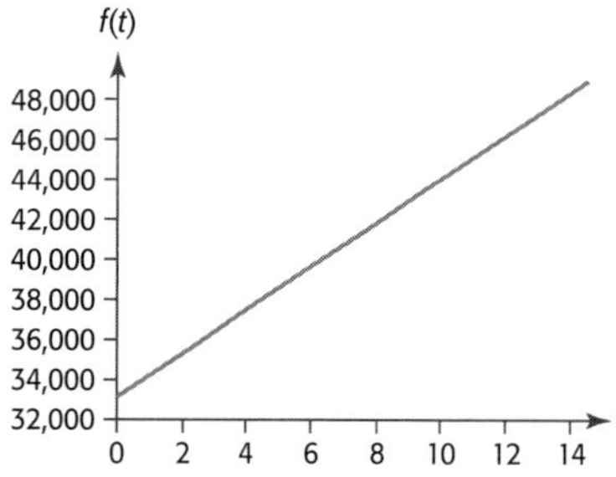

45.

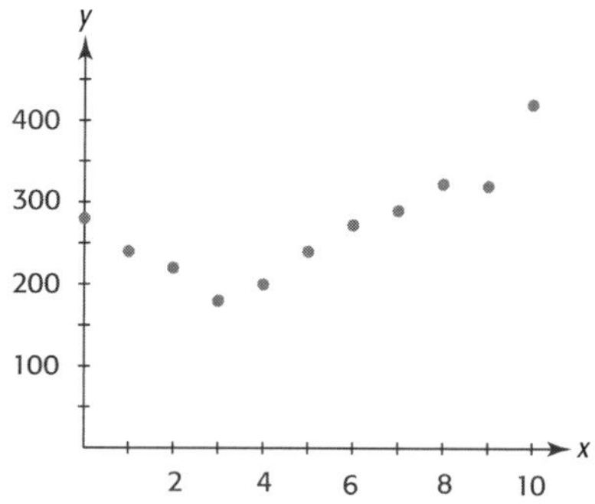

47. a.

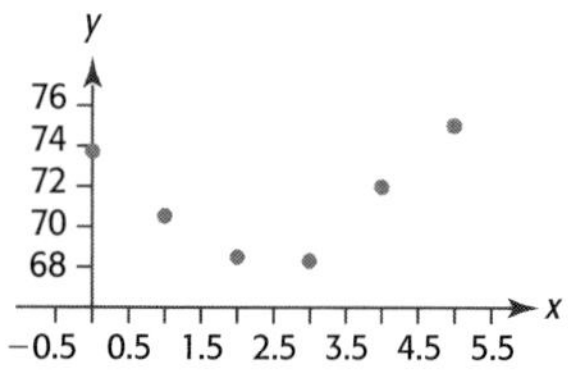

b.

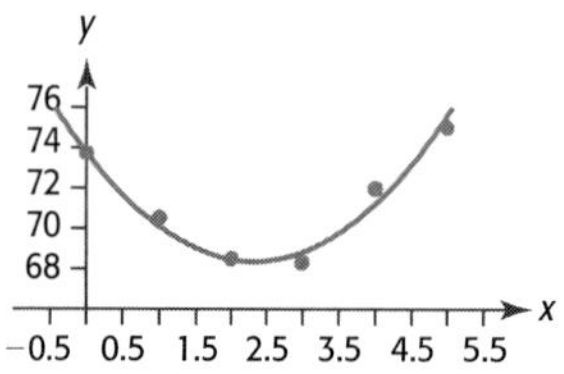

49. a. 3.5% **b.**

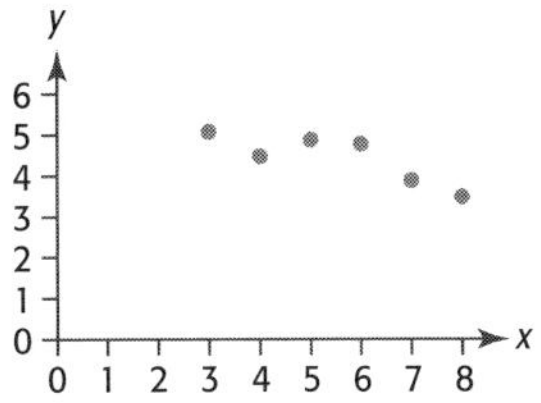

c.

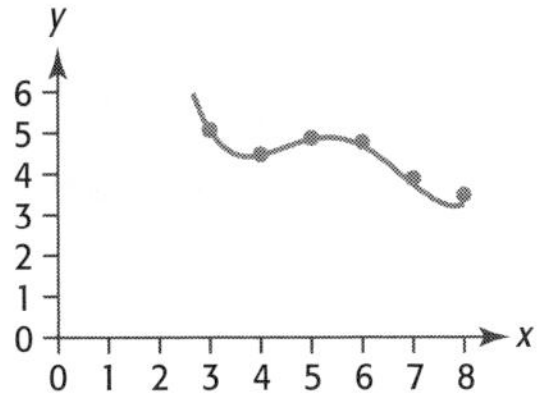

1.3 Skills Check

1. b **3.** Undefined **5.** x: 3; y: -5

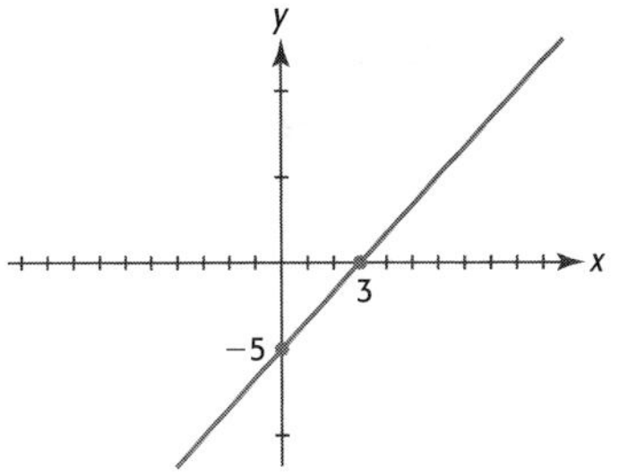

7. x: $\frac{3}{2}$; y: 3

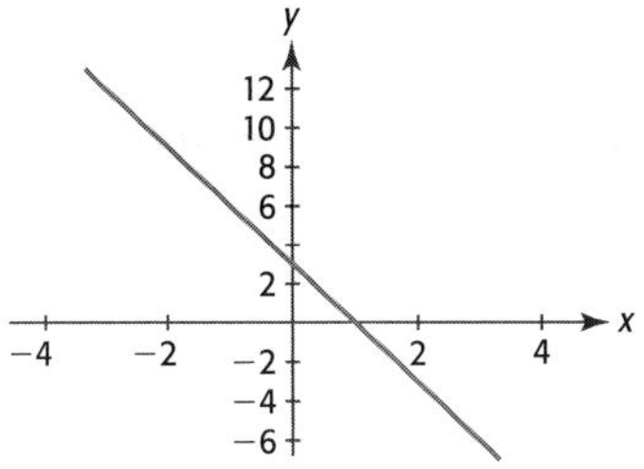

9. 0; undefined **11.** 4; 8 **13.** Slope 0; $\frac{2}{5}$ **15. a.** 4, 5 **b.** Rising **c.**

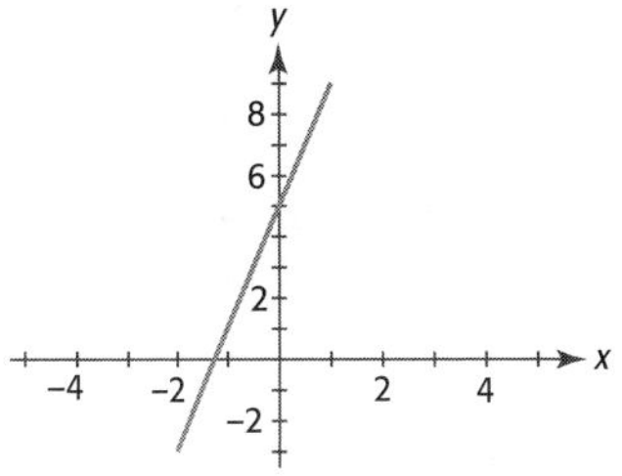

17. a. -100, 50,000 **b.** falling **c.**

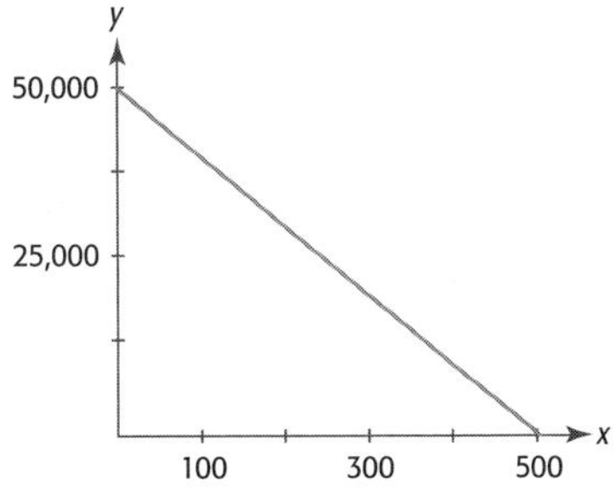

19. 4 **21.** -15 **23.** -2

25. Perpendicular **27. a.** ii **b.** i **29. a.** 0 **b.** 0

1.3 Exercises

31. Linear; rises **33.** Linear; falls **35. a.** $x = -\frac{30}{19}$ **b.** $p = 1$; 1% of high school seniors were using marijuana daily in 1990. **c.** Integer values of $x \geq 0$ on the graph represent years 1990 and after.

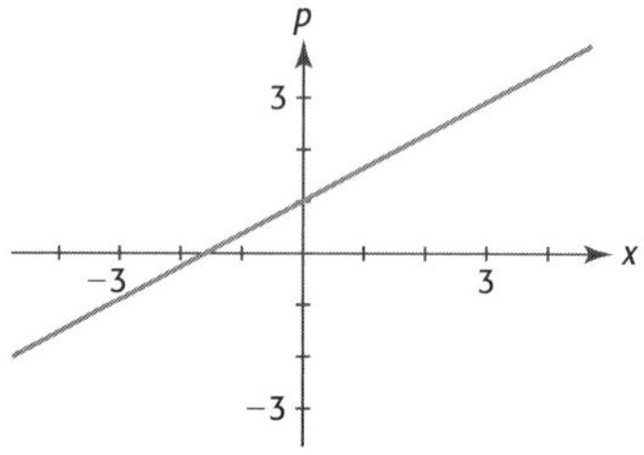

37. a. Constant function **b.** $y = 11.81$ **c.** 0 **d.** 0 **39. a.** 26.5 **b.** The percent of Fortune Global 500 firms that recruited on the Internet increased at a rate of 26.5% per year. **41. a.** Positive **b.** As temperature increases, the number of chirps increases. **43. a.** Yes **b.** .959 **c.** Minority median annual salaries will increase by $.959 for each $1 increase in median annual salaries for whites. **45. a.** $\frac{19}{30}$ **b.** $\frac{19}{30} \approx 63.33\%$ per year

47. x: 50; R: 3500

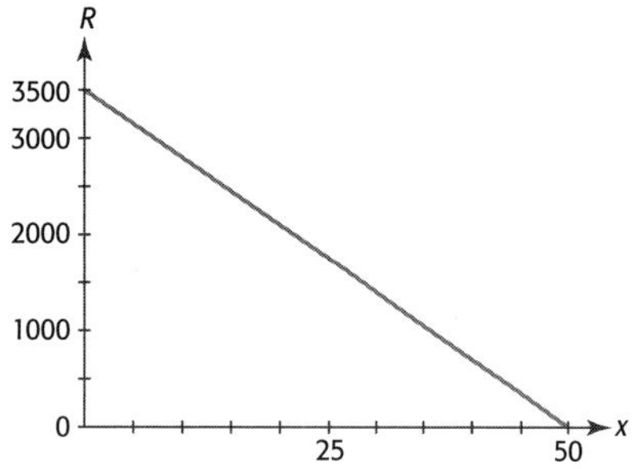

49. a. $m = 5.74$; $b = 14.61$ **b.** In 1995, 14.61% of U.S. population had Internet access. **c.** Percent with Internet access increased by 5.74% each year. **51. a.** $-61{,}000$ **b.** $-\$61{,}000$ per year **53.** $58 per unit **55. a.** 0.56 **b.** $0.56 per ball **c.** The cost will increase by $0.56 for each additional ball produced in a month. **57. a.** 1.60 **b.** $1.60 per ball **c.** The revenue will increase by $1.60 for each additional ball sold in a month. **59.** $19 per unit

1.4 Skills Check

1. $y = 4x + \frac{1}{2}$ **3.** $y = \frac{1}{3}x + 3$ **5.** $y = \frac{-3}{4}x + 2$ **7.** $y = 5x + 9$ **9.** $y = \frac{-3}{4}x - 3$ **11.** $y = 4$ **13.** $x = 9$ **15.** $y = x + 3$ **17.** $y = x + 4$ **19.** $y = 3x + 14$ **21.** $x = 9$ **23.** $y = \frac{4}{5}x + 4$ **25.** $y = 3x + 1$ **27.** $y = -15x + 12$ **29.** 1 **31.** -15 **33.** $4x + 2h$ **35. a.** Yes; a scatterplot shows that data fit along same line. **c.** $y = 3x + 555$

1.4 Exercises

37. $y = 8.95 + 0.0935x$ (dollars) **39.** $y = -3600t + 36{,}000$ **41. a.** $P = 705x + 198$ **b.** 4428 **c.** $x > 6$; after 2002 **43. a.** $25,000 **b.** $5000 **c.** $s = 26{,}000 - 5000t$ **45.** $y = x$; $x =$ deputies; $y =$ cars **47.** $P(x) = 58x - 12{,}750$; $x =$ number of units **49.** $V = -61{,}000x + 1{,}920{,}000$ **51.** $P = -.03t + 32.88$ **53. a.** .05 **b.** $f(x) = 0.05x$ **55. a.** $y = 0.504x - 934.46$ **b.** Yes **c.** They are the same. **57. a.** 3.42 **b.** 3.42 **c.** No **d.** No

59. a. $\frac{61}{36} \approx 1.69$ **b.** 1.69 **c.** The percent is increasing at a rate of approximately 1.69 per year. **d.** $p = 1.69x - 1.94$

61. a. 26,612 **b.** 26,612 **c.** $y = 26{,}612x - 51{,}956{,}196$ **d.** No **e.** 1990 and 1997 **63. a.** No **b.** Yes, they appear to lie on a line. **c.** -0.085 **d.** $y = -0.085x + 174.75$ **65. a.** Increasing at a rate of approximately 2462 thousand people, or approximately 2.462 million people per year **b.** $y = 2462x + 152{,}271$ thousand **c.** 213,821 thousand, or 213,821,000; no **d.** The line does not model the data exactly. **67. a.** Yes **b.** \$0.28 **c.** $y = 0.28x - 3197$ **d.** Evaluating the function at $x = 30{,}100$ gives y = 5231, and evaluating it at $x = 30{,}300$ gives 5287. These values agree with the income taxes due in the table.

1.5 Skills Check

1. $x = -\frac{37}{2}$ **3.** $x = 10$ **5.** $x = -\frac{13}{24}$ **7.** $x = -63$ **9.** $t = -1$ **11.** $\frac{25}{36}$ **13. a.** -20 **b.** -20 **c.** -20 **15. a.** 4 **b.** 4 **c.** 4 **17. a.** 2 **b.** -34 **c.** 2 **19. a.** 40 **b.** 40 **21.** $x = 3$ **23.** $x = -5$ **25.** $t = -4$ **27.** $t = \frac{17}{4}$ **29. a.** $t = \frac{A - P}{Pr}$ **b.** $P = \frac{A}{1 + rt}$ **31.** $F = \frac{9}{5}C + 32$ **33.** $n = \frac{5m}{2} - \frac{P}{4} - \frac{A}{2}$

35. $y = \frac{5x - 5}{3}$

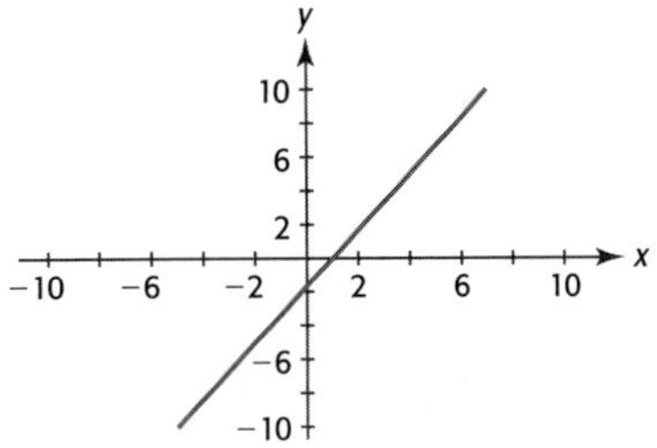

37. $y = \frac{6 - x^2}{2}$

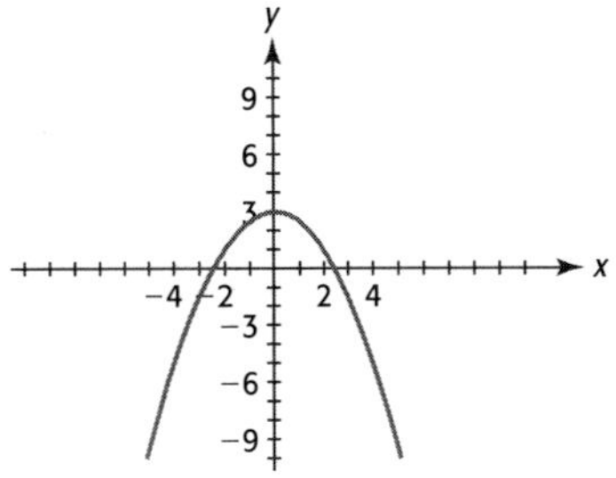

1.5 Exercises

39. a. 60 months, or 5 years **b.** $0 < x < 60$; less than 60 months **41.** \$6000 **43.** \$54,000 **45.** $-40°$ **47.** 26 years from 1970, or 1996 **49.** 1999 **51.** $x = 24$, so 1994 **53.** $x = 1$, so in 1991 **55. a.** $x \approx 46.2$, so in 1996 **b.** No; changing conditions can make models inaccurate **57.** 1997 **59.** 97 **61.** \$78.723 billion **63.** 25% **65.** \$1698

1.6 Skills Check

1. Should not be modeled by a linear function; data points do not lie close to a line.

3.

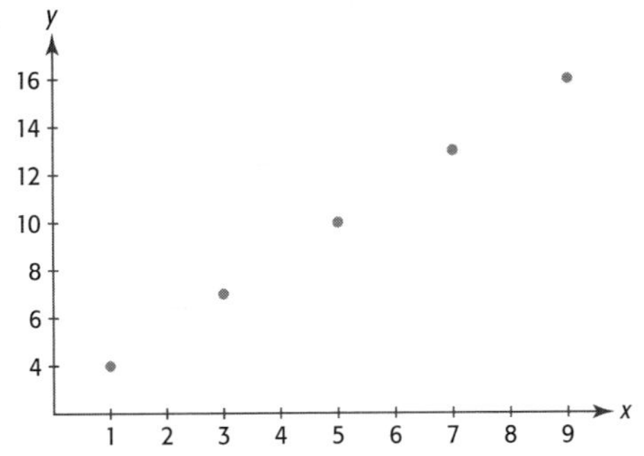

5. Exactly; first differences of the outputs are equal.

7. $y = 1.5x + 2.5$

9.

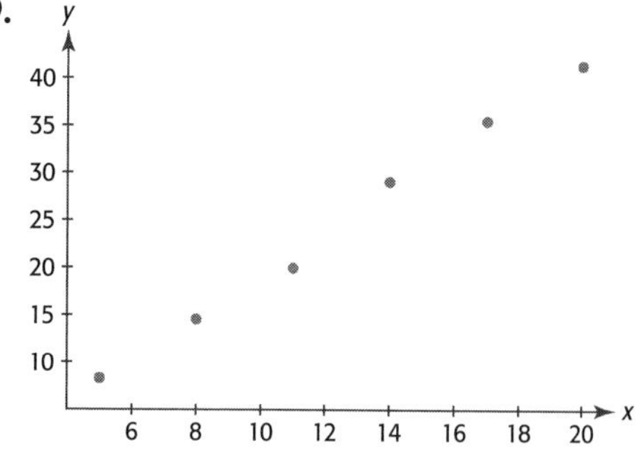

11. $y = 2.419x - 5.571$

13.

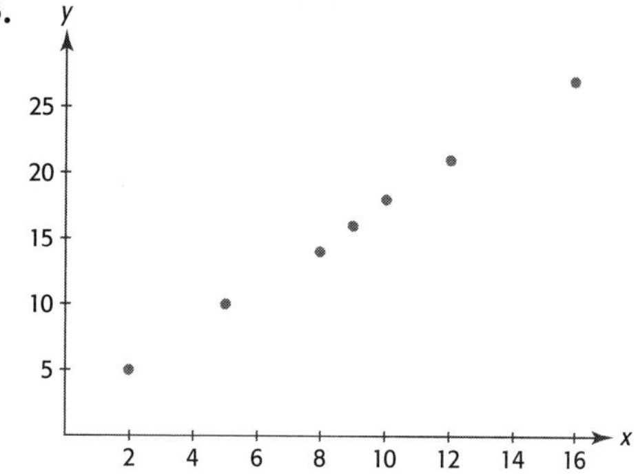

15. $y = 1.577x + 1.892$ **17.** $y = -1.5x + 8$ **19. a.** Exactly **b.** Nonlinear **c.** Approximately

1.6 Exercises

21. a. Discrete **b.** No; the graph of the data has some curvature **c.** Yes **23. a.** Yes. **b.** Yes. **c.** $S = 60t + 1000$ **25. a.** 608,688 employees; interpolation **b.** 720,058 employees; extrapolation **27. a.** $y = 0.723x + 25.631$ **b.** Sometime during the year 1978 **c.** Because the model is only an approximate fit to the data, the answer to part (b) is only an approximation. **29. a.** $y = 0.039x + 1.082$ **b.** Very well; the outputs are very close to the data. **31. a.** \$1.71 **b.** 12 lb **c.** 3.9 cents **d.** 4 cents **33. a.** $y = -0.762x + 85.284$ **b.** 58.6 **c.** 1997 **d.** The answer is an approximation; the actual data set indicates that the marriage rate fell below 50 in 1996. **35. a.** $y = 145x + 3230.667$ **b.** 145,000 children per year **c.** 6,855,667; extrapolation

37. a.

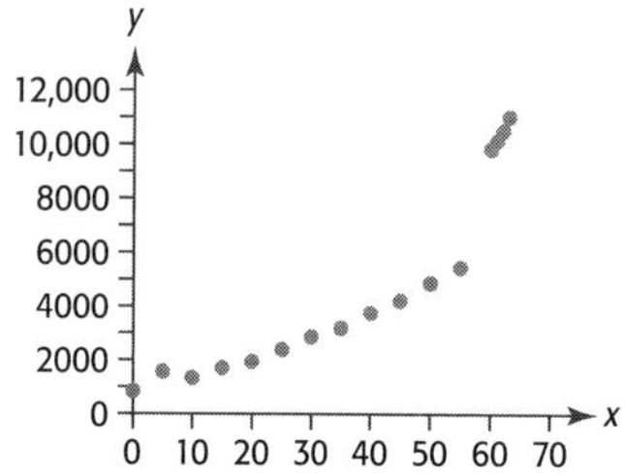

b. $y = 149.931x - 689.033$ **c.** No; see scatter plot

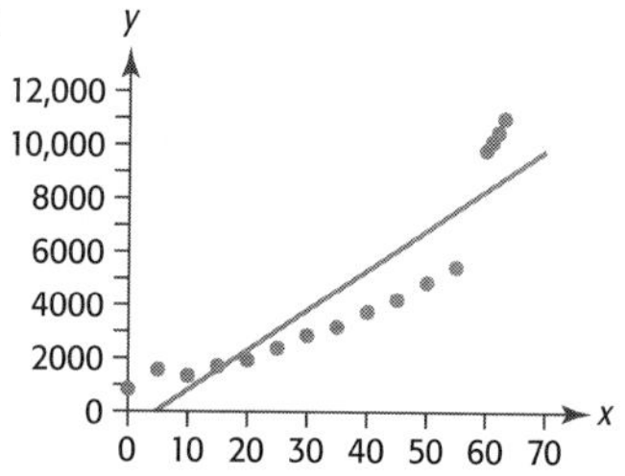

39. a.

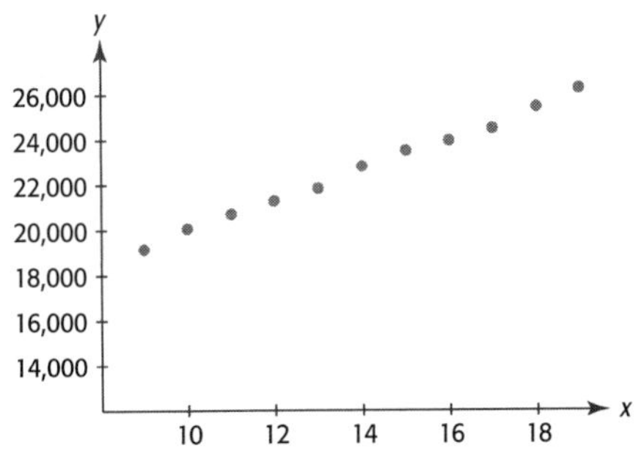

b. $y = 1280.891x + 2096.255$

c.

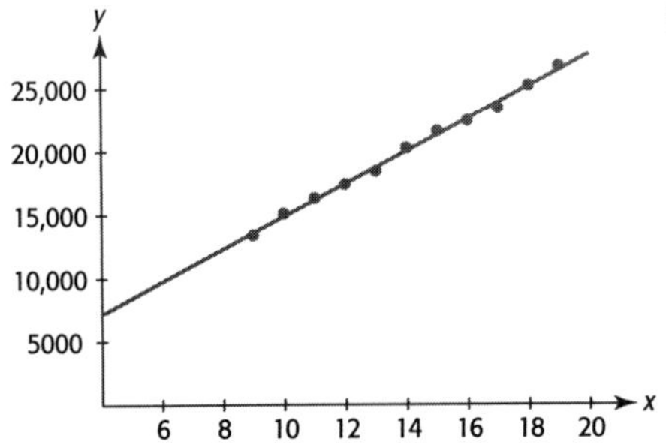

; the function models the data well

d. i. Discrete
ii. Discrete
iii. Continuous

41. a. $f(x) = 2.920x + 265.864$ **b.** 455.6, projected population is 455.6 million in 2065. **c.** 499.4 million; fairly close **43. a.** $y = 3.303x - 18.874$ **b.** Approximately 3.3 million **c.** $x = 11.77$; 2002 **d.** 9/11 terrrorist attacks **45. a.** $y = 0.461x + 3.487$ **b.** 10.4 days

1.7 Skills Check

1. (1, 1) **3.** $x = -14, y = -54$ **5.** $x = \frac{-4}{7}, y = \frac{4}{7}$ **7.** Many solutions; the two equations have the same graph. **9.** $x = 2, y = -2$ **11.** $x = 1, y = -1$ **13.** $x = 2, y = 1$ **15.** $x = \frac{4}{7}, y = \frac{12}{7}$, **17.** $x = 4; y = 3$ **19.** Dependent; many solutions **21.** No solution **23.** $x = 2, y = 4$ **25.** $x = \frac{32}{11}, y = \frac{-14}{11}$ **27.** $x = 3, y = -2$ **29.** $x = 2, y = 1$ **31.** No solution

1.7 Exercises

33. 60 units **35. a.** 8 supplied, 17 demanded **b.** \$80; 12 units **37. a.** 1987 **b.** About 521,787 **c.** $x > 27$;after 1987 **39. a.** $x = 37$ **b.** 2027 **c.** Yes, the scales are different **41.** High: \$52.69; low: \$30.81 **43. a.** $x + y = 2400$ **b.** $30x$ **c.** $45y$ **d.** $30x + 45y = 84{,}000$ **e.** 1600 \$30 tickets and 800 \$45 tickets **45. a.** \$75,000 at 8% and \$25,000 at 12% **b.** 12% account is more risky; might lose money **47.** 3 glasses of milk and 2 servings of meat **49.** 8 cc of the 5% solution and 12 cc of the 10% solution **51. a.** $p = -\frac{1}{2}q + 155$ **b.** $p = \frac{1}{4}q + 50$ **c.** \$85 **53.** No solution **55. a.** $300x + 200y = 100{,}000$ **b.** 250 in the first group and 125 in the second group **57.** 700 units at \$30. When the price is \$30, the amount demanded equals the amount supplied equals 700.

Chapter 1 Skills Check

1. Each value of x is assigned exactly one value of y.

2. **Domain:** $\{-3, -1, 1, 3, 5, 7, 9, 11, 13\}$
Range: $\{9, 6, 3, 0, -3, -6, -9, -12, -15\}$

3. $f(3) = 0$ **4.** Yes; $y = -\frac{3}{2}x + \frac{9}{2}$ **5. a.** -2 **b.** 8 **c.** 14 **6. a.** 1 **b.** -10

7.

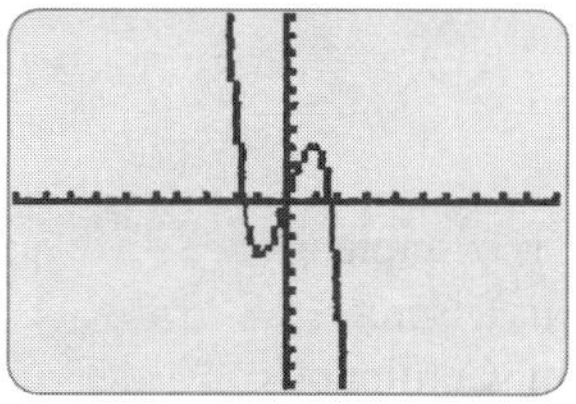

8.

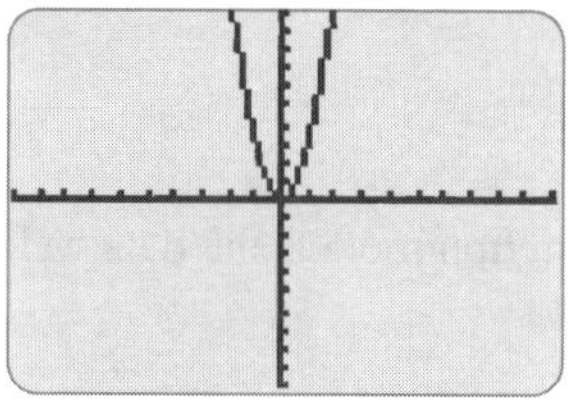

[−10, 10] by [−10, 10]

9. standard window:

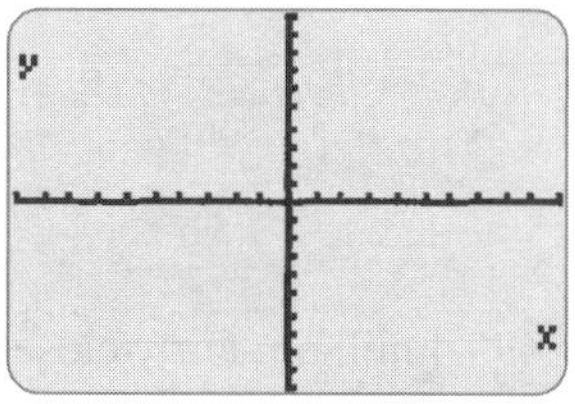

[−10, 10] by [−10, 10]

[0, 40] by [0, 5000] gives a better graph:

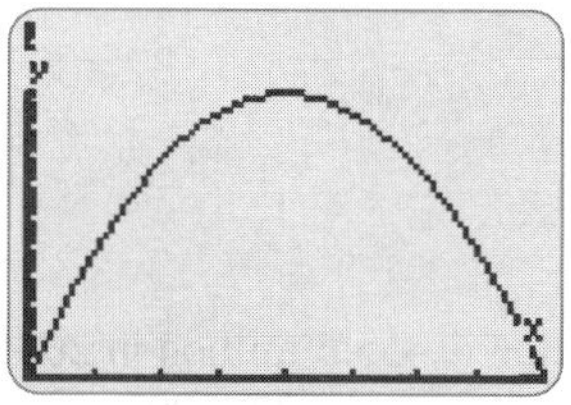

10.

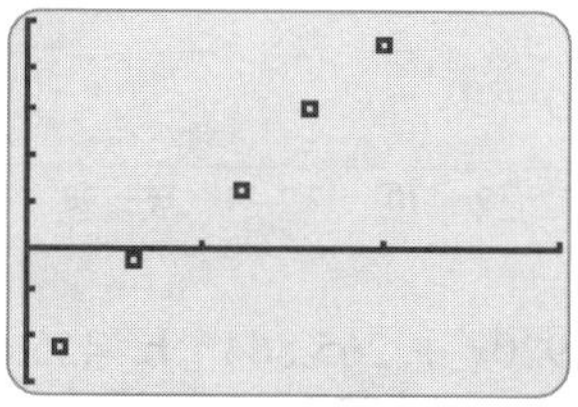

[0, 15] by [−12, 20]

11. $y = 2.8947x - 11.211$

12.

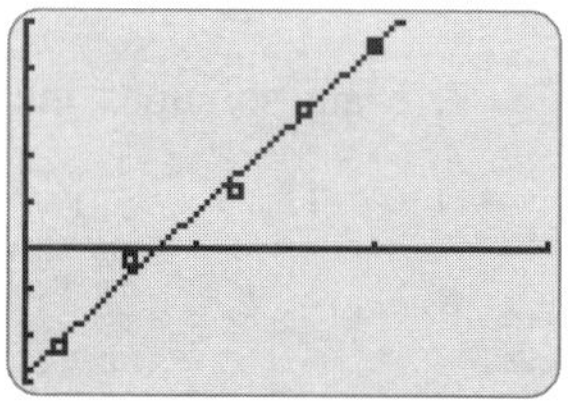

[0, 15] by [−12, 20]

13. No; no, inputs do not have a constant change

14. $-\frac{11}{6}$ **15. a.** $(0, -4)$ and $(6, 0)$ **b.**

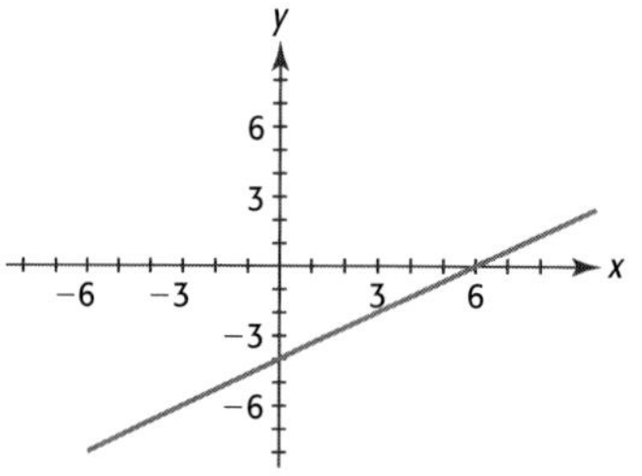

16. $\frac{2}{3}$

17. slope -6, y-intercept 3 **18.** -6 **19.** $y = \frac{1}{3}x + 3$ **20.** $y = -\frac{3}{4}x - 3$ **21.** $y = x + 4$

22. $x = 2, y = -3$ **23.** $x = \frac{-3}{13}, y = \frac{-15}{13}$ **24.** Many solutions **25.** No solution **26.** $x = -3, y = 5$

27. $x = \frac{1}{2}, y = -4$ **28. a.** $5 - 4x - 4h$ **b.** $-4h$ **c.** -4 **29. a.** $10x + 10h - 50$ **b.** $10h$ **c.** 10

30. $x = \frac{34}{5}$ **31.** $x = -138$ **32.** 3 **33.** $y = \frac{4x - 6}{3}$

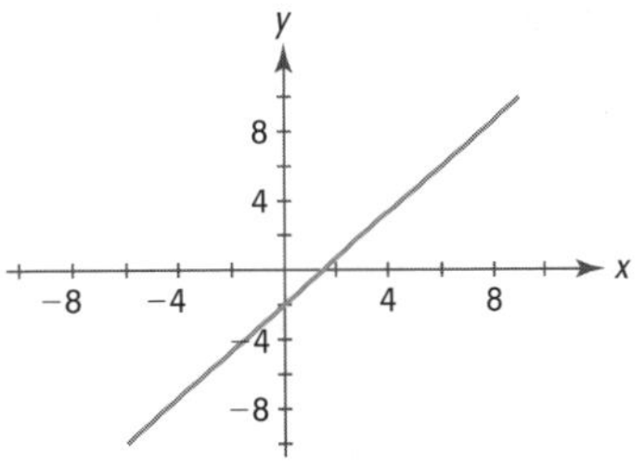

34. $x < -\frac{4}{5}$ **35.** $x \le \frac{25}{28}$ **36.** $6 < x < 18$

Chapter 1 Review

37. a. yes **b.** $f(1992) = 82$; in 1992, 82% of African American voters supported Democratic candidates for president. **c.** 1964; in 1964, 94% of African American voters supported Democratic candidates for president. **38. a.** {1960, 1964, 1968, 1972, 1976, 1980, 1984, 1992, 1996} **b.** No; 1982 was not a presidential election year. **c.** Discrete **39.**

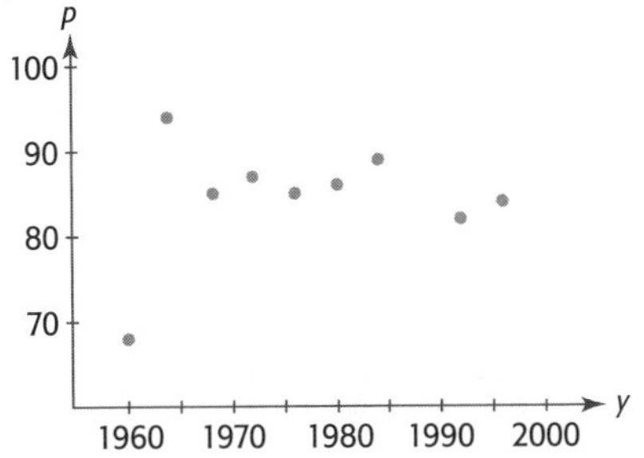

40. a. −0.357 **b.** −0.357 **c.** No **d.** No

41. a. As each input value changes by 5000, the output value changes by 89.62. **b.** $f(25) = 448.11$; the monthly payment to borrow $25,000 is $448.11. **c.** $A = 20{,}000$ **42. a.** D: {10,000, 15,000, 20,000, 25,000, 30,000}; R: {179.25, 268.87, 358.49, 448.11, 537.73} **b.** No **c.** Discrete **43. a.** Yes **b.** Yes **44. a.** $p = f(A) = 0.018A + 0.010$ **b.** 501.882; The predicted monthly payment on a car loan of $28,000 is $501.88. **c.** Yes **d.** $27,895 **45. a.** $f(1960) = 15.9$; in 1960 the average woman was expected to live 15.9 years past the age of 65, or 80.9 years. **b.** 65 + 19.4 = 84.4 years **c.** 1990 **46. a.** 16.9; a 65-year-old man in 2020 is expected to live 16.9 more years, or 81.9 years **b.** 77.8 years **c.** $f(1990) = 15$ **47. a.** $42,724.37; $f(10) = 42{,}724.37$ **b.** $f(15) = 47{,}634.67$; the average is $47,634.67 in 2005. **c.** Increasing

48. a.

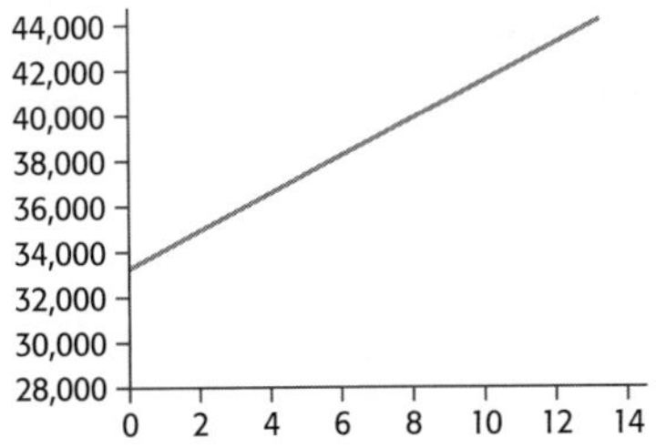

b. Years from 1990 to 2005

49. a. $m = .357$ **b.** Average rate of change = 0.357 million users per year **50.** $f(x) = 4500$ **51. a.** $f(x) = 34.6$ **b.** It is a constant function. **52. a.** $22,000 **b.** More than $22,000 per month **53.** $25,440 each

54. $300,000 in the safe account and $120,000 in the risky account **55.** 1994 **56. a.** $67,680 **b.** $47,680
c. $\overline{MR}$: 564; $\overline{MC}$: 64 **d.** $m = 64$ **e.**

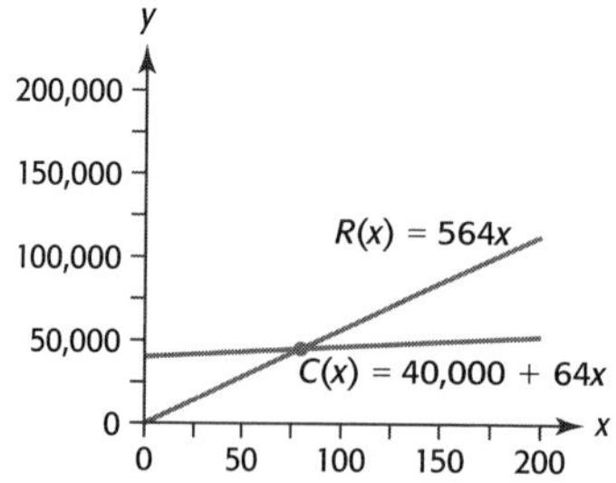

57. a. $p(x) = 500x - 40{,}000$ **b.** $20,000 **c.** 80 **d.** 500 **e.** Marginal revenue minus marginal cost
58. a. $y = 300{,}000$; the initial value of the property is $300,000. **b.** $x = 100$; the value of the property after 100 years is zero dollars **59. a.** Average rate of change = $4.4 per unit **b.** Slope = 4.4 **c.** $4.4x - 205$ **d.** $4.4 per unit
e. Approximately 47 **60. a.** $y = 0.0638x + 15.702$ **b.**

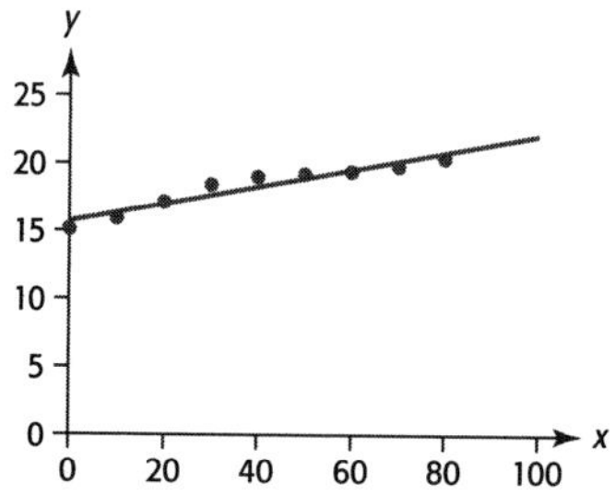

c. 22.34; in 2054 the average woman is expected to live 22.34 years beyond age 65. Her life expectancy is 87.34 years.
d. $x \geq 51.7$; 2002 and beyond **61. a.** $y = 0.0655x + 12.3244$ **b.**

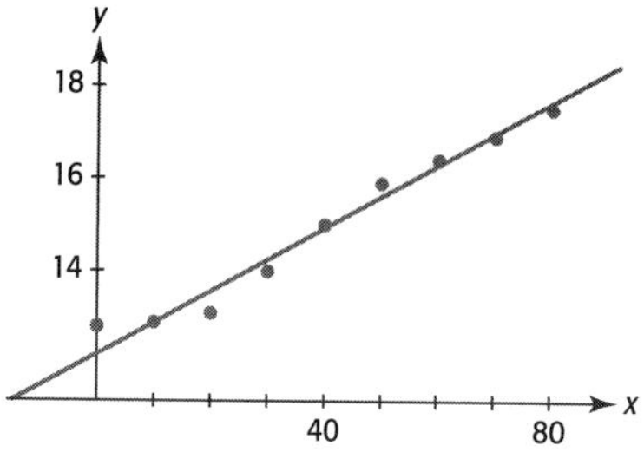

c. $g(130) \approx 20.8$; in the year 2080 (1950 + 130), the average man is expected to live about 85.8 years **d.** 2143
e. $x \leq 56.1$; until 2006 **62. a.** A linear equation is a reasonable model. **b.** $y = 3.317x + 3.254$ **c.** Approximately $43.1 billion **63. a.** A linear equation is a reasonable model. **b.** $y = 285.269x + 9875.17$, where x is the number of years past 1980 and y is population in thousands **c.** 16,151 thousand
64. a. A linear equation is a reasonable model. **b.** $y = 0.301x + 0.477$
c.

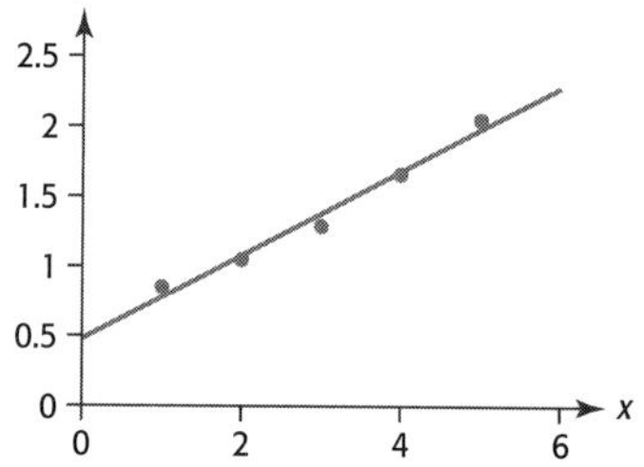

The line seems to fit the data points very well.

65. a. Before 1998 **b.** After 2007 **66.** 1995 through 2001 **67.** Between 17 and 31 months
68. $140,000 at 8% and $100,000 at 12% **69.** 25 **70.** Each dosage of medication A contains 2.8 mg, and each dosage of medication B contains 4.2 mg. **71.** $p = \$100$, $q = 80$ pairs **72.** $p = \$79$; $q = 710$ units **73. a.** $x + y = 2600$ **b.** $40x$ **c.** $60y$ **d.** $40x + 60y = 120{,}000$ **e.** 1800 $40 tickets and 800 $60 tickets **74. a.** $x + y = 500{,}000$
b. $0.12x$ **c.** $0.15y$ **d.** $0.12x + 0.15y = 64{,}500$ **e.** $350,000 in the 12% property and $150,000 in the 15% property

Chapter 2 Quadratic and Other Nonlinear Functions

2.1 Skills Check

1. a. Quadratic **b.** Up **c.** Minimum **3.** Not quadratic **5. a.** Quadratic **b.** Down **c.** Maximum

7. a.

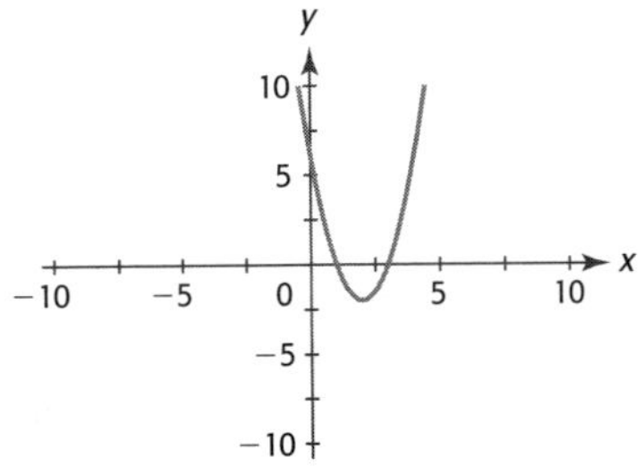

b. Yes

9. a.

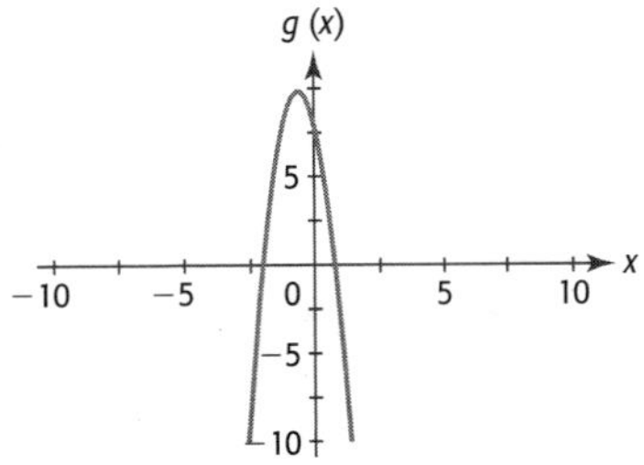

b. Yes

11. a.

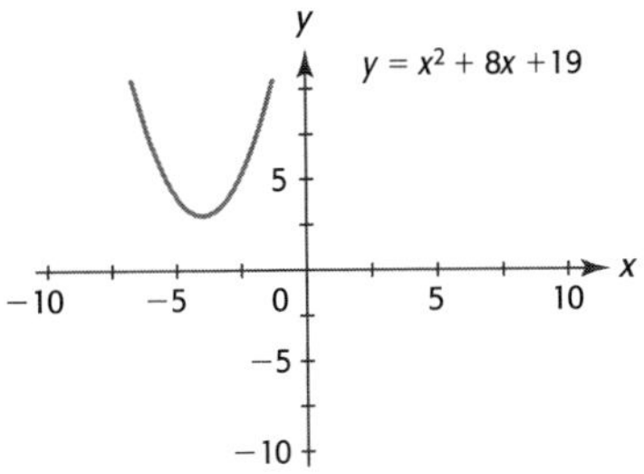

b. Yes

13. a.

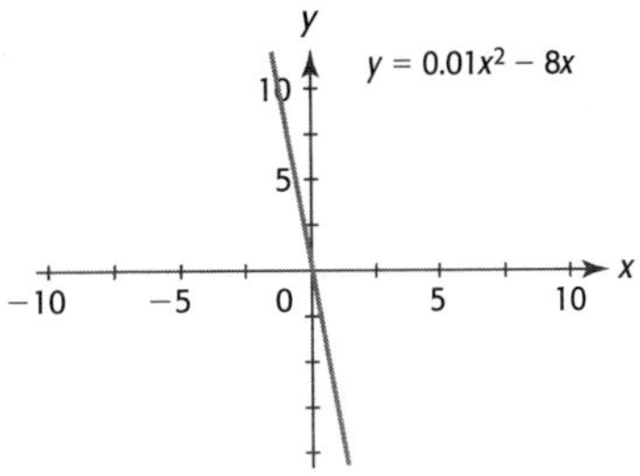

b. No

15. a. (1, 3) **b.**

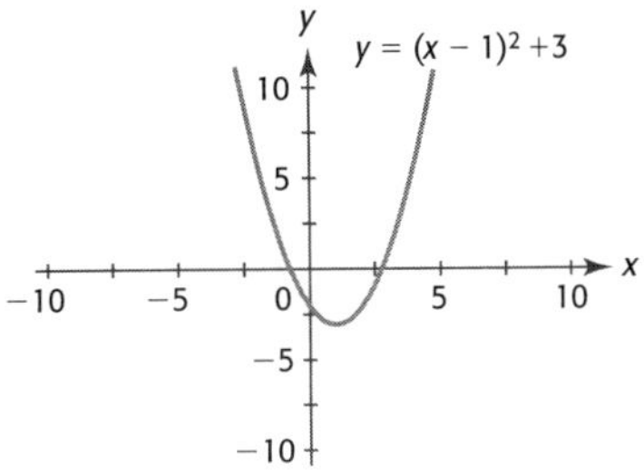

17. a. $(-8, 8)$ **b.**

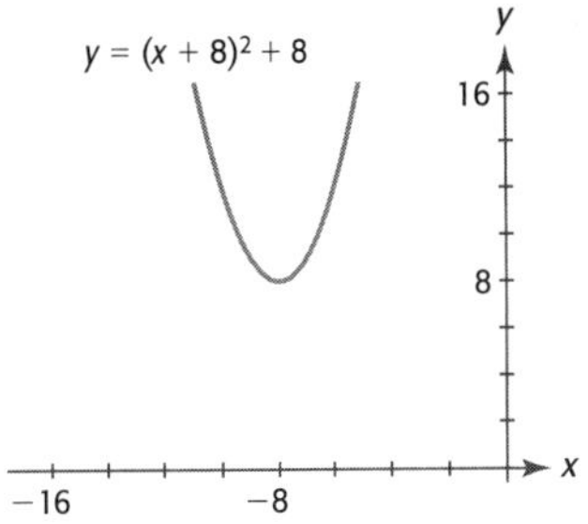

19. a. (2, 12) **b.**

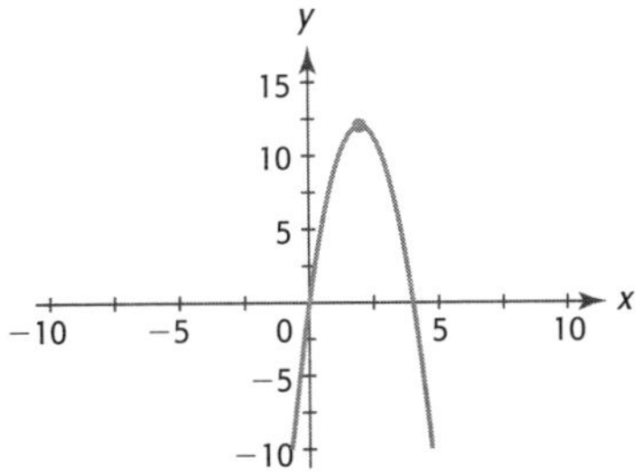

21. a. $(-3, -30)$ **b.**

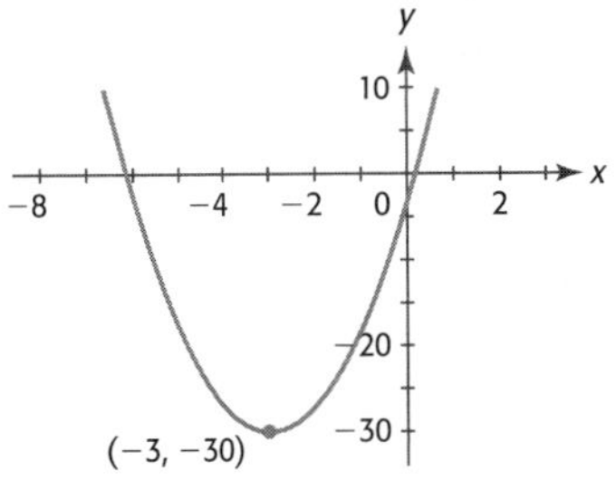

23. a. Vertex $(10, -190)$ **b.**

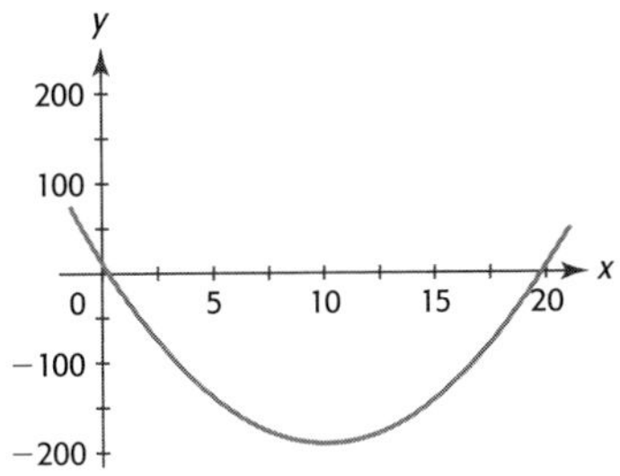

25. a. Vertex $(-80, 1282)$

b.

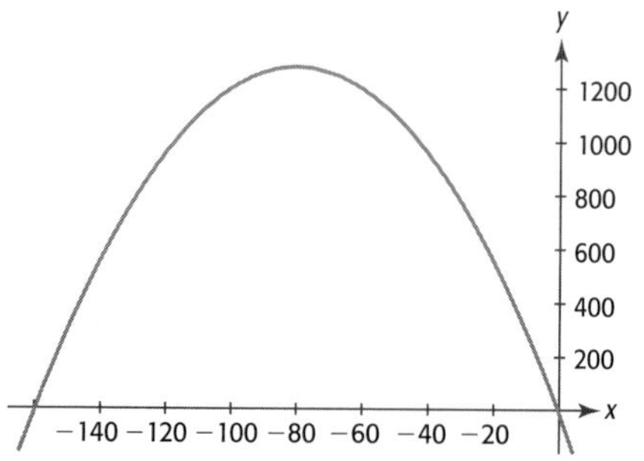

27.

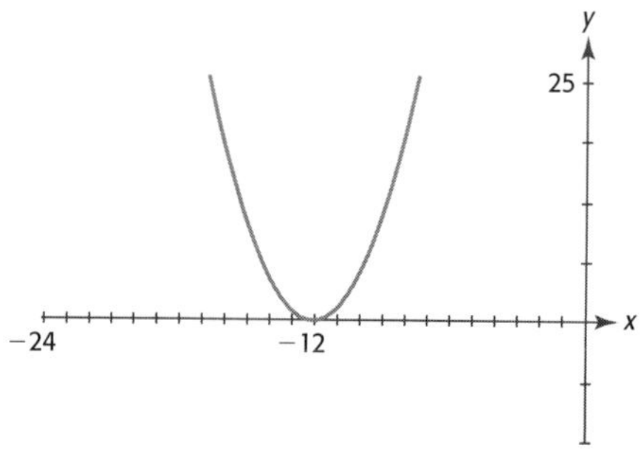

29.

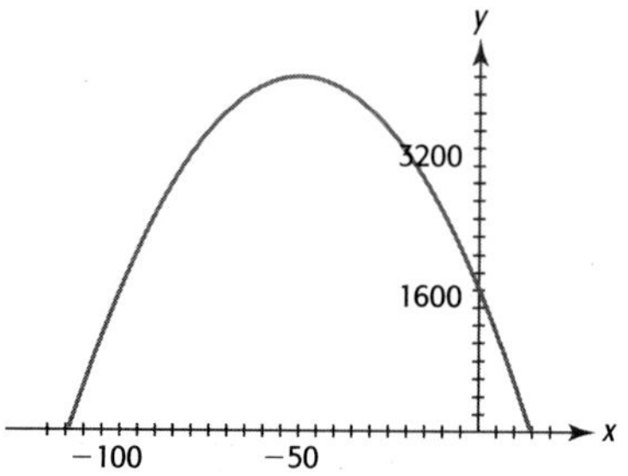

31.

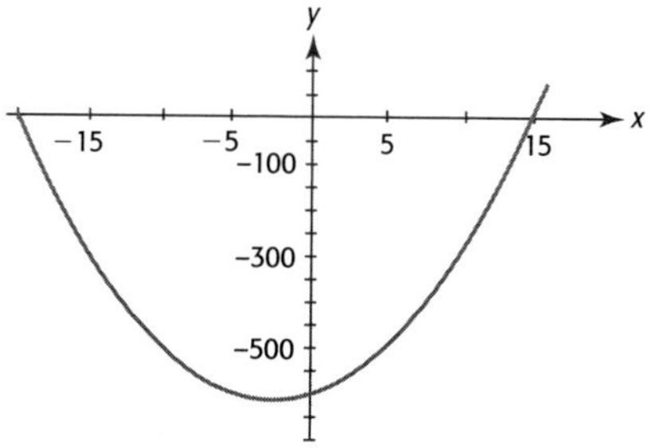

33. $x = 1, x = 3$

35. $x = -10, x = 11$ **37.** $x = -2, x = 0.8$

2.1 Exercises

39. a.

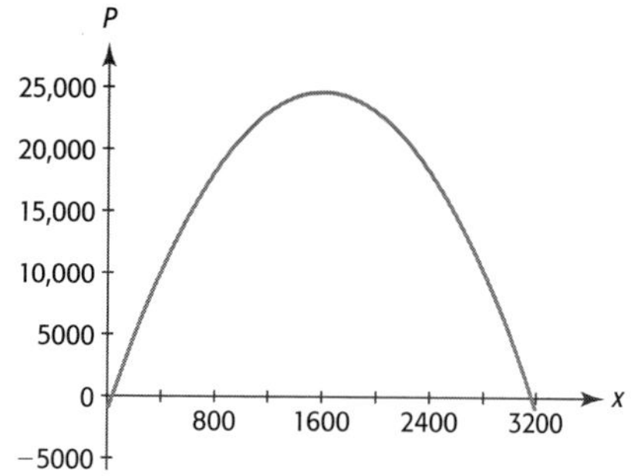

b. It increases. **c.** Profit decreases.

41. a.

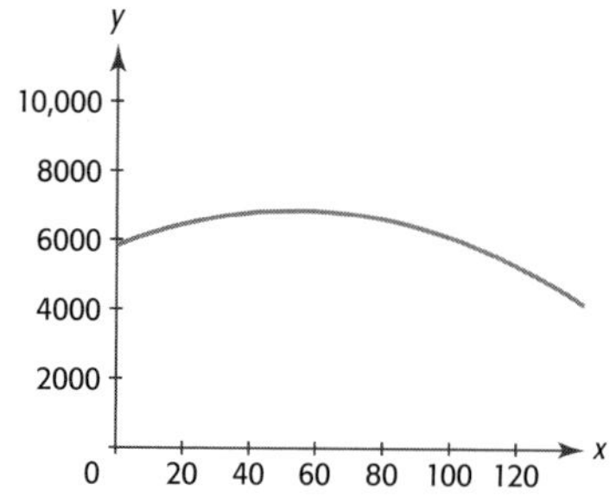

b. 6449.3 million, or 6.4493 billion

43. a.

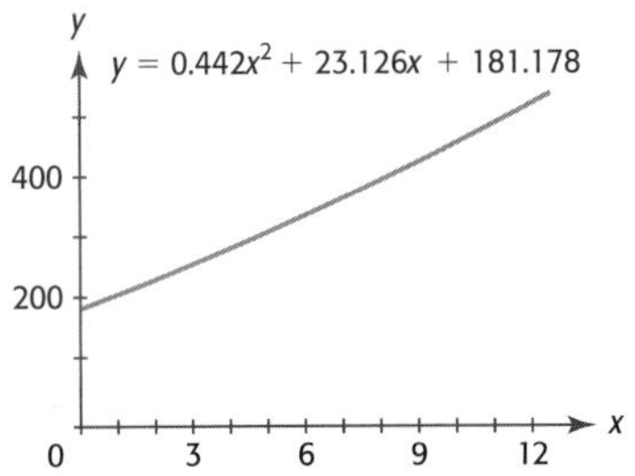

b. $627.52 billion **c.** Extrapolation **45. a.** A parabola opening down **b.** $t = 3, S = 244$ **c.** The ball reaches its maximum height of 244 feet in 3 seconds.

47. a.

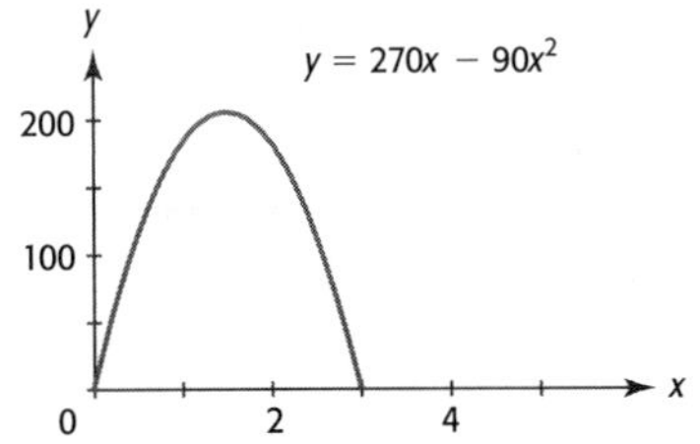

b. 1.5 lumens **49. a.** 2000 units **b.** $37,000 **51. a.** 37,500 **b.** $28,125,000 **53. a.** Yes **b.** $A = 100x - x^2$; maximum area of 2500 sq ft when x is 50 ft **55.** Maximum of ≈16,623 dishes in 1996 **57. a.** Minimum; positive coefficient of x^2 **b.** $x = 69.8, y = 5.2$ **c.** The percent of the U.S. population that is foreign born was at minimum of 5.2 in 1970. **d.**

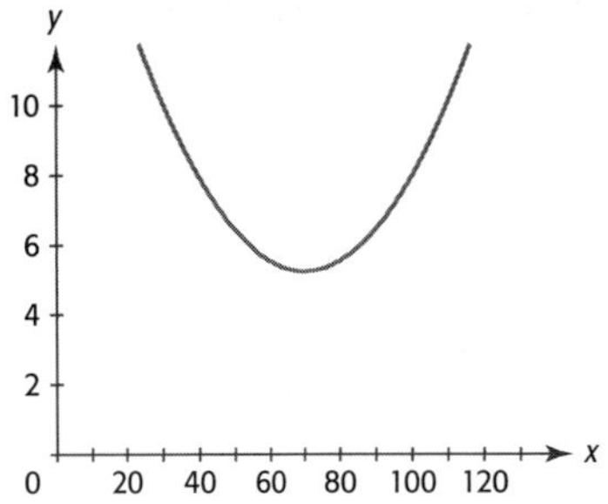

59. a.

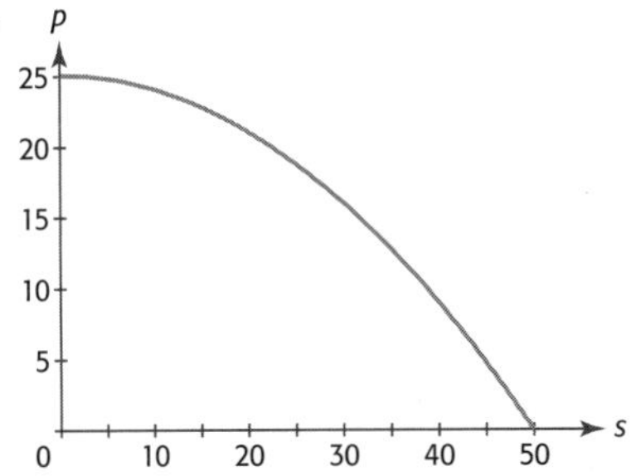

b. Decreasing **c.** 25 **d.** When wind speed is 0 mph, amount of pollution is 25 oz per cubic yard.

61. The t-intercepts are 3.5 and −3.75. This means that the ball will strike the pool in 3.5 seconds; −3.75 is meaningless.

63. a.

Rent ($)	Number of Apartments	Revenue ($)
1200	100	120,000
1240	98	121,520
1280	96	122,880
1320	94	124,080

b. Yes **c.** $1600 **65. a.** $x = 53.5, y = 6853.3$ **b.** World population will be maximized at 6,853,300,000 in 2044. **c.** Until 2044

2.2 Skills Check

1. 5, −2 **3.** 8, 3 **5.** −3, 2 **7.** $\frac{3}{2}$, 4 **9.** $x = -2, x = \frac{1}{3}$ **11.** $x = -2, x = 5$ **13.** $\frac{2}{3}$, 2 **15.** $\frac{1}{2}$, −4 **17.** $-\frac{1}{2}$, 3 **19.** $x = 8, x = 32$ **21.** $s = 50, s = -15$ **23.** $x = \pm\frac{3}{2}$ **25.** $x = \pm 4\sqrt{2}$ **27.** $2 \pm \sqrt{13}$ **29.** 1, 2 **31.** $x = \frac{5 \pm \sqrt{17}}{2}$ **33.** $x = 1, x = -\frac{8}{3}$ **35.** −3, 2 **37.** $\frac{2}{3} \approx 0.667, -\frac{3}{2} = -1.5$ **39.** $x = \frac{-1}{2} = -0.5$ $x = \frac{2}{3}, \approx 0.667$ **41.** $x = \pm 5i$ **43.** $x = 1 \pm 2i$ **45.** $x = -2 \pm 2i$ **47.** $x = \frac{4 \pm i\sqrt{2}}{2}$

2.2 Exercises

49. a. $t = 2$ sec and 4 sec **b.** $2 < t < 4$; between 2 and 4 seconds **51.** 20 units and 90 units **53. a.** $P = 520x - 10{,}000 - x^2$ **b.** −$964; loss of $964 **c.** $5616; profit of $5616 **d.** 20 units or 500 units **55. a.** $s = 50, s = -50$ **b.** There is no particulate pollution. **c.** $s = 50$; speed is positive. **57. a.** $r = 0$ **b.** $r = 0.05$ **c.** $r = 0.1$; against the wall of the artery **59.** $100 is the price that gives demand = supply = 97 trees **61. a.**

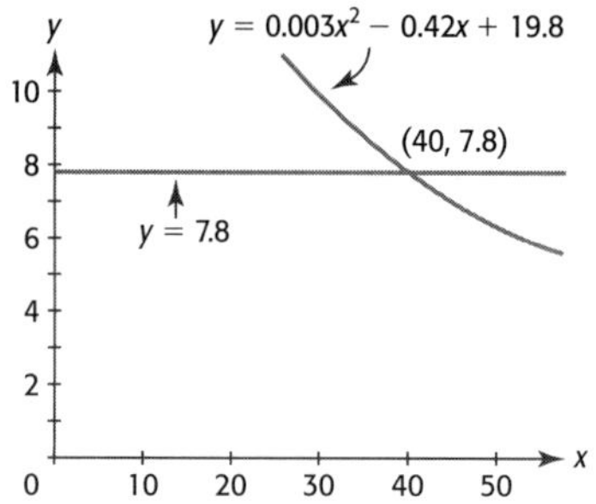

b. $x = 40, x = 100$; 1940 and 2000 **c.** $40 < x < 100$; between 1940 and 2000

63. a. Set the x-values from 0 to 10, so it includes x-values representing 1989 to 1997; use TRACE or TABLE to find y-values for these x-values.

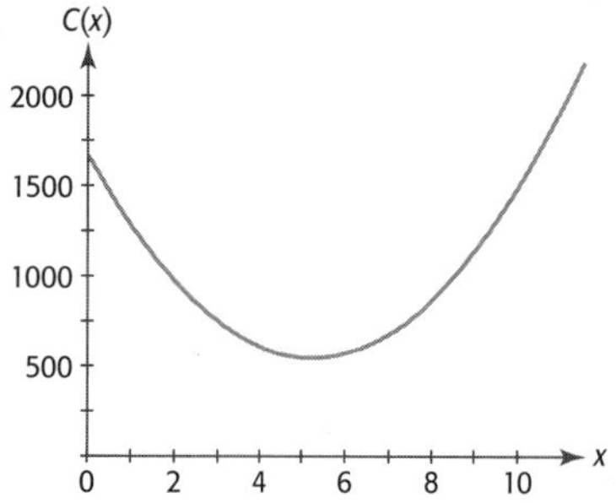

b. $x = 5.2$; testing years shows minimum in 1994. **c.** $x = 10$; 1999; **d.** Extrapolation because the model is used for evaluation beyond the given data **65.** $x = 20$; 2007 **67. a.** Declined by 262,000 **b.** 1994 **c.** 37,851,047; that the model continues to apply **69. a.** $3709.3 million increase **b.** 17.1 years after 1980; in 1998 **71. a.**

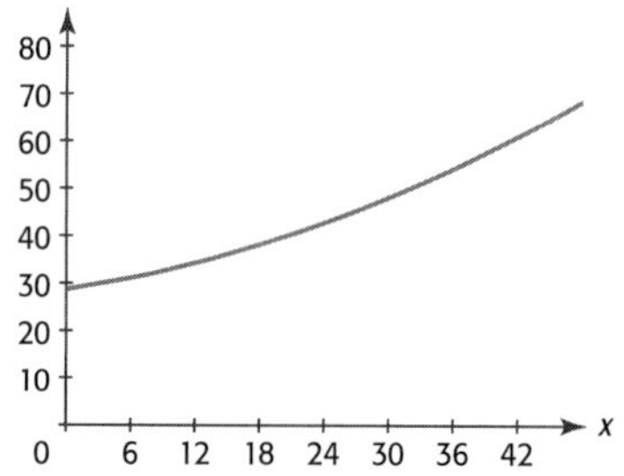

b. 60.8% **c.** 32 years from 1960; in 1992 **73.** 2023

2.3 Skills Check

1.

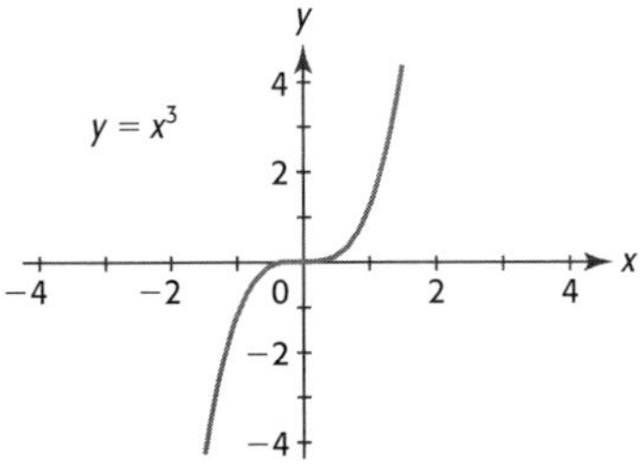

3.

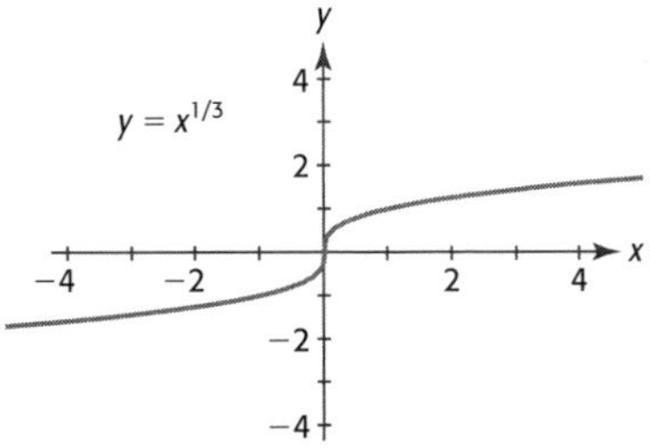

5.

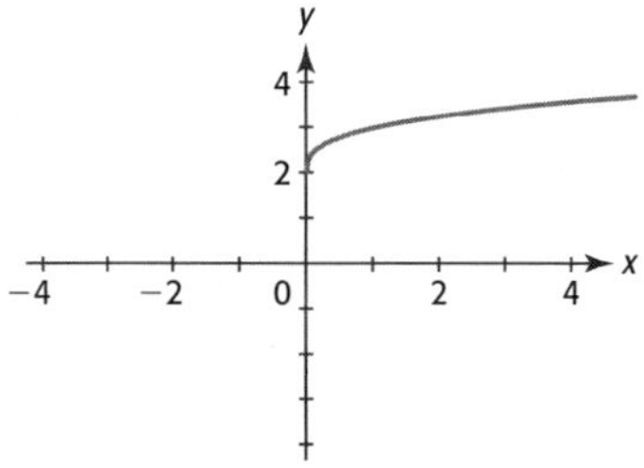

7.

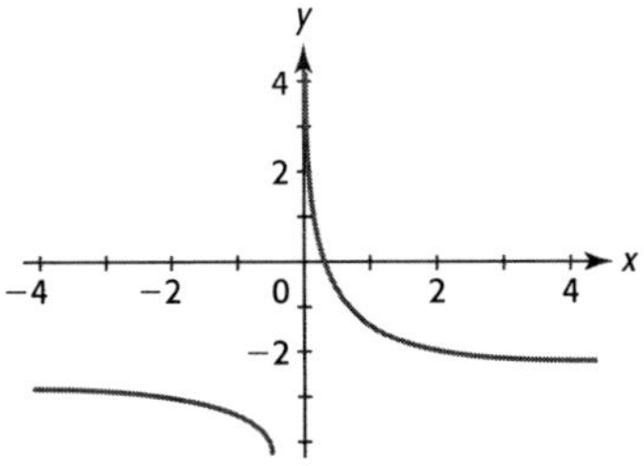

9.

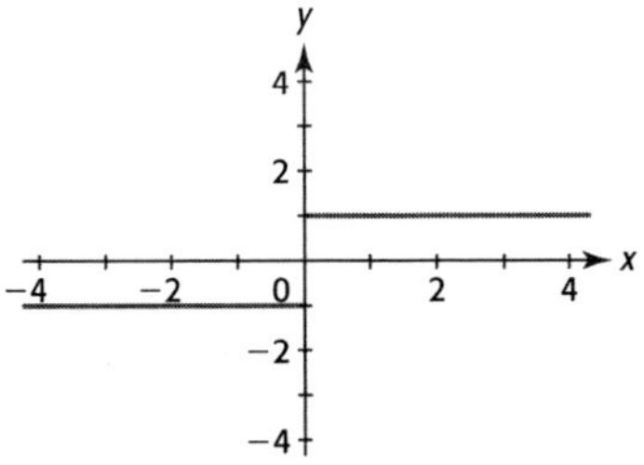

11. a.

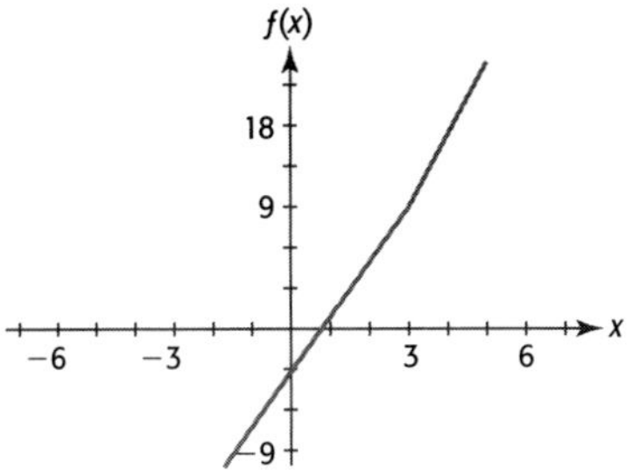

b. 5, 16 **c.** All real numbers **13. a.**

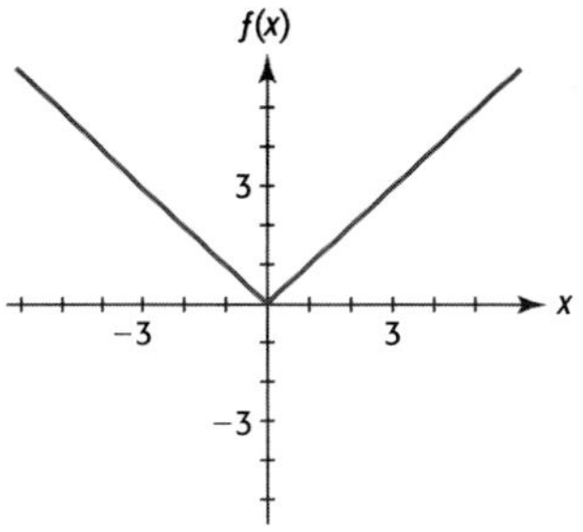

b. 2, 5 **c.** All real numbers **15. a.** -27 **b.** 1 **17. a.** Not possible **b.** 1 **19. a.** $-\frac{1}{3}$ **b.** 1
21. a. -9.2 **b.** -10 **23. a.** 5 **b.** 6 **25. a.** 0 **b.** 29 **27. a.** Increasing **b.** Increasing
29. Concave down **31.** Concave up **33.**

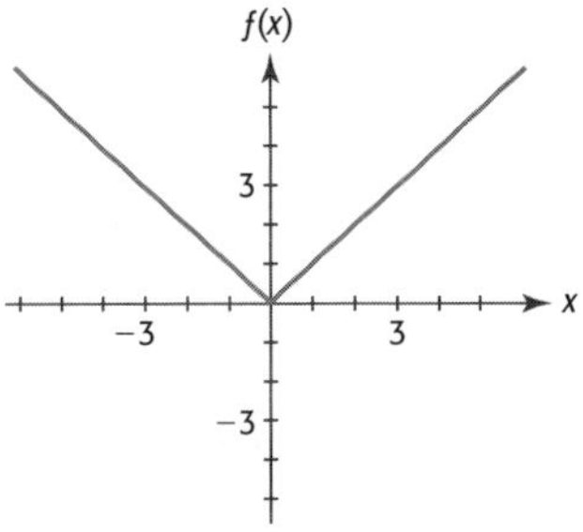

2.3 Exercises

35. a. power **b.** 23,874; 23,874 suicides occurred in 1965 **c.** 28,099

37. a.

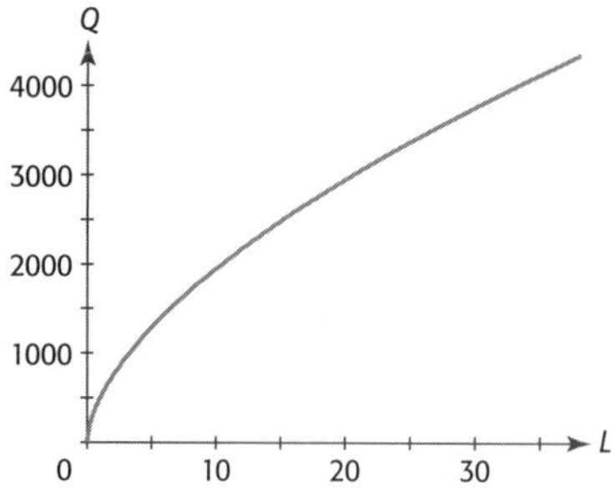

b. 3912 **c.** Increases; yes

39. a. Increasing **b.** Concave up **c.** $x = 10.785$; 2001

41. a.

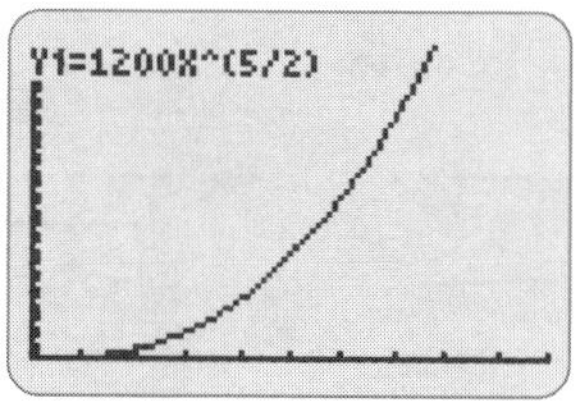

[0, 10] by [0, 200,000]

b. Concave up

43. a.

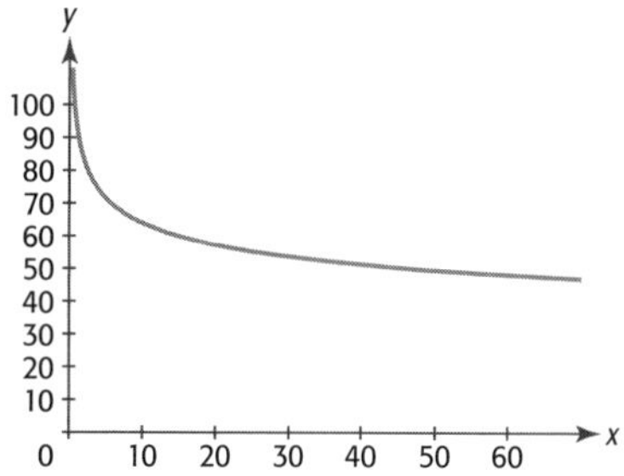

b. Decreasing **c.** 50.3% **d.** 49.6%

e. No, percent was higher **45.** \$130 **47. a.** $P(x) = \begin{cases} 37 & \text{if } 0 < x \le 1 \\ 60 & \text{if } 1 < x \le 2 \\ 83 & \text{if } 2 < x \le 3 \\ 106 & \text{if } 3 < x \le 4 \end{cases}$

b. 60; it costs 60¢ to mail a 1.2-oz letter by first class. **c.** $0 < x \le 4$ **d.** 60, 83 **e.** 60¢, 83¢

49. a. $T(x) = \begin{cases} .15x \text{ if } 0 < x \le 43{,}850 \\ 6{,}577.50 + .28(x - 43{,}850) \quad \text{if } 43{,}850 < x \le 105{,}950 \end{cases}$ **b.** 6300 **c.** \$9699.50 **d.** \$6577.50 and $6577.50 + .28(1) = 6577.78$; 28% is paid only on the amount over \$43,850, not all income

51. a.

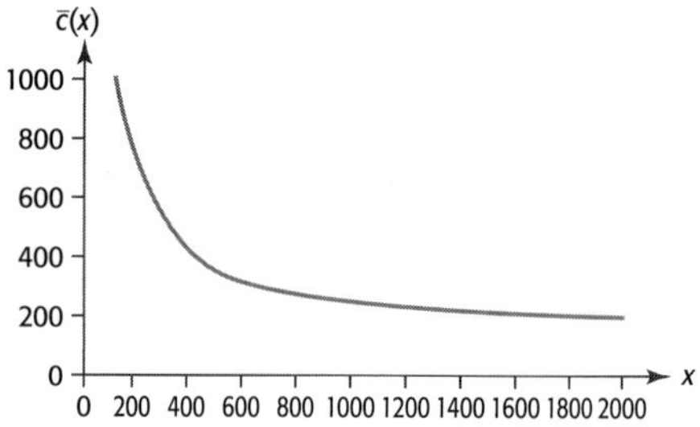

b. Decrease

53. a.

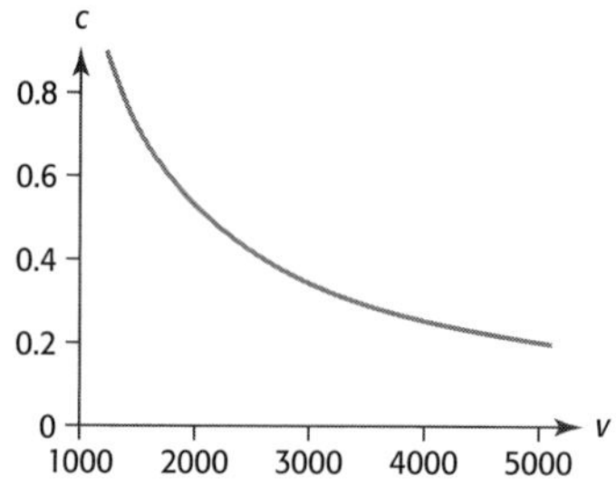

b. Decrease

2.4 Skills Check

1. $f(x) = 3x^2 - 2x$ **3.** The x-values are not equally spaced. **5.** $f(x) = 99.9x^2 + 0.64x - 0.75$
7. a. $y = 3.545x^{1.323}$ **b.** $y = 8.114x - 8.067$ **c.** Power function **9. a.** $y = 1.292x^{1.178}$ **b.** $y = 2.065x - 1.565$
c. They are both good fits. **11.** $y = 2.98x^{0.614}$

2.4 Exercises

Numerical results are computed with the unrounded models.
13. a. $y = -17.57x^2 + 632.35x + 29{,}909$ **b.** 29,256; 25,479 (from unrounded model) **c.** No; extrapolating may not give accurate predictions. **15. a.**

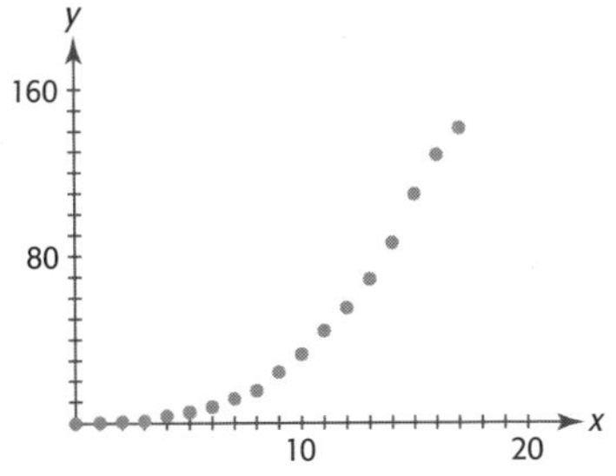

b. $y = 0.731x^2 - 4.346x + 5.478$

c. 211.042 million **d.** Approximately $\frac{2}{3}$ **17. a.**

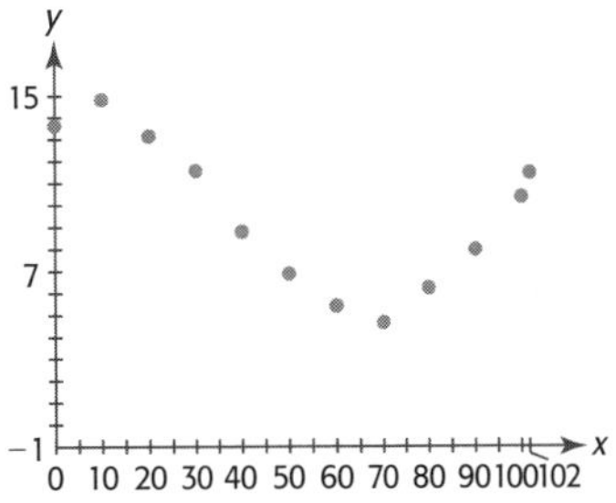

b. Yes

c. $y = 0.0024x^2 - 0.307x + 16.486$ **d.** 10.7% **19. a.** $y = .0134x^2 - .054x + 1.564$
b. $y = .013x^2 + .080x + 1.630$ **c.** 22.1 years after 1965; 1988 **d.** The results would be equivalent.
21. a.

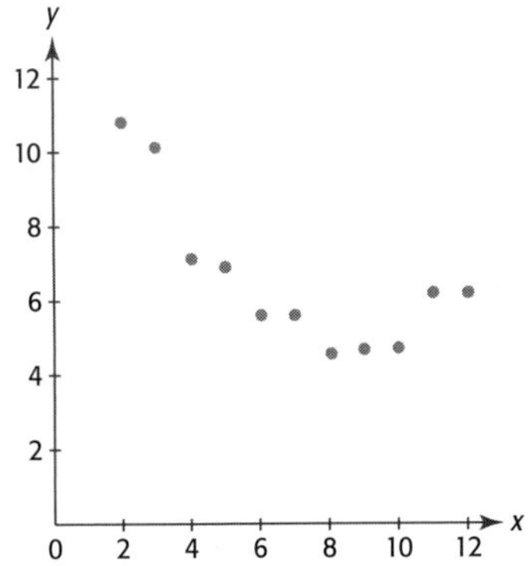

b. Yes; $y = 0.144x^2 - 2.48x + 15.47$

c. The y-intercept is (0, 15.47). This means that unemployment in South Carolina was 15.47% in 1980 according to the model. **23. a.** No **b.** Power function; $y = \frac{1}{3}x^3$ **25. a.** $y = -0.36x^2 + 38.5x + 5822.9$
b. $x \approx 191$, or the year 2181 **c.** After 2181 **27.** $y = 4700\sqrt{110 - x}$ **29. a.** $y = 154.13x^{-.49}$ **b.** 25%
31. a. $y = 52.4x^{-0.29}$ **b.** 16.6 **c.** Decreasing **d.** Eventually the number of students will be 0.
33. a. $y = 34.7x^{1.58}$ **b.** \$7523 million **c.** 1990 **d.** Increased

35. a.

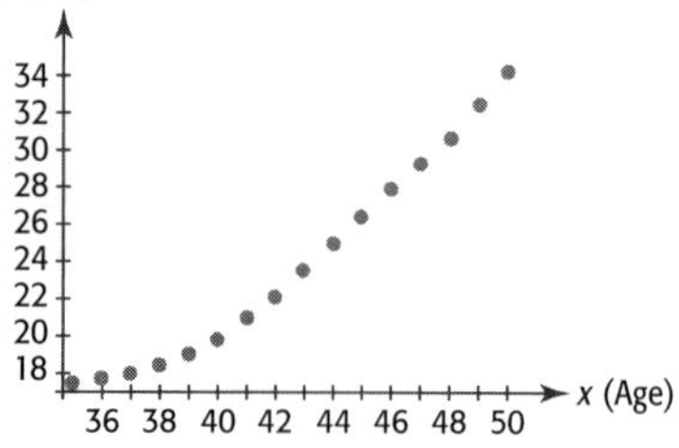

b. Yes; $y = .052x^2 - 3.23x + 66.66$ **c.** $y = .011x^{2.04}$ **d.** Quadratic model

37. a. $y = 0.155x^2 - 9.15x + 235.5$; $y = 474.76x^{-0.473}$; both models fit well. **b.** The discharge rate will be less than the 1996 rate, 102.3. **c.** Quadratic model yields 105.6; power model yields 88.4; power model is a better predictor.

Chapter 2 Skills Check

1. (1, −27) **2.**

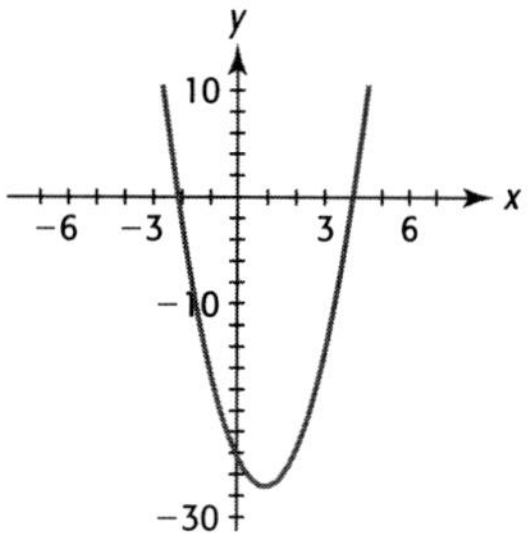

3. −2, 4 **4.** $x = 4, x = -2$ **5.** 4, 1

6. $x = \frac{1}{2}, x = -\frac{2}{3}$ **7.** −4/5, 1 **8.** $x = -2, x = \frac{2}{3}$ **9.** 3, 1 **10.** $x = \frac{1}{2}, x = -\frac{3}{2}$

11. a.

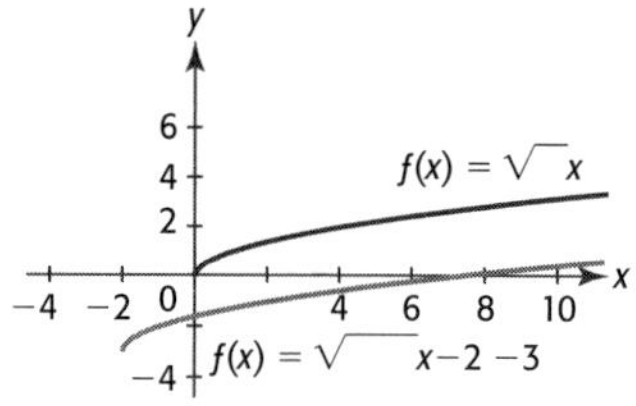

b. The graph of $g(x)$ can be obtained from the graph of $f(x)$ by shifting 3 units down and 2 units to the left.

12. $[-2,\infty)$; real numbers ≥ -2 **13. a.** Increasing **b.** Decreasing **14. a.** Concave up **b.** Concave down **15.** $f^{-1}(x) = \frac{x + 2}{3}$ **16.** $g^{-1}(x) = x^3 + 1$ **17.**

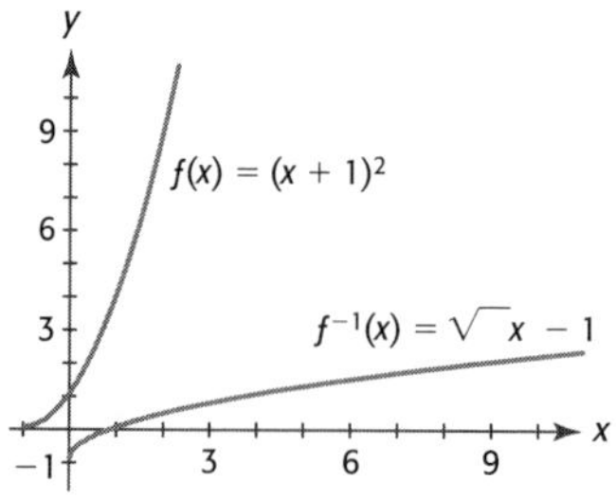

18. Yes

19. $y = 3.5454x^{1.323}$ **20.** $y = 1.04x^2 - 0.51x + 0.977$ **21.** f **22.** c

23. e **24.**

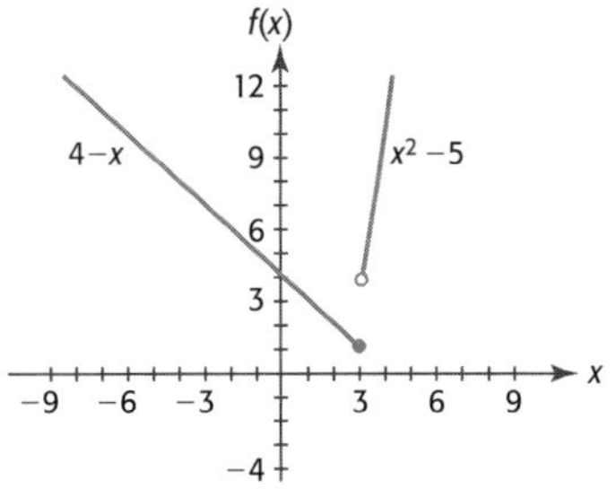

25. $[-2, 9]$ **26.** $x \leq -3$ or $x \geq \frac{1}{2}$ **27.** $z = 2 \pm i\sqrt{2}$

28. $w = 2 \pm i$ **29.** $x = \frac{5 \pm i\sqrt{23}}{8}$ **30.** $x = \frac{-1 \pm i\sqrt{3}}{4}$ **31.** $x < 20$

32. $x \geq 16\sqrt{2} - 2$ or $x \leq -16\sqrt{2} - 2$

33. $x = -\frac{3}{8}$ **34.** $x = -2$ **35.** $x = 10, x = -6$ **36.** $x = 5, x = -8$ **37.** $-2 \leq x \leq 6$

38. $x \geq 4.5$ or $x \leq -3$

Chapter 2 Review

39. a. 3100 **b.** \$84,100 **40. a.** 4050 **b.** \$312,050 **41. a.** 2 sec **b.** 256 ft **42. a.** 1.5 sec **b.** 82.05 m

43. a.

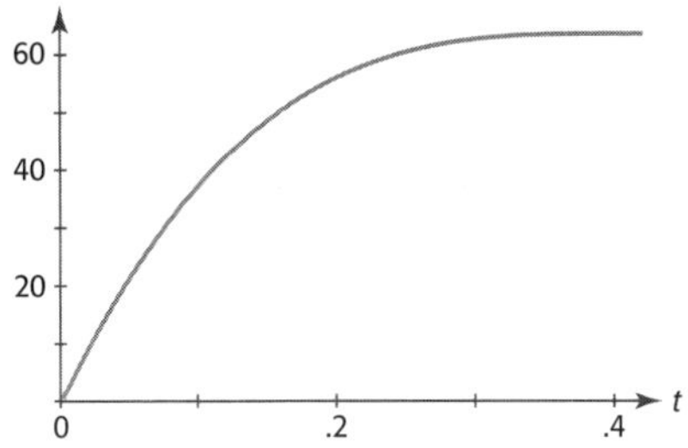

b. 64 in. **44. a.** 4 sec **b.** 380 ft **c.** Shifted 4 units right and 380 units up, stretched by 16, and reflected across the x-axis **45. a.** 1997 **b.** Between 1984 and 2012

46. a. 1989 **b.** 4.79% **c.** 1990 **47.** 30 units and 120 units **48.** 5 **49.** 400 or 3700 units

50. 1990 and 1997 **51.** 1972 through 1994 **52.** 2000 through 2012 **53. a.** 43,905 **b.** 1990 **54. a.** Yes

b.

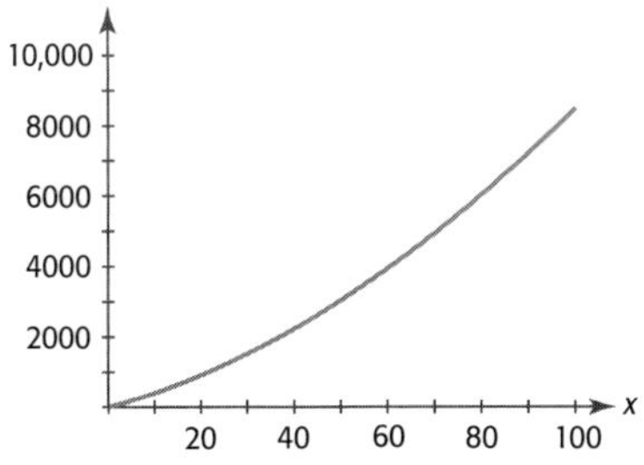

c. Concave up **d.** \$7912.28

55. a.

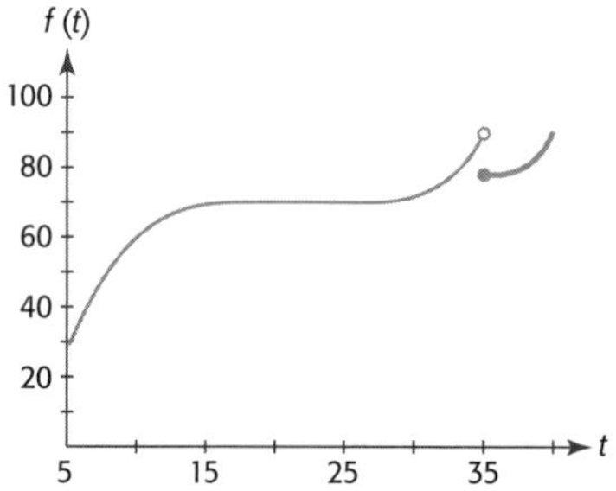

b. \$70.593 million **c.** \$77.486 million

56. a. $\overline{C}(x) = \dfrac{30x + 3150}{x}$ **b.**

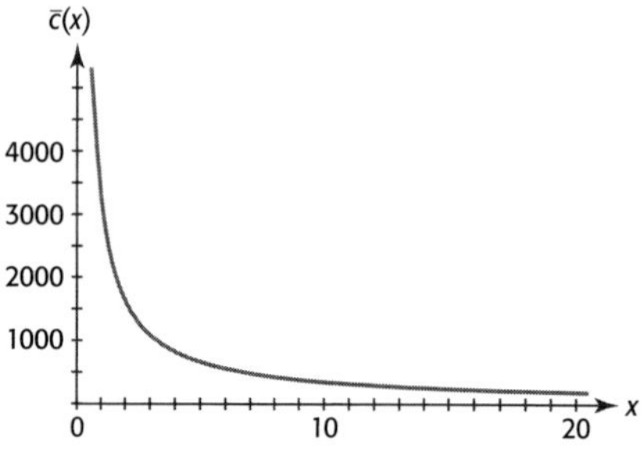

57. a. Demand function **b.**

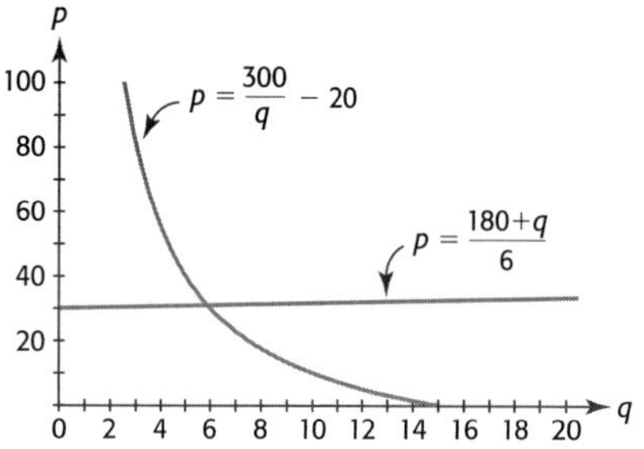

58. a. The demand function is a shifted reciprocal function. **b.**

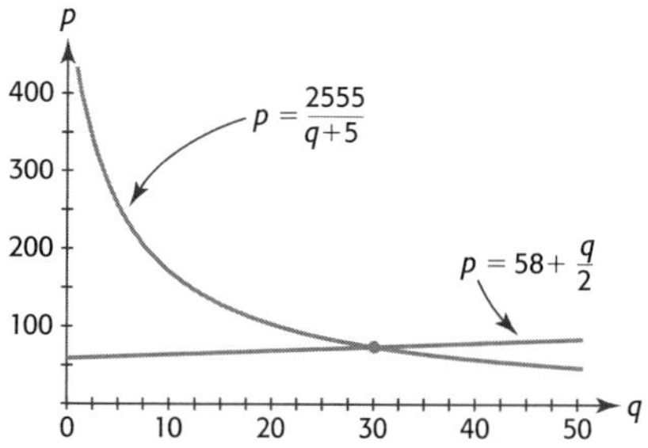

59. a. 5544 thousand **b.** 2000; that the model is still valid **60. a.** For years between 1979 and 1988 inclusive, the function is quadratic. **b.** 12.95, 12.85, 6.05; in 1979 12.95% of people 12 years of age or older used marijuana at least once in the month prior to being surveyed. In 1980 12.85% of people 12 years of age or older used marijuana at least once in the month prior to being surveyed. In 1988 6.05% of people 12 years of age or older used marijuana at least once in the month prior to being surveyed. **c.**

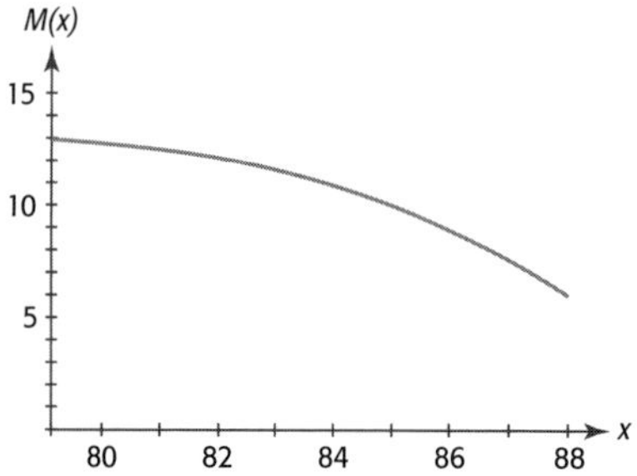

d. The domain is $\{79, 80, 81, 82, \ldots, 87, 88\}$.

e. The maximum occurs in 1979. Marijuana use decreased until 1988.
61. a. $T(x) = E(x) - I(x) = -191.73x^2 - 21604.55x - 57756.99$ **b.** \$292,975.49 million deficit
62. a. $R(E(t)) = 0.165\,(0.017t^2 + 2164t + 8.061) - 0.226 = 0.002805t^2 + 0.35706t + 1.330065 - 0.226 = 0.002805t^2 + 0.35706t + 1.104065$; the function calculates the revenue for Southwest Airlines given the number of years past 1990. **b.** $R(E(3)) = 2.2008995 \approx 2.2$; in 1993 Southwest Airlines had revenue of \$2.2 billion.
c. $E(7) = 24.042$; in 1997 Southwest Airlines had 24,042 employees. **d.** \$3.7 billion **63. a.** $f^{-1}(x) = \dfrac{x + 2.886}{0.554}$
b. The mean sentence length can be found as a function of the mean time in prison. **64. a.** $C^{-1}(x) = \dfrac{9.929 - x}{0.093}$
b. Given the number of cows used for milk production, the inverse function will calculate the number of years past 1990.
c. 8 **65. a.** $y = 11.786x^2 - 142.214x + 493$ **b.**

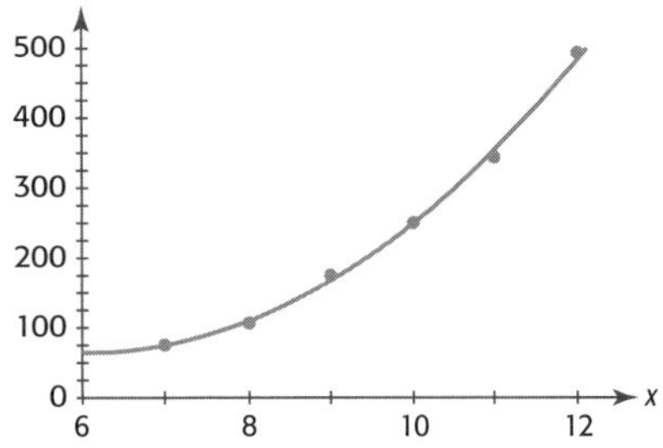

c. Yes d. 2002

66. a. $y = 14.118x^2 + 198.336x + 4795.263$ **b.** 1993 **c.** 2003
67. Values for table: 10,288, 9120, 8712, 8876, 9343; $y = 14.66x^2 - 308.42x + 10{,}293.89$ where y is the number of males in thousands and x is the number of years past 1980 **68. a.** $y = 0.05143x^2 + 3.18286x + 65.40000$ **b.** 16 yr

69. a. Yes, good fit

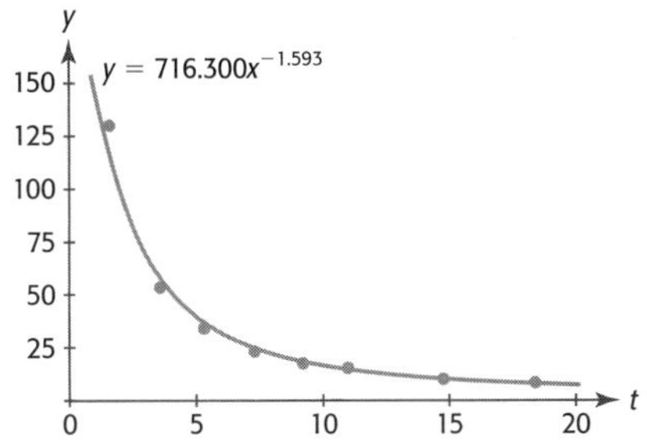

b. Yes, but far in the future, at $x \approx 61.9$ which represents the 2041–2042 school year

c. $C(t) = \dfrac{1}{t}$; stretched by a factor of 380, shifted vertically 15 units down, shifted horizontally 0.3 unit to the left
d. The function in part (a); its function values will never be negative. **e.** $y = C(t)$; no
70. a. $y = 982.66x^{0.0784}$ **b.** 2021 **71. a.** $y = 0.013x^2 - 0.22x + 7.3$ **b.** $y = -0.26x + 9.34$

c. $f(x) = \begin{cases} 0.013x^2 - 0.22x + 7.3 & \text{if } 0 \le x \le 11 \\ -0.26x + 9.34 & \text{if } 12 \le x \le 16 \end{cases}$ **d. i.** 6.4; average stay in 1987 was 6.4 days **ii.** 1992
iii. 6.5 days **72. a.** $y = 14.664x + 169.380$ **b.** $y = -0.141x^2 + 17.475x + 162.351$
c. linear: 535.98, quadratic: 511.38; quadratic is better **d.** 2004 **73.** Between 200 and 6000 units
74. Above 5.5% prior to 1987 and after 1990

Chapter 3 Exponential and Logarithmic Functions

3.1 Skills Check

1. c, e **3. a.**

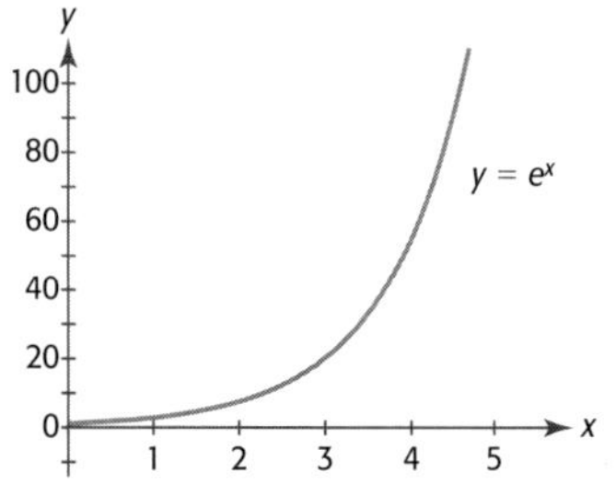

b. 2.718; 0.368; 54.598 **c.** x-axis **d.** 1

5.

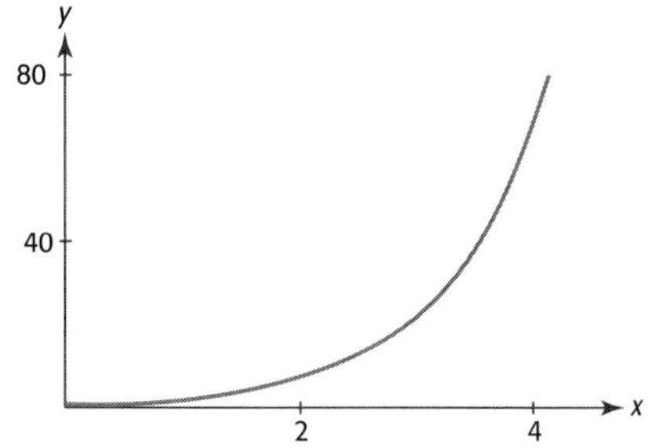

7.

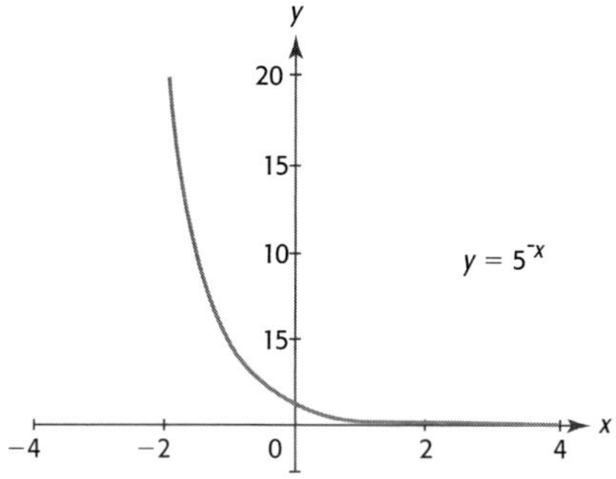

9.

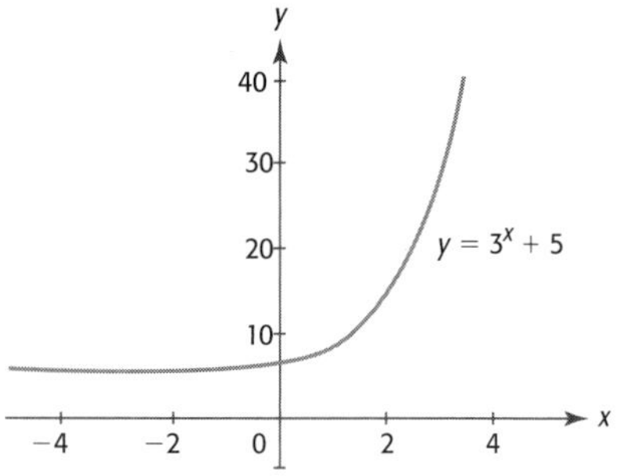

11.

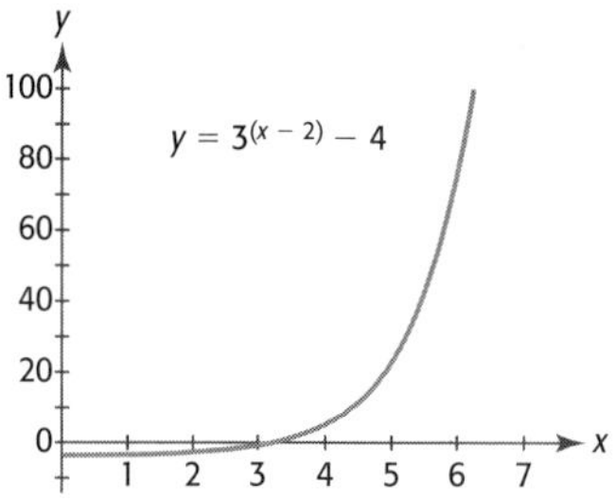

13. Vertical shift, 2 units up

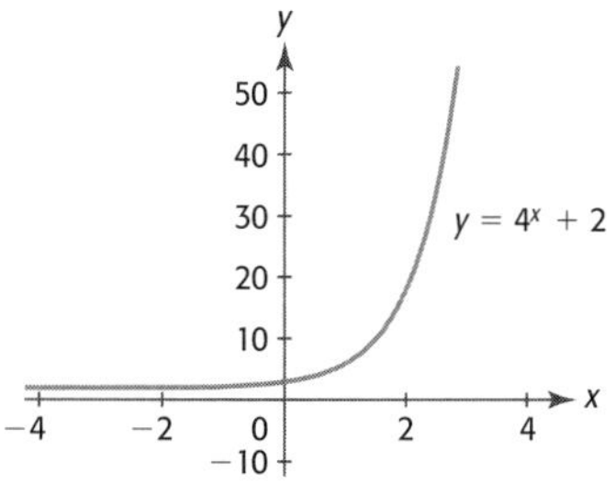

15. Reflection across the y-axis

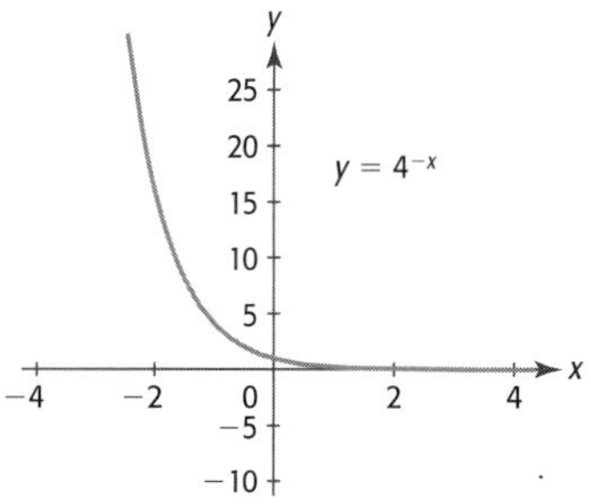

17. Vertical stretch by a factor of 3

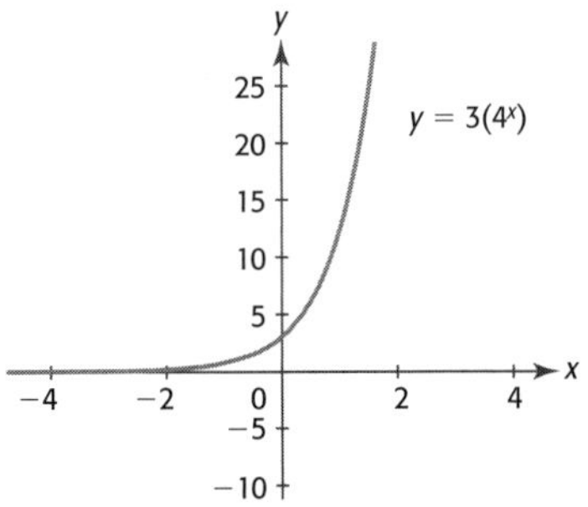

19. The graph of $y = 3 \cdot 4^{(x-2)} - 3$ is the graph of $y = 3(4^x)$ shifted to the right 2 units and shifted down 3 units.

21. a.

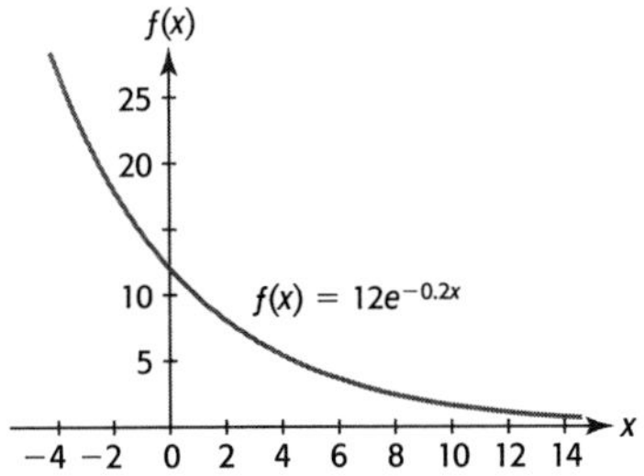

b. $f(10) = 1.624$; $f(-10) = 88.669$ **c.** Decay

3.1 Exercises

23. a. \$12,000 **b.** \$8603.73 **c.** No

25. a.

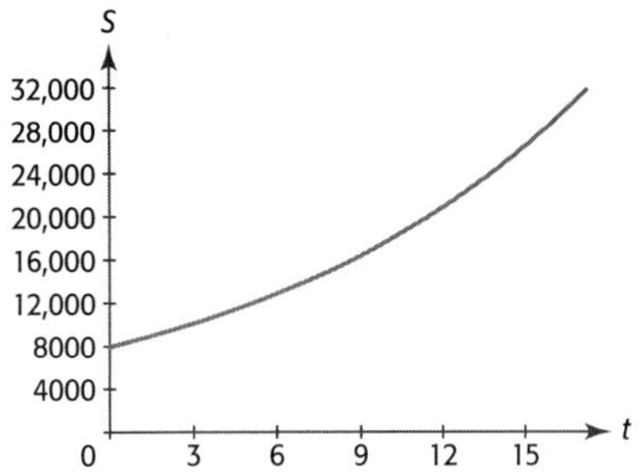

b. $x \approx 11.45$ yr **c.** About 24.5 yr

27. a. 376.84 g **b.**

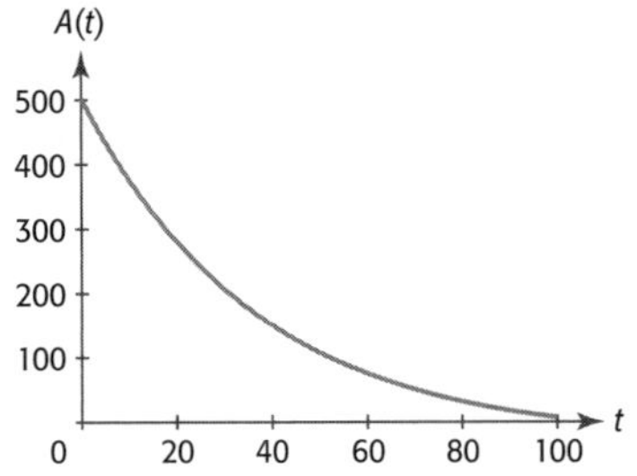

29. a.

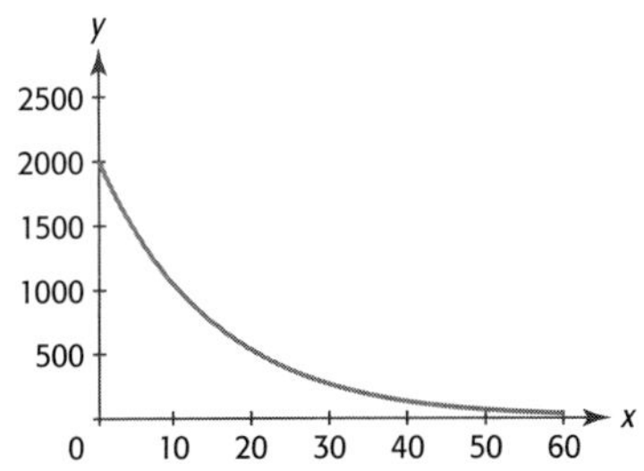

b. \$1000

c. Sales declined drastically after the end of the ad campaign. (Answers may vary.)

31. a. 14,339.44 **b.** Because of future inflation, they should plan to save money. (Answers may vary.)
33. a. \$122,140.28 **b.** In about 14 yr **35. a.** Increasing **b.** 57,128 **c.** 61,577
d. Approximately 858 people per year **37. a.** About 88.4 g **b.** About 18,602 yr
39. a.

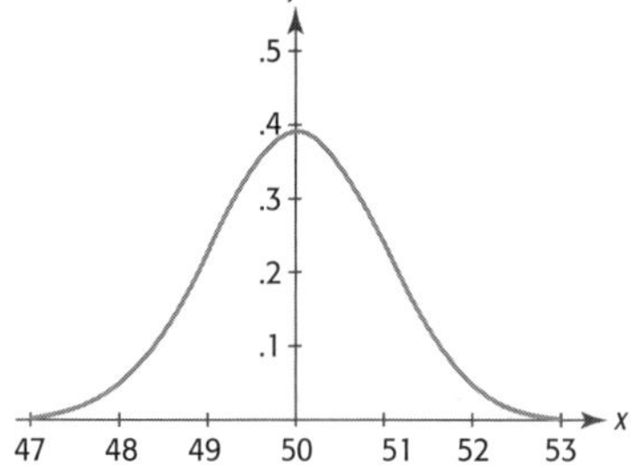

b. 50

3.2 Skills Check

1. $3^y = x$ **3.** $e^y = 2x$ **5.** $\log_4 x = y$ **7.** $\log_2 32 = 5$ **9.** 0.845 **11.** 4.454 **13.** 4.806 **15.** 5
17. 3 **19.** 4 **21.** $\frac{3}{2}$ **23.** 3 **25.** -3 **27.** 1

29.

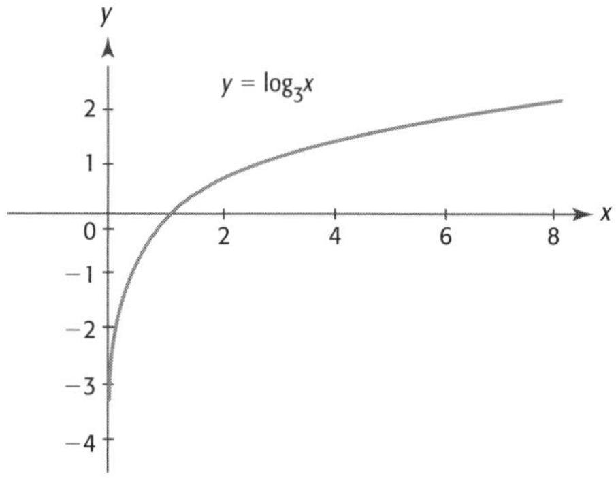

31.

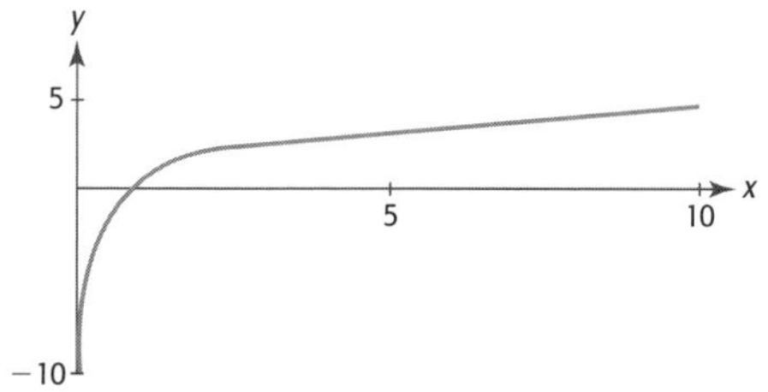

33.

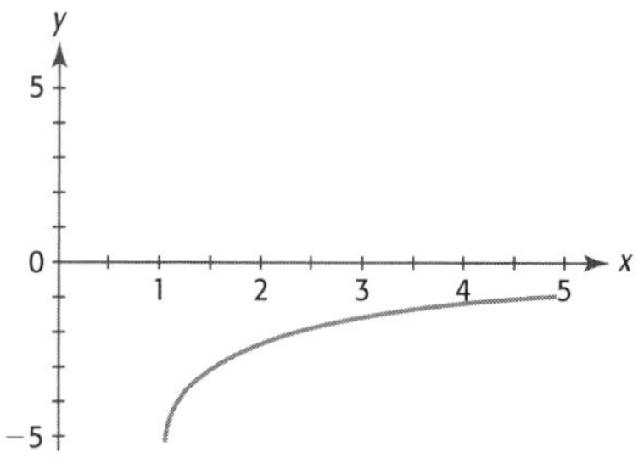

35. a. $y = 4^x, x = 4^y, \log_4 x = y$; therefore, the inverse function is $y = \log_4 x$.
b.

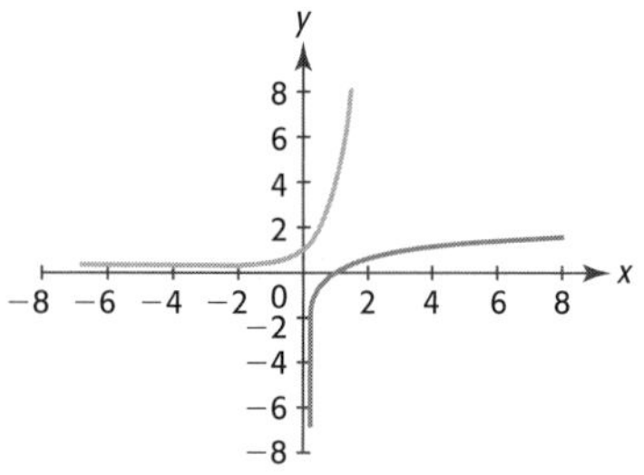

The graphs are symmetric with respect to the line $y = x$.

37. $a^x = a; x = 1$ **39.** $x = 8$ **41.** $x = e^{1.5}$

3.2 Exercises

43. a. ≈59; ≈76 **b.** Improved health care and better diet. (Answers may vary.) **45. a.** 6662.66; 8441.24 **b.** Increasing **c.** Increased cost of living, raise in minimum wage. (Answers may vary.) **47.** ≈6.9 yr

49. b. 35 quarters, or $8\frac{3}{4}$ yr **51.** 7 units **53. a.** $\ln x = \frac{y - 17.875}{12.734}$ **b.** $x = e^{\frac{y-17.875}{12.734}}$ **c.** $x = 88.77$; 1989 **d.** 1989; answers agree **55.** 4.2 **57.** $10^{-14} \leq H^{+} \leq 10^{-1}$

3.3 Skills Check

1. 3.204 **3.** 7.824 **5.** $x \approx 1.819$ **7.** 0.374 **9.** $x \approx 1.204$ **11.** 14 **13.** 12 **15.** 2.2146 **17.** 2.322 **19.** $\ln(3x - 2) - \ln(x + 1)$ **21.** $\frac{1}{4}\log_3(4x + 1) - \log_3(4) - 2\log_3(x)$ **23.** $\log_2 x^3 y$ **25.** $\ln\frac{(2a)^4}{b}$ **27.** 1.6131 **29.** 0.1667 **31.** $x = \frac{10}{3}$ **33.** $x = 2$ **35.** $x \approx 3.5$ **37.** $x \approx 0.769$ **39.** $x = \frac{e^6}{2} \approx 201.71$ **41.** $x = 50$ **43.** $x < 5$ **45.** $x \geq 9$

3.3 Exercises

47. 5 **49. a.** $\ln\left(\frac{S}{25{,}000}\right) = -.072x$ **b.** 6 **51. a.** 3200 **b.** 9 days **53. a.** $1755.358 million **b.** 1996 **55.** 13.86 yr **57.** 23.10 yr **59.** $\frac{\ln 2}{r} = t$ **61. a.** ≈35 **b.** 8.75 yr **63.** $t \approx 14.2$; in 15 yr **65.** $t \approx 17.39$; 17 yrs, 5 mo **67.** Approximately 3 h

69. a.

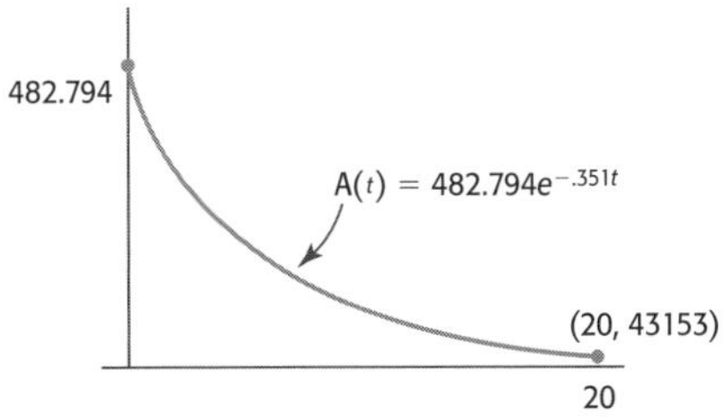

b. 0 **c.** 1993 **d.** LP and EP sales have declined due to the invention of other forms of music such as CD's and mp3 files.

71. a. 16 **b.**

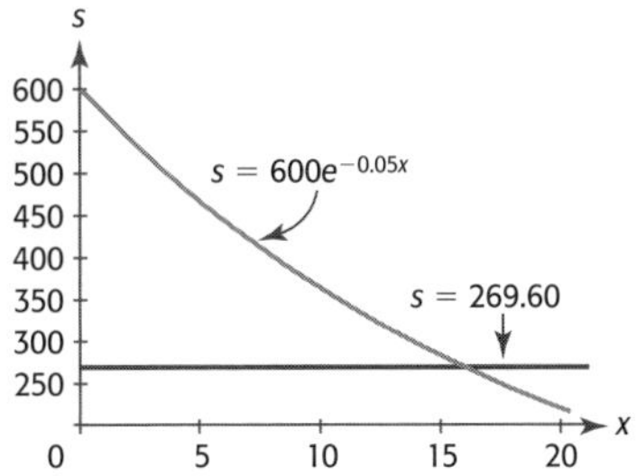

73. After 12 years and 1 quarter **75.** $2P = P(1.10)^n \Rightarrow 2 = 1.10^n \Rightarrow n = \log_{1.10} 2$

3.4 Skills Check

1. $y = 2(3^x)$ **3.** $f(x) = 4^x$ **5. a.**

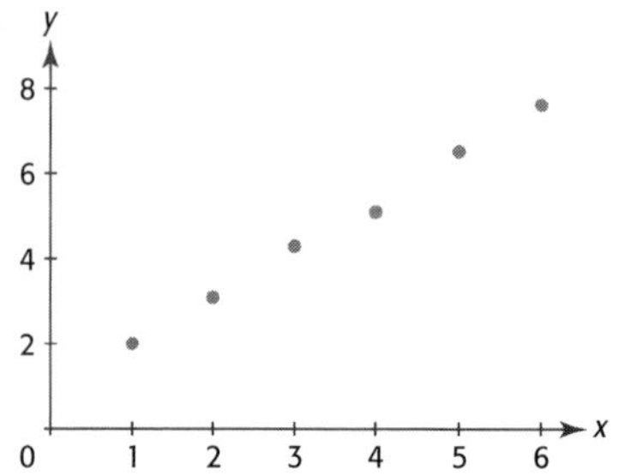

b. Linear **7.** Exponential

9. $y = 0.876(2.494^x)$ **11. a.**

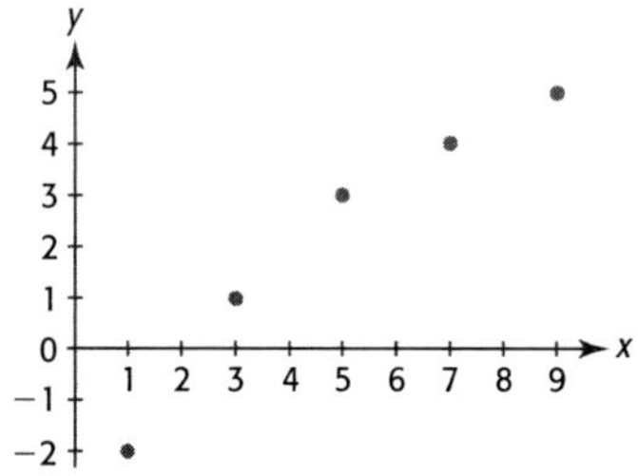

b. Logarithmic

3.4 Exercises

13. a. $y = 30{,}000(1.04^t)$ **b.** \$54,028
15. a. $y = 20{,}000(.98^x)$ **b.** \$18,078
17. a. $y = 238.4752(1.0075^x)$ **b.**

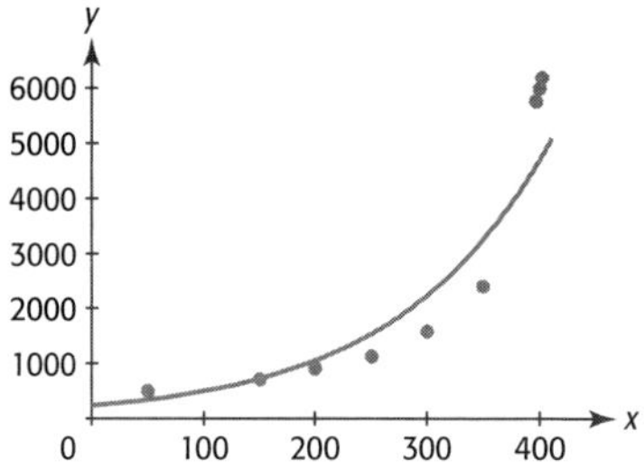

19. a. $y = 4.304(1.096^x)$ **b.** $y = 6.182x^2 - 565.948x + 12810.482$
c. The exponential function appears to be the better fit.

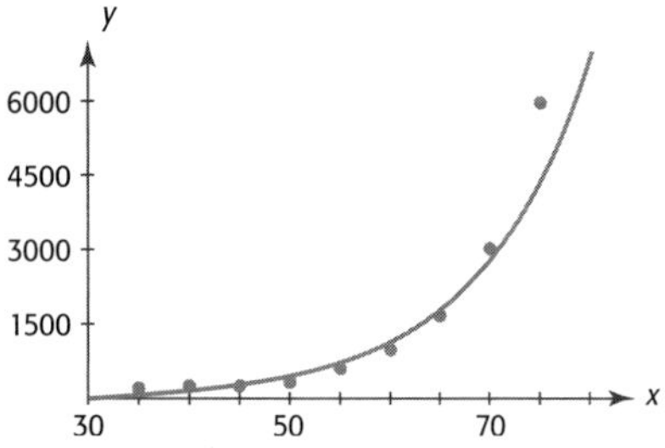

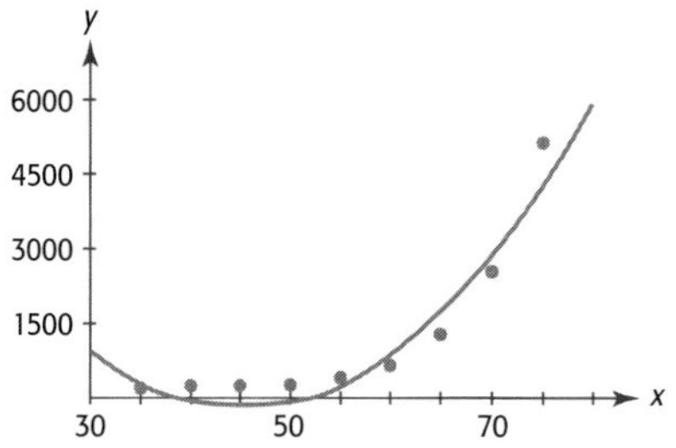

21. a.

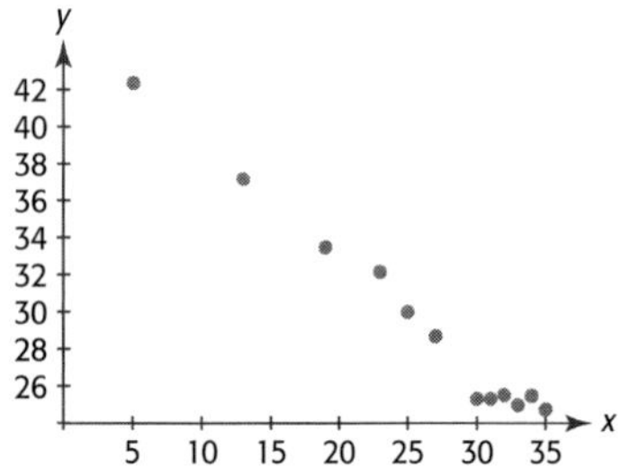

b. yes **c.** $y = 47.818(0.981^x)$ **d.** $y = 74.575x^{-.298}$ **e.** Exponential

23. a. $y = 17.875 + 12.734 \ln x$ **b.** $y = -0.00249x^2 + 0.568x + 44.472$ **c.** Quadratic function looks like a slightly better fit. **d.** Using the logarithmic function: 77.7; using the quadratic function: 76.8; the logarithmic model is the better predictor because the quadratic model falls after 2014. **25. a.** $y = -681.976 + 251.829 \ln x$ **b.** 31.5% **c.** $y = 0.627x^2 - 7.400x - 26.675$ **d.** Quadratic

3.5 Skills Check

Approximate answers to two decimal places.

1. 49,801.75 **3.** 17,230.47 **5.** 26,445.08 **7.** 12,311.80 **9.** 1,723,331.03

11. 1123.60; 1191.00; 1191.00 **13.** $P = \dfrac{S}{\left(1 + \dfrac{r}{k}\right)^{kn}} = S\left(1 + \dfrac{r}{k}\right)^{-kn}$

3.5 Exercises

15. a. \$16,288.19 **b.** \$88,551.38

17. a.

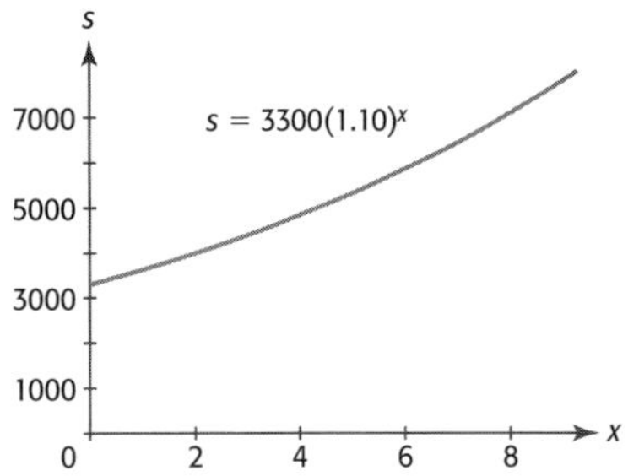

b. $x \approx 7.27$; 8 yr because compounded annually **19.** \$32,620.38 **21. a.** \$33,194.62 **b.** More; compounded more frequently **23.** \$49,958.02 **25. a.** \$20,544.33 **b.** \$29,446.80 **27. a.** \$28,543.39 **b.** \$903.41 more with continuous compounding **29. a.** ≈7.27, 8 yr **b.** ≈6.93 yr **31. a.** \$2954.91 **b.** \$4813.24 **33.** \$6152.25

35. a.

Years	0	7	14	21	28
Future Value (\$)	1000	2000	4000	8000	16,000

b. $y = 1000(1.104)^x$ **c.** \$1640.01, \$2826.02

3.6 Skills Check

1. $P = \dfrac{S}{(1 + i)^n}$ **3.** $A = R\left[\dfrac{1 - (1 + i)^{-n}}{i}\right]$ **5.** $R = A\left[\dfrac{i}{1 - (1 + i)^{-n}}\right]$

3.6 Exercises

7. \$5583.95 **9.** \$4996.09 **11.** \$7023.58 **13.** \$530,179.96 **15.** \$372,845.60 **17. a.** \$702,600.91 **b.** \$121,548.38 **c.** The \$100,000 plus the annuity gives approximately \$452 more per year. **19. a.** \$198,850.99 **b.** \$576,000 **c.** \$377,149.01 **21. a.** 2% **b.** 16 **c.** \$736.50 **23. a.** \$1498.88 **b.** \$639,596.80 **c.** \$289,596.80

Chapter 3 Skills Check

1. a.

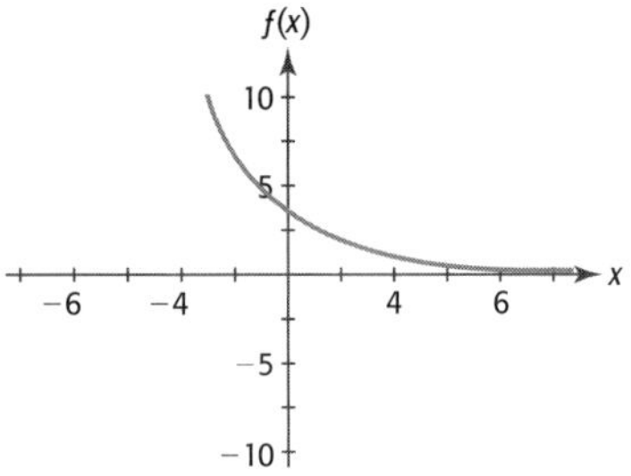

b. ≈ 80.342; ≈ 0.19915 **2.** The function in problem 1 is decreasing.

3.

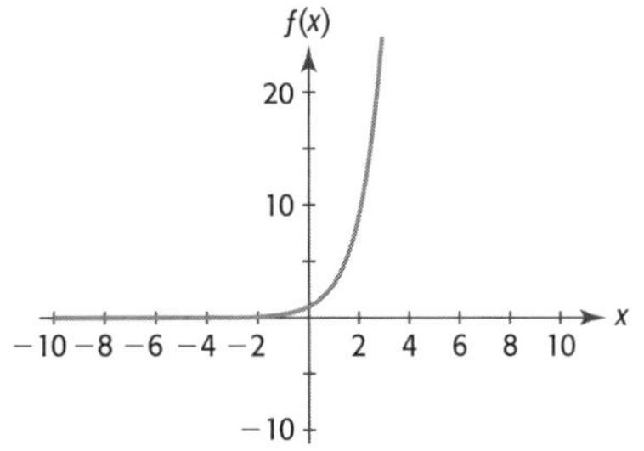

4.

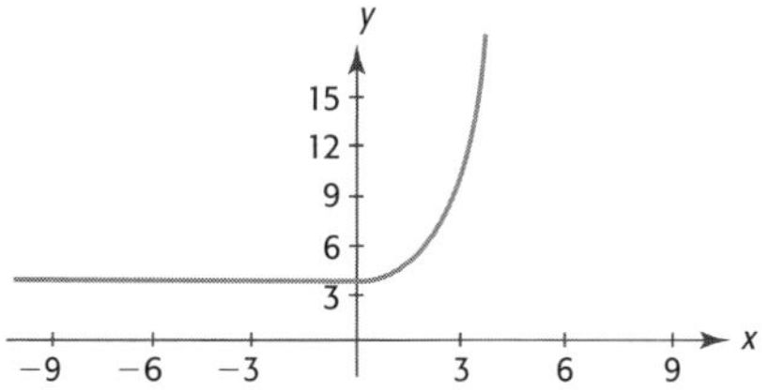

5. The graph of $y = 3^{(x-1)} + 4$ is the graph of $f(x) = 3^x$ shifted to the right 1 unit and up 4 units.
6. The function in problem 4 is increasing. **7. a.** 500 **b.** $x = 200$ **8.** $y = \log_6 x$ **9.** $3x = \log_7 y$
10. $x = 4^y$ **11.** $x = 10^y$ **12.** $x = e^y$ **13.** $y = \log_4 x$ **14.** 1.3424 **15.** 4.0254 **16.** 1 **17.** 4
18. 4 **19.** -3 **20.** 3.6039 **21.** 1.9358

22.

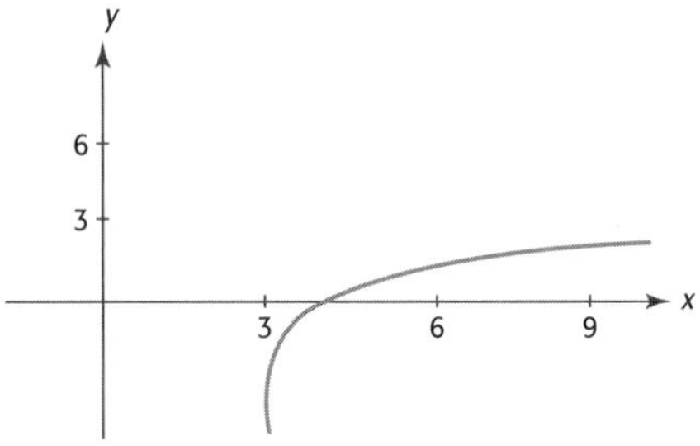

23.

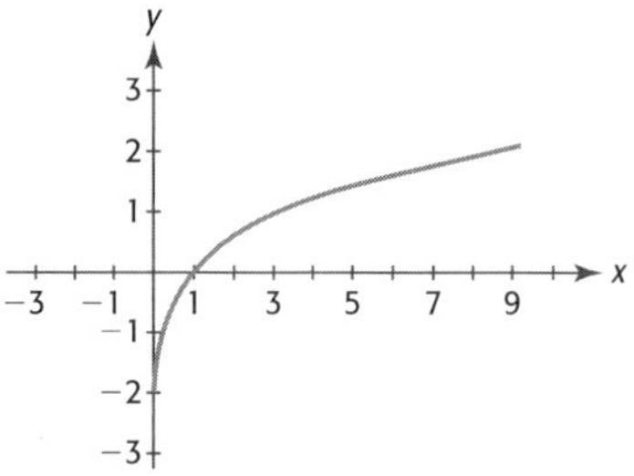

24. ≈ 5.8289 **25.** 0.2012 **26.** ≈ 2.8813 **27.** 2

28. $\ln\left[\dfrac{(2x-5)^3}{x-3}\right]$

$\ln(2x-5)^3 - \ln(x-3)$

$3\ln(2x-5) - \ln(x-3)$ **29.** $\log_4 \dfrac{x^6}{y^2}$

30.

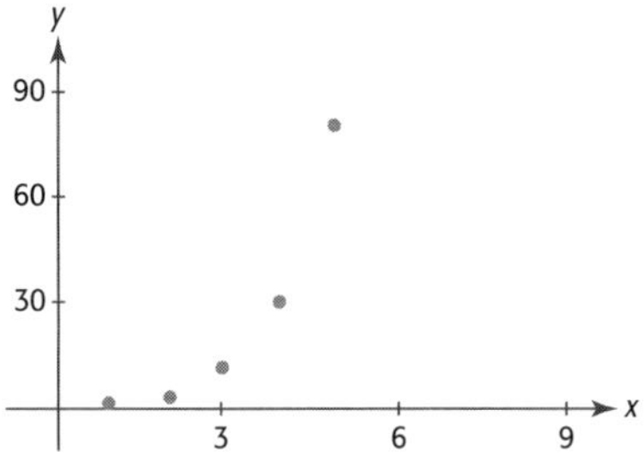

Clearly the data are best modeled by an exponential function. $y = 0.810(2.470)^x$

31. 4926.80 **32.** 32,373.02

33. a.

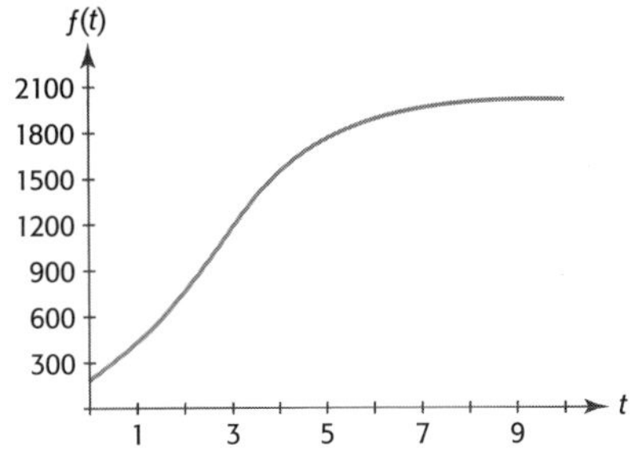

b. 222.22; 1973.76 **c.** 2000

34. a.

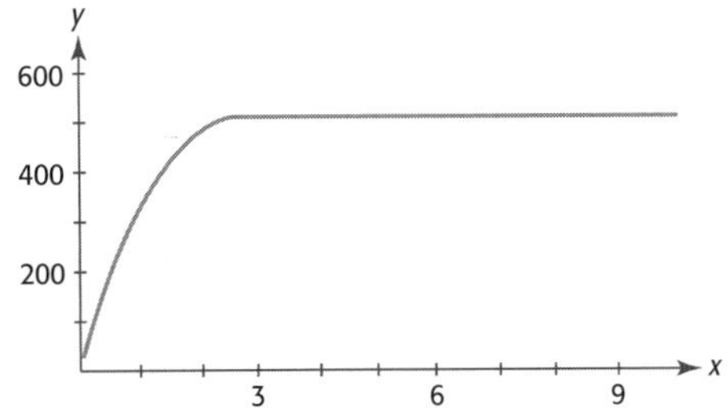

b. $y = 500(0.1)^{0.2^0} = 500(0.1)^1 = 500(0.1) = 50$

c. The limiting value is 500.

Chapter 3 Review

35. 773,760 **36.** \$1515.72 **37.** $x = 2.802$; 1988 **38.** 10 yr **39. a.** $x = 7\log_2 \frac{S}{1000}$ **b.** 30 yr

40. 10 weeks **41.** $x = 29$; 1989

42. a. $P(x) = R(x) - C(x)$

$P(x) = 10(1.26)^x - (2x + 50)$

$= 10(1.26)^x - 2x - 50$

b. Applying the intersection of graphs method, selling at least 10 mobile homes produces a profit of at least \$30,000.

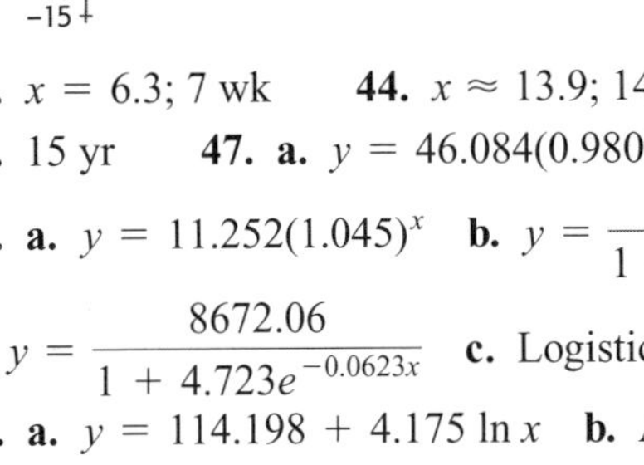

43. $x = 6.3$; 7 wk **44.** $x \approx 13.9$; 14 yr **45.** $\approx$4451.08; 10 yr

46. 15 yr **47. a.** $y = 46.084(0.980^x)$ **b.** 36.3%; 22.5% **48. a.** $y = 57.66(0.978)^x$ **b.** 21.77%

49. a. $y = 11.252(1.045)^x$ **b.** $y = \frac{96.364}{1 + 12.331e^{-0.084x}}$ **c.** Logistic **50. a.** $y = 2102.96(1.0267)^x$

b. $y = \frac{8672.06}{1 + 4.723e^{-0.0623x}}$ **c.** Logistic **51.** $y = 27.945 + 8.84\ln x$

52. a. $y = 114.198 + 4.175\ln x$ **b.** Approximately 127,466,000

53. a. $y = -19.936\ln(x) + 162.148$ **b.**

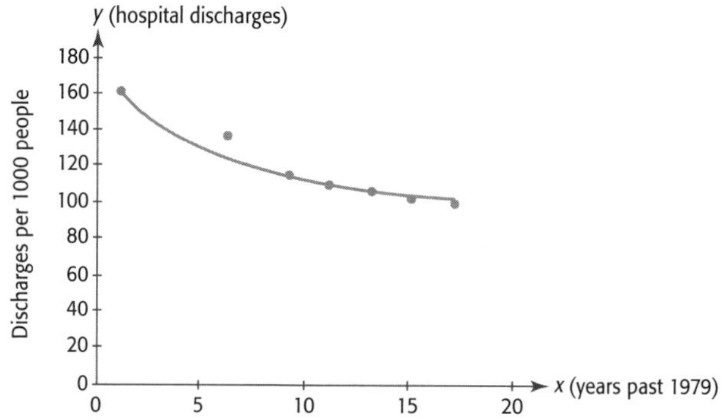

c. Approximately 97 **54.** \$20,609.02 **55.** \$30,072.61 **56.** \$34,426.47 **57.** \$274,419.05 **58.** \$209,281.18 **59.** \$6899.32 **60.** \$66.43 **61.** \$773.16 **62. a.** 67.79%; 76.84% **b.** 96.3641% **63. a.** Approximately 1184 students **b.** 16 days **64. a.** 1298 **b.** 3997 **c.** 4000 **65. a.** Approximately 17,993 Units **b.** 18,000 **66. a.** $C(t) = 5.521(1.349)^t$ **b.** $E(C) = -18.653C^2 + 3226.250C + 3245.657$ **c.** $E(C(t)) = -18.653(5.521(1.349)^t)^2 + 3226.250(5.521(1.349)^t) + 3245.657$ It is the number of employees E as a function of time t. **d.** 110,476.85 **e.** 110,477 employees

Appendix A Toolbox Exercises

Toolbox Exercises

1. $\{1, 2, 3, 4, 5, 6, 7, 8\}$; $\{x{:}x$ is a natural number, $x < 9\}$ **2.** Yes **3.** No **4.** No **5.** Yes **6.** Yes **7.** Integers, rational numbers **8.** Rational **9.** Irrational **10.** $x > -3$ **11.** $-3 \le x \le 3$ **12.** $x \le 3$ **13.** $(-\infty, 7]$ **14.** $(3, 7]$, **15.** $(-\infty, 4)$ **16.**

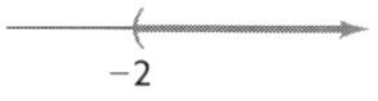

17.

18.

19.

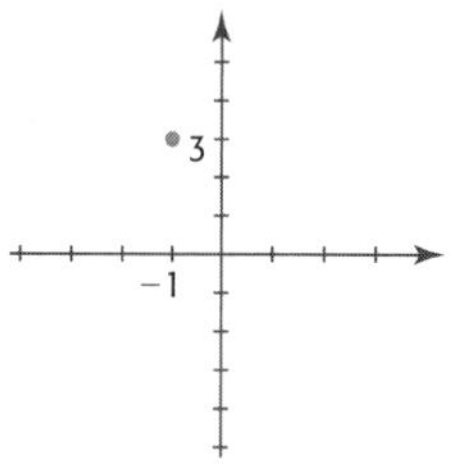

20.

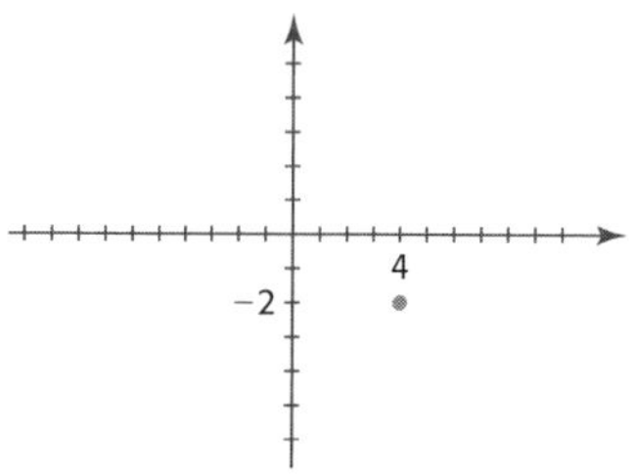

21.

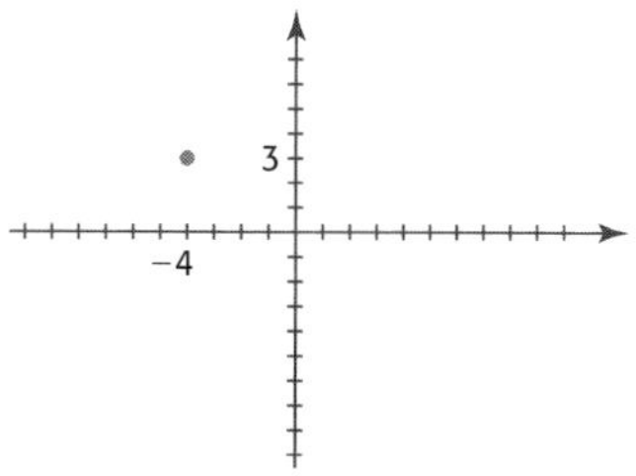

22. 6 **23.** 4 **24.** $-3x^2, -3; -4x^2, -4$; constant 8

25. $5x^4, 5; 7x^3, 7$; constant -3 **26.** $3z^4 + 4z^3 - 27z^2 + 20z - 11$ **27.** $7y^4 - 2x^3y^4 - 3y^2 - 2x - 9$ **28.** $4p + 4d$ **29.** $-6x + 14y$ **30.** $-ab - 8ac$ **31.** $x - 6y$ **32.** $6x + 9xy - 5y$ **33.** $3xyz - 5x$ **34.** $x = 2$ **35.** $x = 18$ **36.** $x = 3$ **37.** $x = 3$ **38.** $x = 3/5$ **39.** $x = 16$ **40.** $x = -10$

Appendix B Toolbox Exercises

Toolbox Exercises

1. a. x^7 **b.** x^5 **c.** $4^4a^4y^4 = 256a^4y^4$ **d.** $\frac{3^4}{z^4} = \frac{81}{z^4}$ **e.** $2^5 = 32$ **f.** x^8 **2. a.** y^6 **b.** w^6 **c.** $216b^3x^3$ **d.** $\frac{125z^3}{8}$ **e.** 3^5 or 243 **f.** $16y^{12}$ **3. a.** 4 **b.** -4 **c.** Not a real number **d.** 3 **e.** 2 **4. a.** $3\sqrt{2}$ **b.** $6\sqrt{6}$ **c.** $2\sqrt{42}$ **d.** $3a^2b\sqrt[3]{9ab}$ **5. a.** $x^{3/2}$ **b.** $x^{3/4}$ **c.** $x^{3/5}$ **d.** $(3^3)^{1/6}y^{9/6} = 3^{1/2}y^{3/2}$ **e.** $27y^{3/2}$ **6. a.** $\sqrt[4]{a^3}$ **b.** $-15\sqrt[8]{x^5}$ **c.** $\sqrt[8]{(-15x)^5}$ **7. a.** 2 **b.** $-3, 8$, 1 variable **8. a.** 4 **b.** $5, -3$, 1 variable **9. a.** 7 **b.** $-2, -119$, several variables **10.** $-12a^2x^5y^3$ **11.** $4x^3y^4 + 8x^2y^3z - 6xy^3z^2$ **12.** $2x^2 - 11x - 21$ **13.** $k^2 - 6k + 9$ **14.** $16x^2 - 49y^2$ **15.** $3x(x - 4)$ **16.** $12x^3(x^2 - 2)$ **17.** $(3x - 5m)(3x + 5m)$ **18.** $(x - 3)(x - 5)$ **19.** $(x - 7)(x + 5)$ **20.** $(x - 2)(3x + 1)$ **21.** $(2x - 5)(4x - 1)$ **22.** $3(2n + 1)(n + 6)$ **23.** $3(y - 2)(y + 2)(y^2 + 3)$ **24.** $(3p + 2)(6p - 1)$ **25.** $y = \frac{5x}{4x + 2}$ **26.** $x = \frac{18}{6y^2 - y}$ **27. a.** Imaginary **b.** Pure imaginary **c.** Real **d.** Real **28. a.** Imaginary **b.** Real **c.** Pure imaginary **d.** Imaginary **29.** $a = 4, b = 0$ **30.** $a = 15, b = -3$ **31.** $a = 2, b = 4$ **32.** $-1 + 9i$ **33.** $2 + i$ **34.** -1 **35.** 1 **36.** 5 **37.** $39 + 18i$ **38.** $-3 - 4i$ **39.** $\frac{7}{6} + \frac{1}{6}i$ **40.** $\frac{\sqrt{15} - 2}{6} - \frac{\sqrt{3} + 2\sqrt{5}}{6}i$ **41.** $6i$ **42.** $7i\sqrt{2}$ **43.** -4 **44.** $-3\sqrt{10}$ **45.** $5 - i$ **46.** $-1 + 3i\sqrt{2}$ **47.** $8 + i$ **48.** $-5 - 12i$ **49.** $1 - i$

Appendix C Toolbox Exercises

Toolbox Exercises

1. $\frac{1}{9}$ **2.** $\frac{1}{8}$ **3.** $-\frac{1}{64}$ **4.** $-\frac{1}{16}$ **5.** 27 **6.** 25 **7.** $\frac{9}{4}$ **8.** $\frac{8}{27}$ **9.** $\frac{1}{100}$ **10.** $\frac{1}{64}$ **11.** $\frac{1}{8}$ **12.** $\frac{1}{256}$ **13.** 10 **14.** 256 **15.** $\frac{1}{x^7}$ **16.** $\frac{1}{y^8}$ **17.** $\frac{1}{c^{18}}$ **18.** $\frac{1}{x^8}$ **19.** a **20.** b^2 **21.** $x^{1/6}$ **22.** $y^{1/15}$ **23.** $\frac{6}{ab^2}$ **24.** $\frac{-8a^2}{b^2}$ **25.** $\frac{x^{10}}{4}$ **26.** $\frac{8y^{18}}{27}$ **27.** $\frac{-7}{a^2b}$ **28.** $\frac{-6y^2}{x}$ **29.** 4.6×10^7 **30.** 8.62×10^{11} **31.** 9.4×10^{-5} **32.** 2.78×10^{-6} **33.** 437,200 **34.** 7,910,000 **35.** 0.00056294 **36.** 0.0063478 **37.** 3.708125×10^6 **38.** 6.460833515×10^1 **39.** \$600; \$2600 **40.** \$6375; \$14,875 **41.** \$210; \$3710 **42.** \$168; \$5768

Index

Note: Information presented in tables and figures is denoted by *t* and *f* respectively.

Index of Applications

Finance

Geometry

Government

Health

Labor

Law Enforcement

Medical

Miscellaneous

Physics

Sports/Entertainment

Statistics/Demographics